Fireground Size-Up

Fireground Size-Up

Michael A. Terpak

PennWell®

Disclaimer

The recommendations, advice, descriptions, and methods in this book are presented solely for educational purposes. The authors and publisher assume no liability whatsoever for any loss or damage that results from the use of any of the material in this book. Use of the material in this book is solely at the risk of the user.

PennWell Corporation
1421 South Sheridan
Tulsa, Oklahoma 74112
800-752-9764
sales@pennwell.com
www.FireEngineeringBooks.com
www.pennwell-store.com
www.pennwell.com

Marketing Manager: Julie Simmons
National Account Executive: Francie Halcomb

Edited by Jared Wicklund
Cover design by Clark Bell
Book design and layout by Brigitte Pumford-Coffman

ISBN 10: 0-91221-299-3
ISBN 13: 978-0-912212-99-9

Printed in the United States of America

5 6 7 8 9 11 10 09 08 07

Dedication

This was a difficult section of the book for me to put together. During my 25 years in the fire service I have met so many inspirational and influential people, that if I listed them all, their names would occupy more pages than the text of this book. However, there are a few that I must mention.

I was a young kid who couldn't wait to get involved with the fire service. I was sworn in as a junior firefighter by Captain John Porter of the Singac Fire Company in Little Falls, NJ on my 16th birthday. As my first fire officer, Captain Porter taught me what a young, new firefighter needed to know, while at the same time looked out for my safety. He put my mom and dad at ease with my new and exciting hobby.

As my ambitions with the fire service grew, my career choice was to be a firefighter with Jersey City. It was as a new probationary firefighter assigned to Engine Co. 10 where I met a man who continues to inspire me today. Lieutenant (now Deputy Chief) William Sinnott remains as my teacher, my leader, and a great friend.

Special Thanks

I have to pay a special thanks to the following members who have assisted me in more ways than I could imagine.

Chief Paul Dansbach of the Rutherford, NJ Fire Department.
Deputy Chief James Smith of the Philadelphia, PA Fire Department.
Both men gave me great advice and technical support during the final stages of the book.

Chief William Peters of the Jersey City, NJ Fire Department.
When frustration would set in, a quick call to his office would help.

To the Officers and men of the 2nd Battalion
It's always best to surround yourself with the best and the brightest.

Photo Thanks

Ron Jeffers, New Jersey Metro Fire Photographers
Bill Noonan, Boston Fire Department
Joe Lovero, Jersey City Fire Communications, a good friend lost while operating at the WTC, 9-11-01
Ira Rubin, Jersey City Gong Club
Chris Fink, Jersey City EMS
Paul Dansbach, Rutherford, N.J. Fire Department
Steven Woodworth, Atlanta Fire Department
Paul Schaetzle, Jersey City Gong Club
Steven Dill, North Plainfield Fire Department

Special Dedication

To my wife Elizabeth, and to my daughters Ashley and Sara. It is also important to surround yourself with people whom you couldn't live without. Without their love and support, nothing would be possible.

Thank you!

Fire Ground Size-Up

Introduction

"Size-up" is the process of gathering information by firefighters and fire officers that allows them to make efficient, effective, and safe decisions on the fireground. Due to the urgency of responding and going to work at an emergency incident, decisions on the fireground must be made quickly, with a continued emphasis placed on their efficiency, effectiveness, and overall safety. In order to meet this demand over the years, fire officers, as well as those aspiring to be fire officers, have developed and used different means of gathering information about a potential or actual incident. The areas in which they sought the information would often be the multiple sources that were available to them. In many of those areas the information that could be gathered would be limited only by the officer's ability and enthusiasm to research it.

As experience often demonstrates, it becomes evident early on in this business that waiting to gather any information until the fire department's arrival at an incident, often proves to be reactive, unprofessional, and at times very unproductive. It was quickly realized that seeking information about a potential or actual incident had to be done as soon as possible in order to have an effective and safe outcome. The question then is, "How and where do we get information?"

Pre-incident size-up

The concept of size-up can take on many different forms for the fire officer if he/she allows it. Ideally, to aid in the decision-making process a fire officer at the earliest possible opportunity needs to have accurate, relevant, and useful information, which should be in place well before the alarm. The information-gathering process that takes shape before the alarm is transmitted is referred to as "pre-incident information" or "pre-fire planning". It is in this area of information gathering where the fire officer seeks data about the building and the occupancy in question from any current information available. Whether it is from a formal pre-fire plan and inspection, or from information noted from a previous incident, by reviewing all of the relevant size-up points, officers can compile a comprehensive list of information that can aid them in their decision-making well before an actual incident occurs.

Once the information is gathered about the occupancy being reviewed, it can then be documented and listed in a binder which can be carried in the cab of the apparatus, or be added into the fire department's dispatch and communication system were it can be reviewed prior to arriving at an incident. In many of the departments researched for this book, specific incident information (SII) is stored on a computer where it can be listed on an alarm sheet printout in the firehouse, or printed from a terminal in the cab of the fire apparatus. In either form, having information about the building you are responding to greatly enhances your ability for sound decision-making. Waiting until you arrive to gather information, limits your ability in making well-informed decisions.

Alarm size-up

When the "alarm sounds" a Fire Officer has the ability to further his/her information gathering relevant to the incident. As the alarm is being transmitted, information such as the time of the alarm, the address of the building, the type of occupancy within the building, and the current weather conditions will influence some of the officer's decision-making. Depending upon your department's method of communicating the alarm, information may also become available about the building's occupancy load, the water supply availability in the area, the possible presence of auxiliary appliances in the building, and any hazardous materials that may be housed within the building. During this phase of gathering more specific information about the location where you are responding, an officer will be able to start factoring the concerns listed, as well as any possible actions he/she might take well before the building comes into view.

On scene size-up

When arriving on the scene of an incident, the opportunity to gather more detailed information will obviously arise. Such specifics as the height of the building, the class of construction, the occupancy type, the location and extent of the fire, and any exposure concerns quickly become evident.

As firefighters and fire officers are assigned and go to work in their specific areas, the information gathering will increase further from such resources as company reporting, or division and sector officers who will provide progress and status reports to the incident commander from their assigned areas. Receiving additional, as well as ongoing information from geographical areas assigned within and around the building allows the incident commander to prioritize decisions, allowing for an efficient, effective, and more importantly, a safe outcome. To make this happen, the on-scene phase of size-up doesn't end until the incident is terminated, and everybody goes home.

Post-incident size-up

Did you think we were done? The process of gathering information can go one step further. After the incident is terminated, a post-incident analysis can be conducted to further the information exchange. It is here where we review what we observed, what we did, the result of our action, and what we learned so we may better ourselves for future incidents. This can be as simple as a review conducted at the kitchen table in the firehouse, or as detailed as a formal review, documentation, and illustration of the incident.

It is important for the firefighter and fire officer to realize that size-up of a potential or actual incident scene is not a concept that you save until the alarm is transmitted. Beginning the information gathering process early, continuing it throughout the incident, and then seeking additional information after the incident is over, continues to be our best weapon in dealing with the unexpected.

How to use this book

What follows in Chapter 1 is a comprehensive look at 15 size-up points that we will use throughout this book. It is the intent of the first chapter to give an overview of the points listed, with each factor covered in detail as we discuss the size-up concerns covered in subsequent chapters.

For easy reference, the points listed in each chapter have been placed in the form of an acronym. Anyone who has studied for high school, college, and fire department promotional exams knows that the use of acronyms can aid an individual's memory to recall and reference when it becomes necessary. This is a useful study tool that many continue to use. The acronym being used is "COAL TWAS WEALTHS". (This was adapted from the traditional East Coast acronym, "COAL WAS WEALTH".)

For each letter listed, there is a size-up point.

C - Construction
O - Occupancy
A - Apparatus and Staffing
L - Life Hazard

T - Terrain
W - Water Supply
A - Auxiliary Appliances and Aides
S - Street Conditions

W - Weather
E - Exposures
A - Area
L - Location and Extent of Fire
T - Time
H - Height
S - Special Considerations

The size-up factors listed throughout this book reflect general thoughts, concerns, and ideas to consider when operating at an incident in the different types of occupancies I have outlined.

The research for the following has been compiled from years of preparing for promotional exams, personal experiences from working in a congested urban city, as well as from listening to those who have been there before. My hope is for those who read this book, that the information-gathering process is only enhanced by their own education and experience.

Michael A. Terpak
Battalion Chief
City of Jersey City

1 The Fifteen Points of Size-Up

Construction: Size-Up

***C**OAL TWAS WEALTHS*

As we respond to incidents in the 21st century there should be no doubt in anyone's mind that the firefighter and fire officer will be continually plagued by newer, lighter, and mixed types of construction. The only sensible way to combat these concerns is by continuing to educate ourselves through pre-fire planning, as well as by enhancing our knowledge in the recognized classes of construction, and the building construction methods and materials used.

Throughout the years we have all been taught that there are five different types of construction the fire service may encounter. It is becoming increasingly more and more difficult to recognize them. Even in cities as old as Jersey City, a chief officer could be standing in front of what looks like a Class 3 constructed building, only to find that behind the brick the building contains all light-weight constructional components. Newer, lighter, or (as we will refer to in this book as) "hybrid" construction methods, are increasingly making their way into every town and city. The construction methods and the materials used with this association are forcing us to change the rules of engagement, and for a very valid reason.

So where do we start to gather relevant and useful information about our construction concerns? It would be dangerous not to review the inherent dangers of the classes that are still present in many towns and cities. These buildings, even if altered, will still contain the original design concerns associated with the different classes. If we use this premise as a basis of our thinking in this size-up category, we need to continue reviewing the existing concerns, adding the "hybrid" methods of construction when present, so we can remain current on the old, as well as new dangers we may face.

The classes of construction

Class 1: Fire-Resistive Construction
Class 2: Noncombustible/Limited Combustible Construction
Class 3: Ordinary Construction/Brick and Joist Construction
Class 4: Heavy Timber Construction
Class 5: Wood Frame
Hybrid Construction

Class 1: Fire-resistive construction

This class of construction refers to a building whose structural components are designed and protected to resist the maximum severity of fire expected within the structure. Buildings that contain this combination of masonry and steel can be vast in design and presence. They may be constructed to fit the needs of a low-rise housing complex or structures up to the size of a large commercial office high-rise building. Although they are different in appearance and in size, each will present many of the same concerns.

Fig. 1-1 *Commercial office building*

Favorable characteristics of Class 1 construction

Of all the types of construction firefighters respond to, Class 1 construction is thought of as being the most resistant to fire spread, as well as the most resistant to significant collapse. If we take an initial look at this type of construction to determine if there any favorable characteristics that can aid the fire service in their efforts, we will find that there are a few. The most significant is that the building's structural components do not add to the fire spread. Buildings built of masonry and steel, where the steel is protected by encasement, or by a membrane covering (as it is in this class of construction) will protect the building's key structural components. Barring any imperfections or alterations to the original work, this factor encompassed with the concept of "compartmentalization" is looked upon as being a plus from a fire control concern. Compartmentalization, which is a significant element in restricting the fire spread in a Class 1 design, divides the floor space into sections or areas that can be closed off from each other. This ability to separate areas creates resistance for fire to spread beyond the area of origin.

Fig. 1-2 *Low-rise residential housing complex*

Negative characteristics of Class 1 construction

Heat. Buildings of this type of construction, however, do not come without some significant concerns to the firefighter and fire officer. By design they are known to retain large concentrations of heat. Anytime you put a fire in a compartmentalized area of fire resistive construction, heat will become your greatest enemy. This concern is well demonstrated in the residential housing complexes built of Class 1 construction. In this particular type of residential building, room sizes are small, allowing heat to collect and concentrate in specific areas. Even though fire spread will often be limited to the involved apartment, members must expect difficulty from the intense heat as they advance into the fire area.

Forcible Entry. Another related consideration for firefighters in a building of Class 1 design is the forcible entry difficulties from the door, jamb, and wall construction. Placing a steel door within a steel jamb, surrounded by a masonry block wall is going to be a little more of a challenge than you may be accustomed to. Using conventional tools and methods are not going to be your best option with this type of construction.

Large open spaces. Depending upon the occupancy type, some buildings of Class 1 construction may lack the compartmentalization previously mentioned and contain large, open spaces. These building types will create concerns with uninterrupted fire spread. Decisions made by engine company officers relating to hoseline selection must take into consideration hose stream reach and penetration needs. Accordingly, firefighters assigned to search operations must use organized searches and search rope procedures for an accurate as well as a safe search. It is within these space types where organization and control must further influence your decision-making.

Ventilation. Ventilation options in a Class 1 design building can create significant concerns. Not being able to adequately "open up" a building of this design only adds to punishing conditions that members will be forced to face in these structure types. Depending upon the type of Class 1 building you encounter, certain sealed windows cannot be opened. Others may have to be broken to provide ventilation. Roof openings, if any, will be limited in these building types. When present, openings may be available from skylights, scuttles, and bulkhead doors that will provide minimal relief.

Collapse concerns of Class 1 construction

Spalling. Collapse concerns of a Class 1 structure, although localized, must still be viewed by firefighters as a dangerous threat. A concern, although only experienced after prolonged involvement, is spalling of concrete from the floor, wall, and ceiling construction within the structure, as well as from the spalling of concrete and marble facades as fire moves to the exterior. Pieces of masonry weighing ten, twenty, or thirty plus pounds have been known to explode and drop from these structures, severely injuring firefighters. Even though this particular collapse threat is localized, it still must be considered deadly.

Suspended Ceilings. An additional, and more common, collapse concern within a Class 1 building is from a suspended ceiling collapse. In buildings that incorporate this type of ceiling design, any measurable heat from the fire can quickly weaken the ceiling's suspension system, resulting in the possible entanglement and entrapment of a firefighter. This type of ceiling design is becoming a more common concern in all types of construction. With more and more firefighters being caught in this spiderweb of material, your best recourse of action, outside of prevention, is often a good pair of wire cutters.

When we compare Class 1 construction to the other classes of construction in this chapter, we note that the fire spread and building collapse concerns will be less severe. However, this fact should not give anyone a false sense of security. As we start to look deeper into this type of construction and its relationship to the different kinds of occupancies listed in the following chapters, we will find that there are many other size-up factors relating to this building design that will significantly increase our concerns.

Class 2: Noncombustible/limited combustible construction

This class of construction refers to a building whose structural design consists of masonry and steel. These buildings will consist of masonry block or steel exterior walls, with a steel-supported roof system. The steel used within the design in most installations will be exposed throughout, and subject to heat from the fire. Buildings of Class 2 construction will generally range in height from one to two stories, with the majority only being one story. Many buildings built of Class 2 construction that are referred to as one story structures may actually be equal in height to a two-, or three-story building. Many fire officers will refer to this type of building as a "high-ceiling occupancy". This association will come with its own set of unique concerns that will be presented in a later chapter.

Favorable characteristics of Class 2 construction

Yes, there is one. The only favorable construction concern (if we could even call it that) of a Class 2 noncombustible/limited combustible building is that concrete and steel, when viewed by themselves, do not contribute to the building's fire load. From this association we gather the reference of the Class 2 definition as "non-combustible". The fact that the building's construction components do not add to the fire load is the only favorable thought of a Class 2 structure that we could find. Everything else is negative, and for a number of reasons.

Collapse concerns of Class 2 construction

Unprotected Steel. From an engineering concept, concrete and steel are known as strong structural components. When incorporated together within a building design they can withstand significant loads and span great heights and distances. This is evident from the numerous high-rise buildings that appear on the skylines of many cities around the country.

However, the same structural members that are used within the design of a Class 1 building are also incorporated within the design of a Class 2 building with one main difference. The steel used within the building is exposed and unprotected throughout. Steel in the form of I-beams, angle iron, and lightweight steel bar joist trusses will make up the structural framework and roof-supporting system of a Class 2 constructed building. The materials and methods employed in the non-combustible/limited combustible design permit the surface area of the structural steel to be exposed. Any sizable fire exposed to the unprotected steel

Fig. 1-3 *Class 2 constructed building*

will expand the steel support system possibly pushing out the exterior walls. This will, in turn, collapse the roof–bringing down any roof-mounted loads in the process. Depending upon the thickness of the steel, the size of the fire, and the loads the steel is supporting, failure times can be within minutes of your arrival.

Fig. 1-4 *Unprotected steel found within a Class 2 building*

If that isn't enough by itself, another significant factor that helps Class 2 structures receive the distinct title as being the class most prone to collapse is the design of the steel deck roof.

Steel Deck Roofs. As a young firefighter, I had the opportunity to earn extra money on my days off from the firehouse as a roofer in the city. One of the many jobs we did was installing roofs of this type on the different commercial buildings throughout the city. It wasn't until my first fire involving a Class 2 building, that I realized the dangers this type of roof presents.

The design and the materials used with the steel deck roof is where Class 2 buildings receive the "Limited Combustible" term in its definition. Steel deck roof construction uses a design referred to as "built-up roof". In this design, corrugated steel deck sheets, approximately 6' by 20' are placed over a lightweight, steel bar-joist supporting system. Hot liquid asphalt used to waterproof the deck is "mopped" into the channels of the corrugated steel sheets–"Q" decking. While the asphalt is still hot and in its liquid form, foam insulation blocks of varying sizes are placed into the liquid asphalt to adhere to the deck surface. On top of the insulation block, more hot liquid asphalt is applied, allowing the installation of multiple layers of roofing material and more liquid asphalt to give the roof its water shedding capabilities.

Fig. 1-5 *Steel deck roof under construction*

The materials applied in this fashion become the ingredients for a giant frying pan of combustible material. As previous incidents have shown, heat from the content fire below would quickly be con-

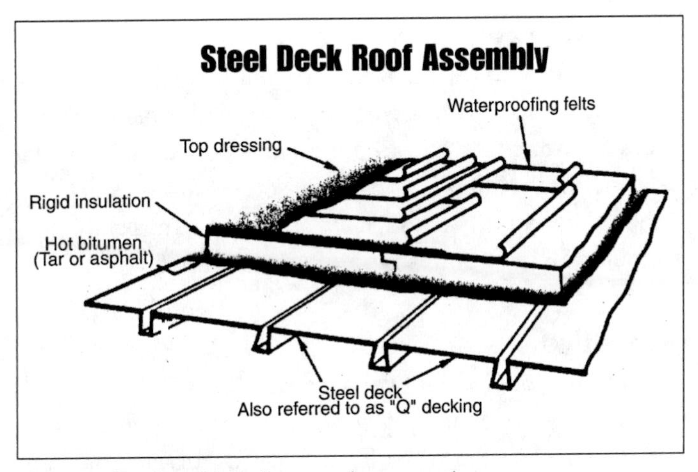

Fig. 1-6 *Combustible/built-up roof construction*

ducted through the metal deck to the now solidified asphalt. The solidified asphalt as it is heated, begins to liquefy and then vaporizes, producing a gas. The combustible gases not being able to escape up through the insulated, sealed roof deck, would then be forced downward through the seams of corrugated steel deck sheets where the gases could be ignited by the heat from the contents fire below. With the possibility of combustible roof deck gases burning below the deck, combustible roofing material burning on top of the deck could also be ignited setting the stage for a rapidly advancing roof deck fire that will soon lead to a collapse.

There's more! Another critical condition produced as a result of the roof fire, is flaming liquefied asphalt can "drip" down from the roof deck during the heating process, possibly causing secondary fires ahead of the primary fire. The "asphalt drips" or "flaming tar balls" should alert the approaching firefighters of the imminent dangers above.

Another and more dangerous fact noted in this roof design stems from the installation of the steel deck sheets. In early installations the steel sheets were tack welded together over the entire deck, regardless of where their seams would end, sometimes in the middle of a roof bay. This practice of tack welding seams, regardless of whether they were supported or not, was the industry practice at one time, but it indirectly added a built-in danger to the firefighter. An unsuspecting firefighter walking on a fire-weakened seam could slide into the fire from a series of failed welds. This installation flaw could actually hinge a sheet downward from the weight of a firefighter dropping them into the fire below. (I found a new part-time job after learning this.)

Class 3: Ordinary construction

This class of construction refers to a structure whose exterior walls are constructed of noncombustible materials such as brick, concrete block, or clay tile with the building's interior floor, wall, and ceiling members constructed of wood. Firefighters also refer to this class as brick and joist construction. Ordinary buildings will range in height from one to seven stories, with two to four stories being the most common.

Favorable characteristics of Class 3 construction

The noncombustible exterior walls, whether of brick or concrete block, are considered by firefighters to be a "moderate" plus factor simply because these structural members will not contribute to the building's fire load. This fact accompanied with the general thought that these types of structures are normally smaller in average square footage when compared to Class 1 or 2 buildings, masks the inherent dangers both on the interior and exterior of this type of construction.

Negative characteristics of Class 3 construction

Alterations. When we reference Class 3 construction we must remember that these structures are generally older class buildings, meaning that they have been around for a while. The fact that they have been occupying the same space for a number of years causes us not to only question their overall integrity, but also to consider the probability that a significant number of alterations may have been made to accommodate the building's changing needs.

As need would dictate change, many have been altered to improve the tenant/owner use of the property. The change might have been to save on heating costs by lowering the ceiling height with a new suspended ceiling, the replacement or addition of new plumbing and electrical fixtures to modernize the structure, or by taking a larger area within the building and dividing it to accommodate a change of occupancy. Regardless, firefighters must be ready to respond and anticipate numerous concealed spaces and voids, with a rapid fire spread in this type of construction.

Concealed Spaces and Voids. Predicting and controlling a fire spread will be a significant challenge for firefighters arriving at a Class 3 constructed building. Fire will have plenty of opportunity to spread throughout the building from the many concealed spaces and voids this particular type of construction presents. Whether the fire originates in an "alteration void" such as a suspended or dropped ceiling, or in a "design void" such as a cockloft or attic, the challenges and concerns will be the same. If you don't get to it quickly, fire will take possession of the building.

Fig. 1-7 *Class 3 constructed buildings*

With the fire's anticipated movement, will come the daunting task of "opening up" above, below, and alongside the fire's origin. Obviously, this is much easier said than done. With the probability of not one, but multiple voids occupying the same space, the tasks become time consuming as well as dangerous.

Depending upon the size of the space, the air within that space, the fire's condition, and the integrity of the structural members that enclose it, concealed spaces and voids can present explosive conditions when penetrated. Conditions ripe for a backdraft will be present within the walls, hanging ceiling space, the cockloft, or the attic of this class of construction. When explored by a firefighter, these areas could violently explode, bringing structural members down with it.

In subsequent chapters, we will explore certain types of occupancies that are characteristic of these concerns, but in the interim what must be noted about any building of Class 3 design, is that these spaces can be numerous and dangerous throughout.

Collapse concerns of Class 3 construction

As one might expect at this point, buildings of Class 3 design will come with significant collapse concerns that stem from fire involvement of those concealed spaces and voids. With the fire's ability to travel and extend throughout the void spaces, failure could occur to any structural members at any given time.

Parapet Wall. The largest and most common concealed space within a building of Class 3 construction will be the undivided cockloft that extends over the entire building. Cockloft spaces within a Class 3 building could range in height from 6" to 6'. Whenever fire involves this space, members must be aware of not only conditions that could present a backdraft, but also conditions that could affect the integrity of the parapet wall. Parapet walls are those walls that extend above the building's roofline. In many cases they are freestanding and only supported at their attachment to the building. Any time a large volume of fire occupies the top floor and cockloft of a Class 3 building, firefighters must

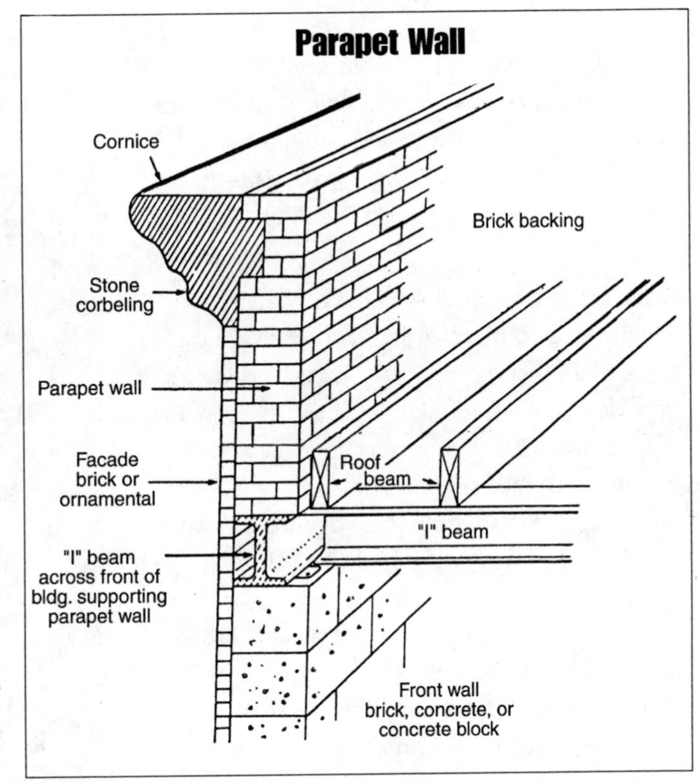

Fig. 1-8 *Class 3 constructed buildings*

expect the integrity of the parapet wall to be jeopardized. These types of walls are known to fail from deteriorated conditions due to their exposure to weather and shoddy workmanship, besides the effects of a fire.

Hanging Ceiling Space/Multiple Ceilings. Class 3 collapse concerns will not be limited to the outside of the structure. When fire involves any hanging ceiling space, there is a potential for a backdraft, as well as ceiling collapse from fire-weakened structural members. It is not uncommon in older Class 3 buildings to find multiple ceilings within a single floor space. The reasons for this can be numerous, but are most commonly associated with the cost to repair the original ceiling, as well as an attempt to save money on heating bills by lowering the ceiling height. Regardless, these alteration voids can breed backdrafts and hidden fire, both of which can collapse a ceiling onto a firefighter.

Class 4: Heavy timber construction

This class of construction refers to a structure that has an average height of 6 or 7 stories, with walls built of brick, block, or stone, and a building interior constructed of large wooden timberwork. Also referred to as "mill construction", wooden columns found within a Class 4 building will be found to be no less than 8" by 8" in thickness, floor and roof girders will be no less than 6" by 10" in diameter, and floor planking approximately 3" in thickness by 12" in width. The timber truss roof design commonly associated with a Class 4 building will generally have top and bottom chords of lumber no smaller than 4" by 6" with smaller dimension lumber used for its web members. This is a very large, and inherently strong building.

Positives of Class 4 construction

Surface-to-mass ratio. The construction features of buildings built of Class 4 construction provide excellent resistance to the early involvement of fire. This reference is not only made from the nature of the building's noncombustible components, but also from the dimensions of the lumber found within the building. Their low surface-to-mass ratio is the key to their early resistance to fire. With most of the lumber's mass found within the structural member, the possibility of early ignition of a structural member is significantly slowed. We can characterize this statement by referencing the difference between a large tree limb placed on a camp fire, as com-

Fig. 1-9 *Class 4 building*

pared to a piece of wood kindling placed on the same fire. The much smaller piece of lumber has a large surface-to-mass ratio allowing for quicker ignition. On the other hand, the larger tree limb, with a much smaller surface-to-mass ratio will take a significant amount of time to be heated in order to produce a vapor that can be ignited.

Lack of Void Spaces. Withstanding early ignition, buildings built of Class 4 construction were originally designed to have negligible void spaces, if any at all. Generally within these type structures, void spaces present themselves only after the building has undergone some type of renovation. This added plus factor only enhances the theory that buildings of Class 4 construction are listed as the second most resistant to collapse when compared to all other classes of construction.

Fig. 1-10 *Columns pictured are 12" by 12"; girders are 12" by 14".*

Negative characteristics of Class 4 construction

What was referred to as a plus during the initial stages of an incident, returns during the later stages of the fire to become the most significant concern to firefighters and fire officers. The large wooden interior members, once ignited, will produce heavy fire conditions with tremendous amounts of heat, making it extremely difficult for fire streams to penetrate into the building. Further fueled by oil-soaked floors, cork-lined wall and ceilings, as well as combustible stock from the occupancy content itself, a fire in a heavy-timber building will tax even the largest fire departments. Fires of this magnitude will produce severe exposure problems from the radiated heat, flying brands, and the flying embers they will produce. Firefighters and fire officers should never underestimate the speed at which these fires can spread once the large timberwork is ignited. The age, dryness, and abuse the structural members have been exposed to through the years will produce a hot and fast moving fire once ignited.

Collapse concerns of Class 4 construction

Involvement of these large structures can produce conflagration conditions making involvement and collapse onto nearby buildings a real possibility. Incident commanders should establish collapse zones for distances equal to twice the height of the building on all four sides. These distances must take into consideration any nearby exposure buildings and any fire department operations involving those areas. Collapse of a heavy-timber building is generally not a small, localized event. The collapse of a building of this magnitude will encompass large areas. In addition to the immediate area of collapse, they have been known to throw structural debris for extended distances when they fall.

This is the reason many incident commanders seek extended collapse zones. The only plus of a heavy fire involvement of a Class 4 building is that the radiant heat that they produce, generally drives fire forces far enough away from any direct exposure to their collapse. However, you should be aware of the secondary collapse potentials of nearby buildings. This is the place where firefighters can get into trouble.

Fig. 1-11 *Class 4 building collapse*

Class 5: Wood frame construction

This class of construction refers to a structure primarily built of wood. Most of the newer and larger buildings of Class 5 construction now contain a number of "hybrids," ranging from the installation of steel girders to accommodate the larger spans, as well the use of truss assemblies in the floor and roof designs. Keeping this in mind as we review Class 5 buildings, the majority of structures of this design will contain walls, floors, ceiling rafters and roof rafters constructed of milled lumber. Since this is the most common type of construction found, you shouldn't be surprised that they also have the most fire activity.

When firefighters consider a Class 5 constructed building, many pictures may come to mind depending upon the area they are assigned to protect. Urban firefighters may picture a structure one to four stories in height, measuring approximately 20' X 60'. Suburban and rural firefighters may picture a structure one to two stories in height encompassing 4000ft^2. Each type of Class 5 building mentioned, although much different in size, will share significant concerns that must be identified by responding firefighters and fire officers.

Fig. 1-12 *Suburban Class 5*

Negative characteristics of Class 5 construction

The most significant fire spread concern for those responding and operating at a multi-story wood frame building fire will be from the open interior stair construction. This large, vertical artery within the building's living space allows the unimpeded movement of fire and its byproducts to invade upper-levels of a structure very quickly. With a vertical opening as large as an interior staircase placed in between floors of a multi-floor building, super-heated gases, smoke, and fire will quickly penetrate the building's upper-levels. This is without a doubt the largest and the most significant fire extension concern within a building of this type.

Fig. 1-13 *Urban Class 5*

Collapse concerns of Class 5 construction

A significant and often overlooked fact of a Class 5 building is that the entire structure is combustible, especially the load-bearing members. Firefighters and fire officers must remember that what holds up the second, third, and even the fourth floor of these buildings is structural members of wood sometimes no larger than $1^1/_2$ x $3^1/_2$ inches. When a fire extends from the building's contents to the building's structure and attacks these structural supports, it could collapse the building.

Framing methods of Class 5 construction

Braced Frame. Two early Class 5 constructed buildings that continue to plague firefighters in many areas of the country are those of the balloon frame and braced frame design. The earlier of these two versions is the braced frame, also known as the "post and beam". This type of wood frame construction

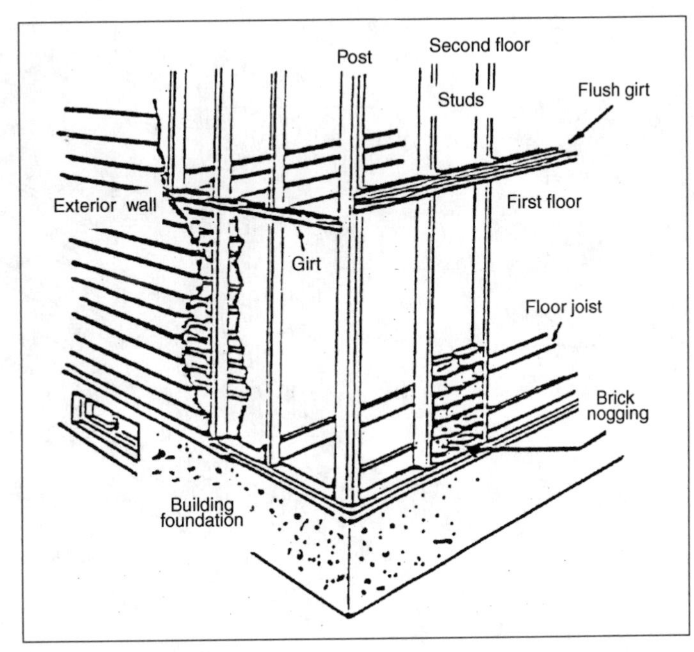

Fig. 1-14 *Braced Framing*

has a braced framework of vertical timbers know as "posts" which are positioned at each of the four corners of the building assembly. Horizontal timbers, known as "girts," will be found attached to these at each floor level. Where they meet, the post and girt are fastened together by use of a mortise and tenon joint. At the end of each girt, the horizontal members will be cut down to form a tenon that is designed to fit into the mortise slot cut into the vertical timbers. The obvious concern to the firefighter and fire officer is the wooden connection points that most definitely have been subject to age, possible rot from weather, any previous fire damage, as well as any current fire involvement. Failure of a connection point on a lower floor of this type of framing can collapse the entire building.

Balloon frame. The next oldest type of wood frame construction that also continues to plague firefighters is that of the balloon frame design. Balloon frame design became

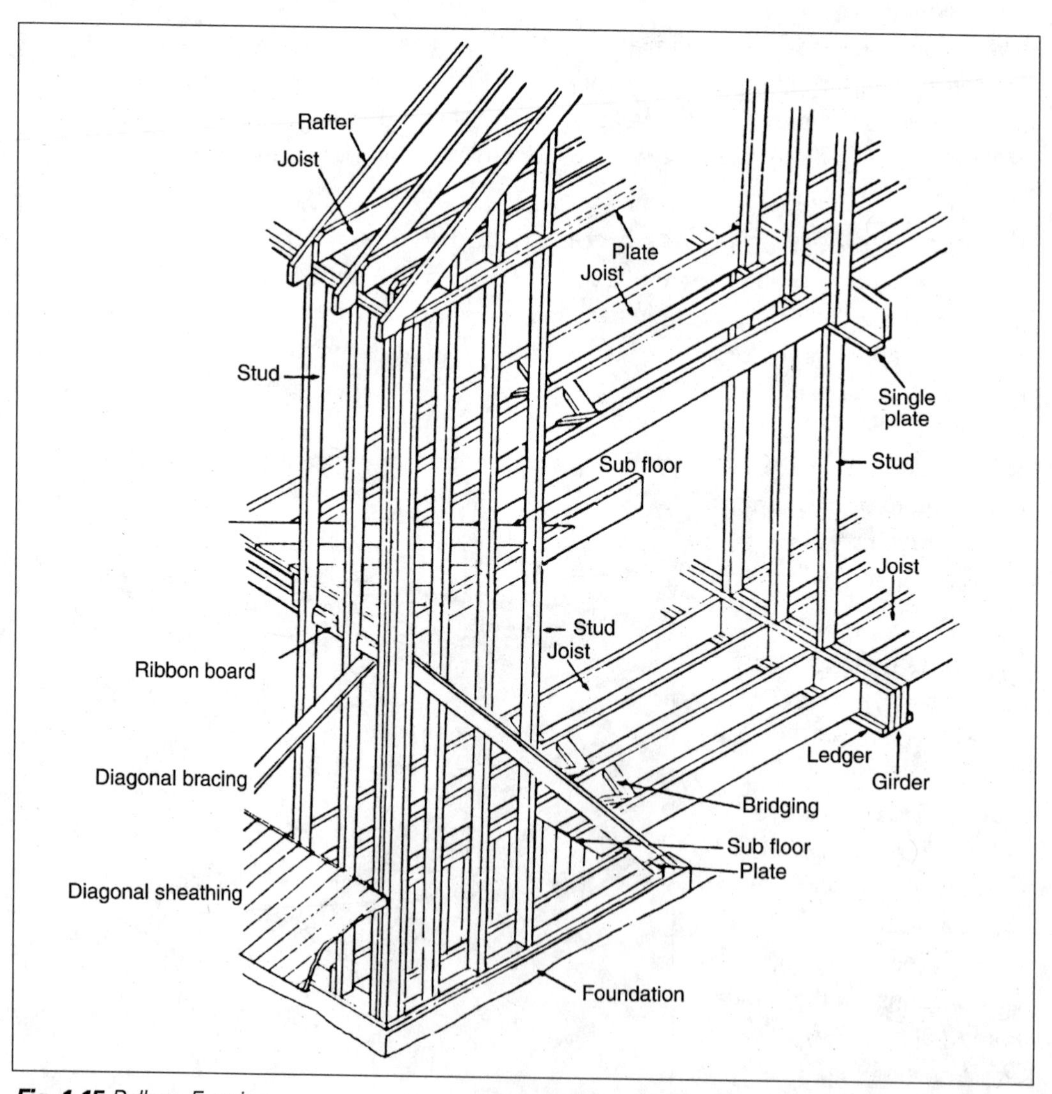

Fig. 1-15 *Balloon Framing*

a product of the braced frame building simply due to its speed in construction. Exterior walls of 2" x 4" studs extending for the entire height of the structure allowed the builders to erect these buildings with speed, but unknowingly allowed fire to travel even faster. In this design, wood studs are erected vertically to run the entire height of the building. At each floor level, ribbon boards are nailed to the vertical studs allowing the building's floor joists to rest on top. From this installation, an unprotected, vertical opening is created for the entire height of the building. The most apparent concern of this type building is the speed at which they allow fire to travel throughout the building.

Platform frame. As the building industry and the fire service learned of the dangers of both methods of framing, we entered into the most often used wood frame design of today; the platform frame. In platform frame design, each floor is erected as one inde-

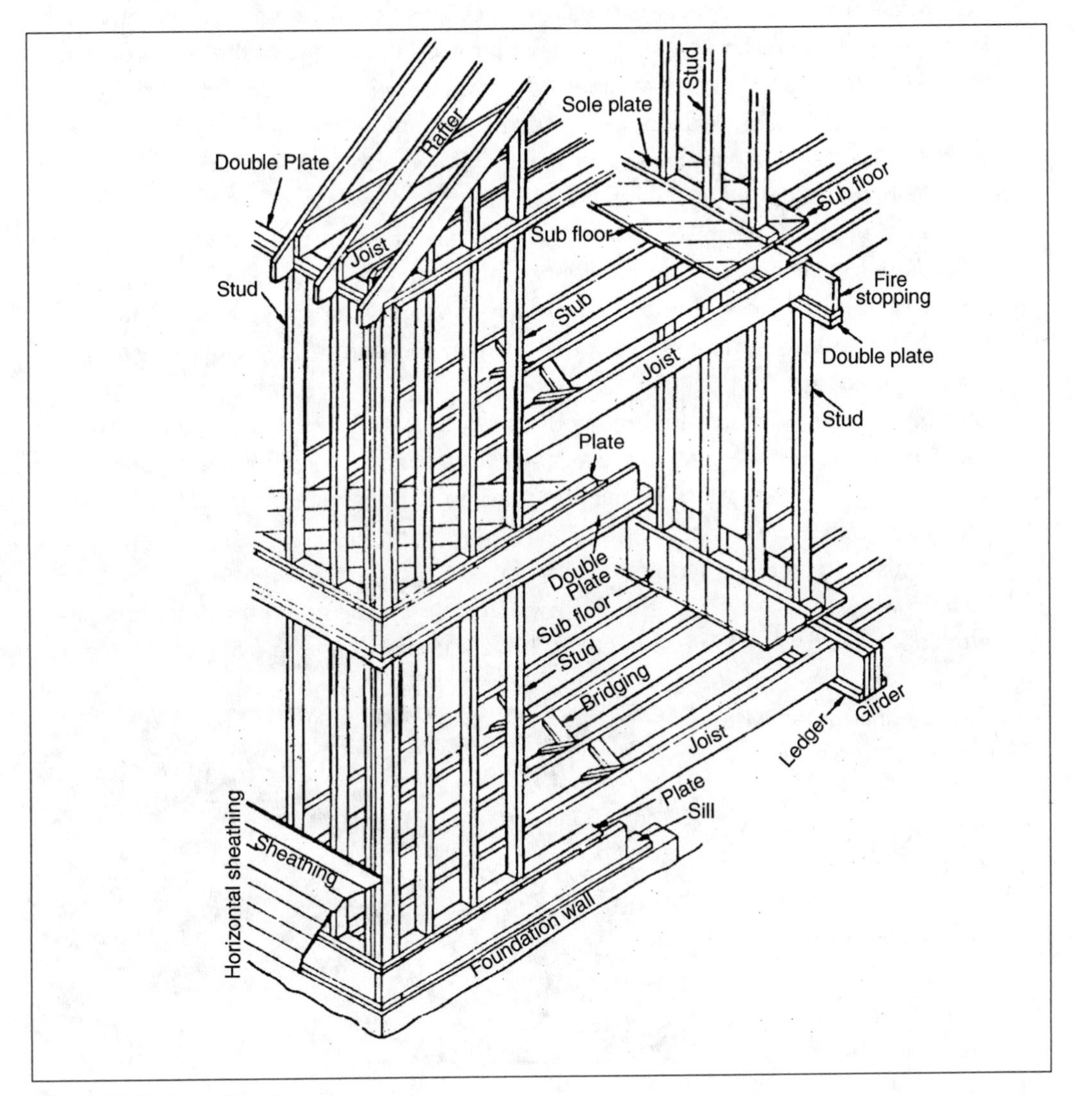

Fig. 1-16 *Platform Framing*

pendent unit eliminating the devastating vertical fire travel found in the balloon frame, as well as the elimination of the use of mortise and tenon connection points used in the braced frame design.

As we consider the three types of Class 5 construction listed previously, we must continue to remember that regardless of when they were built, their load-bearing supports are still made of wood. When a fire goes beyond the contents and attacks the structure, the integrity of the entire building is at risk.

Hybrid construction

Construction features referred to as "hybrids" are those elements made of light-weight or pre-engineered materials that may be mixed within any of the previously mentioned classes of construction. Examples of these materials include light-weight wood trusses, unprotected steel, and wooden I-beams (to name a few). As builders and developers seek newer and cheaper ways to erect buildings, the fire officer's concerns with hybrids will increase. For years we have been taught the five classes of construction and the inherent dangers they present. With hybrid construction materials within a building, or a building totally constructed of lightweight materials, the rules of engagement change. It is no longer possible to put time frames on when buildings will fall down. Lightweight materials and truss construction force us to operate very cautiously and carefully, with time against us.

One of the main concerns that many of us have when discussing these building features is, "how do you know if any of these materials are in the building where you will fight a fire?" Barring any actual visual inspection during the construction or renovation stage, you probably won't know unless the building is marked, or the construction materials used are characteristic to a certain type of occupancy.

To help with this concern, in the State of New Jersey all commercial buildings and some residential structures are required to have a marking on the front of the building indicating whether the building is constructed with floor or roof truss assemblies. As one could imagine, this is an important piece of information upon arrival at the building. The marking eliminates any guesswork related to the possible presence of a truss design within the building. These required markings

Fig. 1-17 *This building contains a truss roof and floor.*

became law in the state after a number of firefighters died in structures with truss roof construction.

As we go to work in the 21st Century there is no doubt that the job of the firefighter and fire officer is becoming increasingly difficult, dangerous, and demanding. It seems that our concerns with hybrid construction methods and materials will be one of the paramount reasons why.

Occupancy: Size-Up

COAL TWAS WEALTHS

When firefighters and fire officers reference the size-up concern termed "occupancy," its definition can come with more than one specific meaning. In the text of this book we will consider its definition to mean the following:

- Occupancy Classification
- Occupant Load and Status
- Occupancy Content
- Occupancy/Construction Associations

Occupancy classification

Initially many will view the occupancy to mean the classification. For example buildings can be referenced as residential, commercial, mixed (combination of residential and commercial), institutional, educational, health care, and manufacturing (to name a few). Undoubtedly, each classification will present their own set of concerns, specifically as it would relate to the life hazard and the fire load.

Occupant load and status

The occupancy classifications mentioned previously can be used by the firefighter and fire officer to gather information about the occupant load and status. Questions that must be researched or considered in this category include:

- How many people are in the building at this time?
- Are they young or old?
- Are they ambulatory or non-ambulatory?
- Are they awake or asleep?

Although this category is cross-referenced in varying forms in our life hazard size-up, it allows the fire department to place additional focus on those individuals not immediately affected by the location and extent of the fire.

Occupancy content

The occupancy of a building can also refer to the contents of the structure–specifically what the building houses. Considerations by the fire officer during a pre-plan size-up or the on-scene size-up should include the following information.

1. Are there hazardous materials in this building? If so, what is their
   - reactivity; together or with water?
   - flammability/combustibility?
   - radioactivity?
   - health concerns?
   - specific hazards?
2. What is the building's fire load? Is it light, medium, or heavy?
3. Are there any specific stock concerns within the building? Examples might include these:
   - Plumbing supply warehouse – Increased collapse potential
   - Carpet and rug store – Hydrogen Cyanide gas from content involvement
   - Toy stores – Hydrogen Chloride gas from plastic involvement
4. Is there any water absorbent stock within the building? Concerns might include these:
   - Increased load concerns within the building
   - Stock swelling could push out walls
5. What is the content value? Examples might include these:
   - Jewelry shops
   - Camera shops
   - Computer stores
   - Museums

Occupancy/construction associations

We know that certain types of commercial occupancies require large open areas simply due to the nature of their business. Structures that house supermarkets, auto dealerships, furniture showrooms, bowling alleys, and movie theaters will all require large open areas for their business needs. This type of occupancy indicates to the firefighter and fire officer that the structure will most likely be constructed with a truss roof assembly and possibly a truss floor assembly.

Fig. 1-18 Movie theater viewed from the "A" side.

Other than being able to view the classic hump associated with the "bowstring truss," officers will be able to sense their presence from this association allowing for a calculated, as well as cautious, approach.

Fig. 1-19 *Same movie theater viewed from the "D" side.*

Apparatus and Staffing: Size-Up

COAL TWAS WEALTHS

When we reference the size-up factor Apparatus and Staffing, a number of areas can be considered to assist with the fire officer's decision-making. In this chapter we will take a brief look at them, and then add additional considerations in the chapters that follow as they relate to engine and ladder company operations.

Staffing

This continues to be one of the controversial issues plaguing today's fire service. When I was doing my research for this text, regardless of the town or city I reviewed, all firefighters complained about the lack of staffing in their respective departments. This continues to be a sad fact of reality. People who don't know or respect the fire service want us to do more with less. This continues to be a dangerous concept that simply doesn't work!

The concerns in this area seem to be no better in the volunteer fire service. As a former member of a volunteer department, staffing availability would always vary between the day and the evening hours. During the workday, there were only a few members who would be able to respond. In the evening hours, we would often have more volunteer members respond than the apparatus could seat. For many volunteer fire departments this is changing. It was quickly becoming obvious for many departments that I was exposed to that staffing shortages were not only becoming critical during the workday, but were also becoming critical in the evening hours as well. This stemmed from the simple lack of people within the community who were able to volunteer their time. Whether it was because of the demands of an individual's personal or professional life, the time required for training, or a simple lack of interest, the staffing of the volunteer fire services across the country is at an all time low. In an attempt to combat this concern, many towns and states are offering monetary incentives in the form of state tax reductions, as well as monthly stipends to increase membership in the volunteer service, but the incentives are being met with limited success.

Is there an answer to this concern in the career fire service? Sure there is; employ more people. The benefits far outweigh the costs. Barring the most obvious and the safest answer, what we are seeing is that many departments continually and increasingly rely on mutual aid from neighboring fire departments. This will undoubtedly provide you with additional staffing, but not during the first few critical minutes of the incident, when the staffing size of your response matters the most.

Response compliment and capabilities

This category within our apparatus and staffing size-up directs our attention to the number and type of apparatus responding to an incident, as well as those same unit's capabilities. Information gathered from this area will allow the incident commander to consider deployment and assignment for those companies, based on identification of these specific areas.

Number and capability. Knowing initial information such as the number of apparatus on the assignment may not only provide information specific to the amount of staffing responding to your incident, but it can also identify the resource capabilities of the individual units responding as well.

Let's take a look at what I mean by this. As an example, an engine company responding to your incident capable of delivering 1000gpm of water is very different in comparison to another engine company capable of delivering 2000gpm of water. This is information you would want to know, especially if you're looking to assign one of them to a water source at the incident.

Resource typing. Expanding on that idea, a company referred to as a "ladder company" is much different from another company referred to as a "tower ladder company," "ladder tower company," or "snorkel company". To identify the differences, a "ladder company" is most commonly associated with an aerial ladder of a specific height. A "tower ladder" company has a telescoping boom with a bucket or platform on the end of it. A "ladder tower" has both an aerial ladder that can be climbed as well as a bucket or platform on the end of it. A "snorkel" has a platform on the end of an articulating boom. Each one can handle specific and defined tasks if you know what they are, as well as their capabilities. In addition we have the following:

- *Quints* – fire apparatus equipped with an aerial ladder as well as a pump, hoses, tank, and compliment of ground ladders.
- *Quad* – a quadruple combination fire apparatus that carries a water tank, compliment of hose, pump, ladders, and the equipment associated with both an engine and a ladder company.
- *Squirt* – an engine company that also has an articulated or telescoping boom used for water application.
- *Tanker/Tender* – an apparatus designed for the transport of water to fires. It usually carries a minimum of 1000 gallons, and may or may not have a pump on board.

Fig. 1-20 *A ladder company provides an uninterrupted flow of people and equipment to a designated area.*

The simple point that needs to be made here is knowing what the apparatus and its personnel can do, by simply calling it what it is. In many departments they refer to this as "resource typing". One of the most commonly used fire department slang terms that forced many fire departments to resource type was the reference to a truck company. The term "truck company" could identify a company that is either an aerial ladder, tower ladder, ladder tower, or snorkel. A chief officer calling for Truck Company 10 because of its capabilities as a tower ladder is all well and good, if that is what he is going to get every time it's needed.

As an example, in Jersey City, all of the city's fleet of aerial and tower ladder companies were once all referenced as truck companies. The department made the decision a number of years ago to resource-type its fleet of apparatus to eliminate this term for a number of reasons. One simple and quickly realized reason came from the transmission of an alarm. If the chief officer read the alarm printout, or heard the radio transmission stating that Ladder Company 12 and Tower Ladder 4 were assigned to his incident, he would immediately know what his responding ladder company capabilities were going to be.

If, on the other hand, he read or heard that Ladder Company 12 and Ladder Company 4 were responding to the incident, he would know based on the resource typing policy used within the department, that Ladder Company 4, normally a tower ladder company, must be using a spare/reserve aerial ladder. This identification change would immediately tell him that there are no tower ladder apparatus responding to that particular incident; therefore, if he wanted one, he would need to specifically request one.

Fig. 1-21 *A tower ladder company was primarily designed to provide for an elevated stream.*

This is a beneficial piece of information, especially when different types of fire apparatus are capable of doing different jobs. If you always call it what it is, the name will tell you what it can do. Otherwise, you have to hope they're sending you something you can use.

Engine and ladder company operations

Although different departments may reference and use different pieces of apparatus as well as have varying staffing levels, subsequent chapters within this book will present operational considerations for both the first arriving engine and ladder company. The suggested tactics that will be presented in each chapter are based on the building and occupancy described, as well as the tasks researched and currently used.

Life Hazard: Size-Up

*COA**L** TWAS WEALTHS*

This size-up factor without question is the most important within the fire officer's size-up. At an emergency incident, it will be influenced by such factors as the time of the day, the occupancy of the building, as well as the location and extent of the fire. To focus our concerns within these factors it is important that we reference the different groups of individuals involved in this aspect of size-up.

Firefighters

Your first obligation as a fire officer is to protect the men and women you are responsible for. A firefighter's dedication to the profession, the citizens they have sworn to protect, as well as their commitment to each other, is stronger than in any other profession. Lives count on your education, training, and decision-making. It is important to always remember that.

Firefighters and fire officers have focused on the safety and rescuing of others for so many years, that many indirectly develop tunnel vision, which could lead to overlooking their own safety by not being aware of the dangers that are present around them. Through the continued use of incident management, firefighter accountability systems, officer training programs, and back-to-the-basics programs, etc., the fire service has

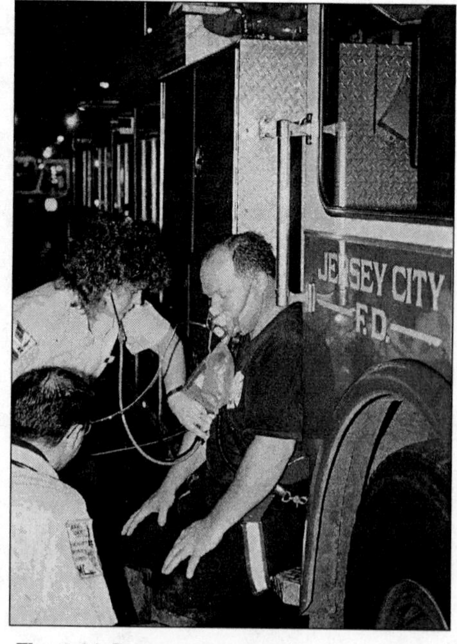

***Fig. 1-22** Protect your people.*

become better prepared to help our people work safely at an incident. Still newer technology in the areas of personnel-protective equipment, self-contained breathing apparatus, and thermal imaging have increased our ability to provide better equipment for our members while operating at an incident. These advances alone are not enough. It is all of our responsibilities, especially the fire officers, to continue to train and educate ourselves in any area that can impact on a firefighter's safety. The worst thing members of the fire service can do after receiving a promotion is to think that their education and learning is now complete. In reality, it has just begun. Your new responsibility requires that you continually add to what you already know.

Occupants

A second group of individuals that we will focus on are the occupants. This group of people includes those individuals within the fire building, as well those within any exposure building. Their inability to exit a building involved in fire requires that the fire department be trained to rescue them. Our ability to safely remove them requires sound decision-making by the firefighter and excellent leadership by the fire officer. Uncoordinated or unplanned attempts will compound an incident. This may not only limit your abilities to save a trapped individual, but it may also affect the life safety of the firefighters involved.

The location and extent of the fire and the building's areas of greatest danger will influence initial occupant life hazard considerations within a building. Obviously people who are trapped above an out-of-control fire are going to be in more danger than those that are below it. Depending upon the building's construction, its size and area, combined with the behavior of the fire, the fire officer may have to consider additional factors.

Our study of fire behavior has always reminded us of how fire travels. Its ability to travel will be influenced by its size, temperature, and any barriers or obstructions within the building. These factors have to be considered when identifying the areas of greatest danger in a building.

Areas of greatest danger in a building

- The fire floor
- The floor above the fire
- The top floor
- Floors in-between
- Floors below the fire

The same factors that we use in identifying the building's areas of greatest danger will also require that specific decisions be made pertaining to the occupants within the building(s).

Occupant life hazard concerns

- The number – How many occupants do we have?
- Their location – Where are they in relation to the fire?
- Their condition – How are they? Can any of them self-evacuate?
- Accessibility – Can we get to them?
- Resources – Are resources assigned and coordinated in their search and removal?
- Care – Are we providing for their treatment and transportation

These questions that may appear simple in thought are often difficult to get answers to while standing in the street. Nevertheless they must be addressed at every incident that we respond to where occupants are involved.

Bystander

A third group of individuals that must be considered in our life hazard size-up are the bystanders. This category would encompass anyone near the fire ground perimeter. People who often fall into this category would include the spectator or the curious onlooker, the media, as well as the fire photographer or fire buff.

Members of the media have been known to be respectful of the boundaries of a fire line or fire ground perimeter. However, there has been an occasion where some of these individuals have been known to ignore fire department restrictions in order to get their story. It is here where the strong and vivid presence of the public information officer has proven valuable.

A more aggressive group, second in nature to the actual firefighters is the fire photographers and the fire buffs; a valuable group of people in the fire service. Their enthusiasm and energy is the driving force for many departments. However, watch for the occasional photographer looking for that unique picture. You may be surprised at the places where some of these individuals will show up.

Other emergency service personnel

The last group identified within our life hazard size-up is referred to as "other emergency service personnel". These people are those who are part of any other agency or service operating in or near the incident scene. Examples of these people include:

- Police Department
- Emergency Medical Service
- Coast Guard
- Hazardous Materials Teams
- Arson Squad
- Marine Units
- Mask/Air Service Units
- Technical Specialists
- Other

Many times these agencies and/or services will have an active role at your incident. If they do, you must be aware of them, account for them, and ensure their safety.

Terrain: Size-Up

COAL TWAS WEALTHS

This size-up factor refers to the topography and/or obstructions that might interfere, delay, or cause concern to fire department operations. By the simple definition presented, this can be considered as a vast area of concern. The first thing we need to do is identify those areas, and then describe how they may impact fire department operations.

Terrain Categories include these:

- Setbacks
- Buildings built on a grade
- General accessibility

Setbacks

As we look at this first category, we can identify and place two similar areas together to emphasize a shared concern. Buildings that are set back from the street or from accessible areas, as well as buildings surrounded by and set back by decorative landscaping, will equally affect fire department operations. Although different in appearance, they both have the same impact on the fire officer's decision-making.

The unique concern with this category is that it could encompass all types of occupancies, with all types of construction. From private dwellings with long winding driveways, to commercial or residential high-rise complexes surrounded by decorative fountains and shrubbery, the effects will be the same.

Concerns that should quickly come to mind with your setback size-up is the immediate effect on apparatus placement and the company's initial operations. As an example, the engine company officer will initially be concerned with hoseline selection, stretch, and place-

Fig. 1-23 *Building set back from the street.*

ment in these situations. Buildings that are easily accessible from the street may only require a hose stretch of three to four lengths to reach the seat of the fire. On the other hand, buildings that are significantly set back from a street or accessible area may require a hose stretch of nine or ten lengths in order to accomplish the same objective. In this example not only does your hose stretch need to be longer, but also depending upon the size hoseline chosen, you may have the added concern of friction loss from the smaller hose line.

Ladder companies are not going to be exempt from a setback size-up either. Ladder company officers will also be concerned with apparatus placement relating to their aerial's or bucket's ability to reach a target spot on the building. Buildings that cannot be reached by an aerial device may require the use of additional ground ladders, which, depending upon the setback, may need to be walked in from extended distances. Decorative landscaping and sloped terrain around a structure will also cause havoc to ladder company personnel who are attempting to place ground ladders.

By what seems aesthetic to some becomes a major concern for the firefighter and fire officer when operating at an incident scene.

Buildings built on a grade

Your terrain concerns increase significantly when operating at buildings that are built on a grade. Structures with this concern could cause serious problems with your fireground management, as well as cause increased life hazard concerns to the firefighter if not preplanned.

With structures that are built on a grade, buildings that may appear to be two stories from the street address side may in fact be three to four stories from the rear. Firefighters who enter from the rear may have a different thought on floor designations as compared to firefighters who only view and enter from the front. This can turn into a logistical night-

Fig. 1-24 *Building viewed from side "A".*

Fig. 1-25 *Building viewed from side "D".*

mare especially when a firefighter reports that he/she is trapped and needs help on floor number three, only to find that firefighters were misdirected in their rescue attempts. If you have buildings in your town or city that are built on grade, it is critical that this piece of size-up information be discussed and planned for prior to the receipt of an alarm. In Jersey City, firefighters and fireofficers pay specific attention to the initial radio report where the first due company is required to give a brief on the scene size-up that includes, the address, the *height* of the building, the class of construction, the occupancy type, any conditions showing, and the company that is establishing command. If upon entering from the rear, there is a difference in the number of floors designated from the initial radio report, the incident commander must be alerted to this grade size-up difference in order to designate any additional areas.

Fig. 1-26 *Look up!*

General accessibility

The final terrain category is defined as "general accessibility". This is a very broad term and can be viewed to mean practically anything that is in the way. Accessibility obstructions at grade level could include parking garages, fences, waterways, etc. Accessibility obstructions from the overhead area could include trees limbs, electrical lines, cables, and telephone wires. The concerns for both engine and ladder company in this category are universal; to find the safest and easiest way around this.

Water Supply: Size-Up

COAL TWAS WEALTHS

This size-up concern will refer to what is available for fire department use, what is needed or required to extinguish the fire, as well as how it is to be delivered to the fire scene, and then to the fire.

Water availability

Water availability will obviously vary from location to location. Urban and suburban areas usually rely upon domestic water supplies from reservoirs through a series of underground piping known as a grid system, before becoming available at a fire hydrant. In addition, some of these same areas may have the added ability to use natural water sources such as lakes, rivers, and streams for drafting purposes.

Rural areas normally rely upon natural sources of water supply as their primary means of extinguishment, and must compensate for this limited access by having engine companies or water tankers/tenders with large water supply capabilities. Whatever the specific situation, water availability concerns in your area or jurisdiction must be viewed well before the alarm is received. It is here where we need to know such specific information as the location of any draft sites, the quantities available, the main sizes, hydrant locations, and their available flows for efficient fire ground operations.

Waiting until you receive the alarm to "seek" out an adequate supply of water is going to allow the fire to extend. Fire will wait for nobody!

Fig. 1-27 *The fire department that services this area can take advantage of the lake water supply.*

Water needed

When we reference water needed for a specific incident, we need to identify factors that will influence our decision-making. The most common factors include these:

- The location and extent of fire.
- The class of construction of the fire building.
- The contents/fire load of the fire building.
- The height and area of the fire building.
- The proximity of exposures, including their construction and features, their contents/fire load, and their height and area.

Many of us throughout the years have grown up on the old premise of "Big fire = Big water". This old saying still holds true today, but there is much more fire personnel need in order to deliver efficient and effective flows. One of the most common ways of determining what will be needed is to pre-plan a specific building or target hazard and obtain information pertaining to the building's square footage and its fire load. By determining the type of fire load and the area's square footage, the fire department can begin to determine the minimum flows they will need to extinguish a particular type and size of fire.

N.F.P.A. FIRE FLOW FORMULAS

10gpm per 100ft^2 for a Light Fire Load

20gpm per 100ft^2 for an Ordinary Fire Load

30-50gpm per 100ft^2 for a High Fire Load

Water delivery

As we review how water can be delivered, remember this reference will also vary from location to location and from department to department. Water delivery to the fire scenes in rural areas may require tanker/tender water shuttle operations as well as delivery from the air in difficult to reach locations. Urban and suburban areas may vary from the concepts of in-line pumping by bringing a water supply from a hydrant to a pumper, while others may use the concept of positive pumping which places a pumper on a hydrant and pushes water to the fire. Whatever option you choose, once a water supply is established you're only half way there. You need to now figure out how to get as much of it as you can on to the fire. What it all boils down to is that water delivery to the fire is based upon:

- Selecting the right size diameter hoseline to handle the job.
- Placing the right type of nozzle on the end to give the best flow.
- Stretching the hoseline to its desired location the quickest, easiest, and shortest way possible.
- Supplying the pressure to meet the required flows.
- Hopefully having enough people to handle the task.

Suggested Pump Pressures For Needed GPM Flows For 1¾" Hose:

G.P.M.	Pump Pressures/200 ft.	Pump Pressures/250 ft.
150	120 SB/170 Fog	135 SB/185 Fog
175	145 SB/195 Fog	165 SB/215 Fog
200	170 SB/220 Fog	200 SB/250 Fog

2" Hose:

G.P.M.	Pump Pressures/200 ft.	Pump Pressures/250 ft.
200	110 SB/160 Fog	125 SB/175 Fog
225	120 SB/170 Fog	140 SB/190 Fog
250	135 SB/185 Fog	160 SB/210 Fog

2½" Hose:

G.P.M.	Pump Pressures/200 ft.	Pump Pressures/250 ft.
225	70 SB/120 Fog	75 SB/125 Fog
250	75 SB/125 Fog	80 SB/130 Fog
275	80 SB/130 Fog	90 SB/140 Fog

Note: SB denotes smooth bore nozzle. Fog denotes fog nozzle.

(Pressures indicated above may vary depending upon fog nozzle design.)

Whatever the situation, it is the fire department's primary responsibility to find the water and determine the needed amount for a specific incident. Then it becomes the responsibility of the engine company officer to ensure it gets where it's supposed to go.

Auxiliary Appliances and Aides: Size-Up

COAL TWAS WEALTHS

This section will review those devices, equipment, and people that could aid the fire department in their efforts. When any information in this particular size-up category is available at a particular building or complex, it should be identified and reviewed by the fire department in order to be fully utilized. They can range from detection devices and suppression equipment, to on-site personnel educated and trained to assist the fire department. Their presence and expertise must be identified and used whenever possible.

Detection equipment

Fire detection systems installed in a building can be numerous in design and in type. By attempting to identify just some of the more common types found, we will be able to get a better understanding of their design, as well as their function. Equipment designed to detect the products of combustion will do so by sensing smoke, heat, or flame.

Smoke detection. Smoke detection equipment and systems can be presented in a variety of devices. The most common methods used are *ionization*, *photoelectric type*, or *projected beam*.

Ionization detectors are designed to sense the invisible products of combustion, sensing a fire in its earliest stage. Photoelectric detectors are designed to sample the visible products of combustion (e.g., smoke particles), indicating a more advanced condition at the sounding of the alarm. Projected beam detectors are similar in design to photoelectric detectors. In this design the detector projects a beam across an open space from a sending device to a receiving device. If smoke penetrates and interrupts the beam, a signal is sent indicating detection.

Heat detection. Heat detectors are the oldest type of automatic fire detection. In addition to being the oldest they are also known to have the lowest false alarm rate of all automatic fire detectors. Unfortunately they are the slowest in detecting fires.

The two most common types are the *rate-of-rise detector* and the *fixed temperature detector*. A rate-of-rise detection system will signal when the temperature in the room or area exceeds a predetermined level, typically around 15 degrees per minute. Fixed temperature detectors are designed to signal when the temperature of the operating element reaches a predetermined number.

Flame detection. Flame detection equipment is designed to sense radiant, infrared, or ultraviolet energy. Their presence will be particular to a certain type of occupancy and its operation. They can be designed to respond to the appearance of radiant energy visible to the human eye, or they can be as sensitive in detecting radiant energy outside the range of human vision. These detectors can be designed to sense glowing embers, coals, or actual flames.

Suppression equipment

Equipment that is designed to aid us in our suppression efforts will generally fall into three types of systems:

1. Sprinkler systems
2. Standpipe systems
3. Special extinguishing systems

Sprinkler systems. Sprinkler systems can be either the wet, dry, pre-action, or deluge type. Use of a building sprinkler system has always been a tremendous asset for firefighters. Having pre-incident information about the system greatly enhances the fire department's ability to fully use the system in an efficient and effective manner. Information the fire officer would like to have prior to the receipt of the alarm, or at the very least prior to arrival on scene, should include the type of system or system used within the building, the location of the fire department connections, and the zones that each will serve. Sprinkler systems have an outstanding performance record for the extinguishment, and at the very least, containment and control of the fire until the arrival of the fire department. The few times that a sprinkler system fails in its suppression efforts, it is normally traced back to human error.

Reasons sprinkler systems fail

- Blocked or damaged fire department connections
- Blocked or obstructed sprinkler heads
- Closed zone and control valves
- Premature closing of zone and control valves
- Failure or inadequate supply to the proper zone or system
- Improper design of the system
- Improper maintenance of the system
- Excessive pressures supplied to the system
- Vandalism

Standpipe systems. Standpipe systems can be categorized as either wet or dry systems with classes ranging in numbers from one to three. Here again, the fire officer needs to know the type of system, specifically if it's wet or dry, the class of system, the location of fire department connections, and the location of zone or control valves.

Standpipe class systems

Class 1 Standpipe – A system designed for fire department use with 2½" hose connections for attack of the fire.

Class 2 Standpipe – A system designed for occupant use only. This system will have a 1½" hose (house line) and nozzle.

Class 3 Standpipe – A system designed for both occupant and fire department use. This system will have both 2½"and 1½" hose fittings for either operation.

Specialized extinguishing equipment

Specialized extinguishing equipment includes those devices that are capable of extinguishing a specific type of fire within a certain type of occupancy. Common extinguishing devices found might include:

- Fixed dry chemical or wet chemical systems found in kitchen hoods in restaurants
- Halon or other clean agent systems found in areas that house delicate electronic equipment
- Fixed foam systems found in flammable liquid occupancies
- Fixed total flooding systems of carbon dioxide for hazards such as high voltage electric transformers

Examples of some of the inherent concerns when responding and operating in these areas might include oxygen depletion from total flooding systems, toxic environments from specific extinguishing agents, as well as member disorientation from total flooding. Pre-incident information about special extinguishing equipment and their affected areas will provide those members responding with enough knowledge to operate more efficiently, and more importantly, to operate safely.

Aides/assistants

A valuable resource that should be noted as an auxiliary aid is those individuals who could provide information and/or assistance at a specific incident. Knowledge of these individuals in many of the instances should be sought out well before the receipt of the alarm, or at least immediately upon our arrival.

Fig. 1-28 *Officers should seek information from on-site personnel.*

Examples of these individuals are these:

- a safety plant manager at a chemical manufacturing facility with significant, useful information pertaining to the complex and its operation.
- a building engineer at a commercial high-rise fire who could pressurize remote HVA.C zones to retard smoke spread.
- the on-site fire brigade or hazardous material response team that could assist with extinguishment or product control.Or they can be as ordinary as:
- a building superintendent at a multiple dwelling attempting to aid you in resetting a fire alarm.

There are many civilian resources available to us. We must remember that we are not the initial on-scene experts at every incident that we respond to. Sometimes you need to check your ego at the door. When available, tap these resources, and tap them early.

Street Conditions: Size-Up

*COAL TWA**S** WEALTHS*

In this section, conditions will be identified that could affect apparatus movement, placement, and operation.

This size-up factor also will vary from city to city and from town to town. However, it seems that the older the community, the more there is an increased awareness of this particular size-up factor.

Street width

As older, congested cities attempted to meet their housing demands, they were forced to continually construct buildings on streets that were originally designed for the horse and buggy. These same narrow streets that lined many of our cities are still there today, but they now contain an asphalt surface, sidewalks for individuals to traverse, and are traveled by vehicles of varying size.

Fig. 1-29 *Narrow car-lined streets will restrict apparatus movement.*

In many older cities with this concern firefighters and fire officers know that there may be streets where they must make sure that a tractor-drawn aerial ladder is on the assignment if they want a ladder company in front of the fire building. Even in certain areas where streets are wide enough for a rear-mount aerial or tower ladder to operate, one car parked on the corner, or one double-parked car will delay, or eliminate critical ladder company placement.

Fig. 1-30 *Narrow car-lined streets will restrict apparatus placement.*

Traffic flow

Two-way traffic in many cities is a luxury only associated with major thoroughfares. Most streets that run perpendicular to these main arteries of traffic have been designed as one-way streets. Besides the concern of a single lane of traffic, many departments deem it critical that the initial responding fire department units proceed into the street with the flow of traffic. The obvious concern with this is to eliminate apparatus coming in from two directions, possibly pinning civilian vehicles in the middle of the street, even in front of the fire building. An additional concern is ensuring critical placement of your first due engine and ladder company in what is anticipated to be a difficult and congested area to operate in.

Ignorance and complacency with this particular size-up factor will guarantee eventual confusion, as well as a crucial delay in your operations, all of which could easily be eliminated with some direction. (Chapters 2 and 3 will give suggested guideline considerations for this size-up concern. Take a look.)

Street surface

Another factor that we must consider within our street size-up is the actual condition of the street. Not all streets that we will be operating on are asphalt or concrete. Conditions that many are forced to deal with include dirt, stone, and gravel. Immediate concerns with a less-than-adequate street surface would include

Fig. 1-30a *Unstable surfaces will require the apparatus to be repositioned.*

water accumulation from the weather, as well as water runoff from fire department operations. As water penetrates the surface area you are working on, concerns with the apparatus sinking as well as with their stability will require your immediate attention.

Fig. 1-30b *Marine company?*

Unusual

Streets affected by difficult or unusual weather conditions will become a major concern with movement, placement, as well as operation of firefighters and equipment. When encountering heavy snow, ice, mud, or flooded streets, the delay concerns become increasingly worse.

Geographical areas prone to heavy snows, mudslides, and water accumulation must be given early consideration with this size-up factor when dangerous and difficult weather is forecast.

Whatever street conditions your department might encounter, if their difficulties are already known or anticipated, members must be prepared to deal with them.

Weather: Size-Up

*COAL TWAS **W**EALTHS*

Weather size-up refers to those elements that can affect fire department response and fire department operations.

Weather is one of the most difficult size-up factors for us to forecast. This simply stems from the fact that it is often unpredictable, and has been known to change hourly. What may have started out as a clear, calm, sunny day, can turn into a windy, snowy day within a few short hours. Modern forecasting equipment has allowed the weather service to become more accurate with their predictions, nevertheless, the weather's unpredictability requires us to review universal concerns regarding how it may impact fire department operations, as well as fire department members.

Wind

Wind continues to cause great difficulties for firefighters operating at any type of fire. Whether fighting a wildland fire in the Midwest, or a structure fire in an inner city, the effects of the wind will seriously hinder the fire department's progress.

What may have started out as a small brush fire in a remote wooded area, can produce blow up fire conditions that can outrun and trap a firefighter once introduced to a significant wind. What may have also appeared to be a small, manageable building fire on arrival, can quickly escalate into a major incident, with that fire's embers causing other fire incidents downwind.

When we gather information about the forecasted wind condition, winds that gust or become sustained at 10 mph should be classed by the fire department as significant, and should become a concern. Wind even at low velocities, can have a major impact on the fire department's operations. From directly affecting the fire conditions within a building, your ability to ventilate a building, as well as hose stream reach and penetration, there should be no doubt in anyone's mind how this particular weather event can affect your operations.

Fig. 1-31 *A wind-fed fire*

Temperature

Temperature is a universal, as well as a serious, concern to all firefighters regardless of the area they are assigned to protect. The emphasis behind this statement is directly related to our most important incident priority; the life safety of the firefighter.

Cold. Temperatures that fall below 40°, or rise above 85° warrant early consideration. When temperatures fall below the 40°-mark the concern for firefighter injury increases. Cold temperatures cause firefighters to fatigue early and will also promote conditions conducive to slips and falls. Long durations in cold temperatures will promote frostbite, as well as dull the senses of a firefighter affecting an individual's ability to think clearly. When temperatures drop, early attention must be given to relief and rehabilitation of operating forces. The number of resources needed to fight a fire in the cold weather usually doubles for the reasons mentioned.

The same concerns extend to the building's occupants as well. Anytime we have a fire in an occupied building during cold weather periods, incident commanders must provide for displacement and temporary relocation of the building's occupants. This assistance may come from additional resources such as the Red Cross, to obtaining temporary relocation in public buildings as churches and schools, as well as bringing shelter to the site in the form of heated transit buses and trains. Whatever the options, the cold weather will force you to explore them.

Cold temperatures will also affect your operations from delays due to frozen hydrants, to possibly ineffective tools and general equipment. Also during the cold weather months we often encounter advanced fire conditions from delays in actually reporting a fire. In extreme cold, people aren't just out and about like they would be during the warmer weather months. People will be inside attempting to stay warm with windows closed to conserve heat, therefore eliminating individuals' ability to see as well as smell smoke early. Extreme cold can add to this delay by creating smoke conditions that appear white in color, actually giving the appearance of steam from a distance. For some of the reasons mentioned, fires often go unnoticed for longer periods of time in colder weather. Usually by the time we get to it, it's going good.

Fig. 1-32 *High temperatures will require early and frequent reliefs.*

Hot. Temperatures that climb above the 85°-mark must also be viewed as dangerous to operating members. Heat exhaustion and heat stroke at this point become another serious concern. Incidents that occur on hot weather days require early and frequent reliefs. During periods of hot weather, incident commanders should not hesitate to call for additional alarms to relieve their people. The point that must be gathered here is, firefighters exposed to extreme temperatures, whether hot or cold, will require incident commanders to pay close attention to their people.

Humidity

Reliefs. Humidity is defined as the amount of atmospheric moisture in the air. Our initial concern with humidity refers to the life safety of the firefighter. High humidity periods result in higher heat indexes, again causing immediate concerns with heat exhaustion and heat stroke. Incidents that occur on days that exhibit this type of condition will also require early and frequent relief and rehabilitation of your firefighters.

Fire spread. Humidity can also affect our fireground operations. Concerns with fire spread during low humidity periods become a much greater concern. Low humidity in the atmosphere results in lower moisture contents, therefore much dryer combustibles. Combustibles, once ignited, allow fire to spread at an accelerated rate of speed.

Ventilation. Fire ground ventilation efforts can also be hampered by high humidity periods. High atmospheric moisture during a summer evening has often caused the products of combustion to become cooled to a point were they would loose their buoyancy, and often at times stay close to the ground. Not only do these types of conditions prevent you from initially finding the actual building that is on fire, but also a lack of visibility in the street or on the roof will cause a safety concern to all firefighters operating. Undoubtedly this will also affect the fire officer's ability to accurately size-up the fire building.

With high humidity conditions, expect fires to be smoky and difficult.

Precipitation

Our last weather concern is always the most obvious, precipitation. Precipitation can be a major influencing factor with fire department response and fireground operations. Heavy snow, rain, and fog will increase the fire department's response times, possibly even limiting apparatus getting to the fire building at all. In heavy snows, many departments send a larger compliment of apparatus and firefighters on the initial alarm with the hope that somebody can make it down heavy snow-covered streets, and for those that can't, they will be able to assist with long hose stretches and equipment that may need be walked in.

Exposures: Size-Up

COAL TWAS WEALTHS

This size-up factor refers to those areas that surround the main body of fire. Exposures can be categorized into two reference areas: *interior exposures* and *exterior exposures*. Interior exposures are those areas that surround the fire area or fire floor. They are identified from the avenue and direction of fire travel from within the building. Direction of fire travel is referenced from the fire's ability to move vertically up through the building, horizontally left or right from the point of origin, as well as any areas below where the fire may be able to drop down. It is from this simple geometric thought that we gather the association that all fires have six sides, and depending on the possible avenues of fire travel within a building, all six sides must be examined for potential extension. Concerns more specific and identifiable to this size-up category are further diagnosed in the "Fire Location and Extent" section of this chapter, and will become more occupancy specific in the chapters that follow.

Fig. 1-33 *Exterior exposure protection*

Exterior exposures are a major concern for many fire departments across the country.

Congested urban and suburban areas must deal with buildings that are literally built right alongside one another. With buildings that are attached to each other, or those that are only separated by a few feet, extension into adjoining or nearby structures has to be considered as an early possibility. In these settings, hesitation can allow fire not only to occupy one, but three buildings in a short amount of time.

In order for incident commanders to direct efficient and effective fire ground operations, officers must identify and prioritize protection of exposure buildings from a number of factors. Those factors should include these:

1. *The life hazard.* Specifically those nearest. Buildings must be prioritized based on "the most severely threatened exposure" verses "the most severely threatened life exposure". There is a difference.
2. *The location and extent of the fire.* The current fire involvement with its extension probability and possibility.
3. *Exposure proximity.* Is it attached or separated, and by how much?
4. *Fire building construction and features.* Examples are exterior sheathing, light and air shafts between the buildings, common cocklofts, cornices, etc.
5. *Exposure building construction and features.* Examples are exterior sheathing, light and air shafts between the buildings, common cocklofts, cornices, etc.
6. *Wind.* Consider its speed and direction.

Exposure/life hazard

Exterior exposure concerns must immediately focus on determining the "most severely threatened exposure" verses the possibility of the "most severely threatened life exposure". The most severely exposed building may not necessarily be the one that receives the incident commander's initial attention. Initial courses of action must be based on any threatened life hazard to nearby or attached properties. The best way for me to identify this concern is to give you an example.

Let's say you arrive at a well-involved structure severely exposing a nearby vacant, two-story frame building on the west side, and a closely spaced two story, Class 5, occupied dwelling on the east side. Now even though the building to the west is referenced as being severely exposed, the occupied building on the east side would warrant the initial exposure protection simply based on the potential life hazard. As simple as this example is to identify with, everything that we do in this business must be based on the life hazard size-up. Additional resources can always be delegated as the incident commander sees fit, but your initial course of action is always based on the life hazard.

Exposure proximity/construction

Buildings that are attached or separated by narrow alleyways are immediate concerns for responding firefighters. Due to their proximity to each other, resources will have to be assigned to check and protect them from any fire extension. Factors that will further influ-

Fig. 1-34 The building's sheathing enhanced this exposures involvement.

ence this area of concern will be the construction or sheathing of the building, the presence of any inherent building features, namely light and air shafts, direct window lines, as well as inherent features in and around the building.

Buildings whose exterior sheathing is combustible can enhance the fire's spread from one building to another. Buildings covered in wood or asphalt siding will warrant earlier consideration than those made of brick or stone. On the other hand, buildings exposing each other with brick or masonry exteriors will still present an extension concern from any direct window lines. Fire venting out of one window of a building across an alleyway in line with a window of an exposure building, will be transmitting its radiant energy to the window and possibly through the glass to the interior curtains, draperies, and combustibles if not protected.

Changing the type of construction of the exposure building to a brick or masonry exterior will definitely buy you some time, but never underestimate the ability of a well-vented fire to transmit its energy to and through anything nearby.

Exposure designation

Exposure designation is a vital part of fireground management. Fire that extends from one structure to another will require you to assign resources and delegate responsibilities to those new areas. When assigning resources to work in other geographical

***Fig. 1-34a** Example of Exposure Designation*

areas, operations must be coordinated to provide for efficient, effective, and safe fireground operations. This coordination begins with the designation of exposures.

There are different forms of exposure designation being used out there. Whether your department uses letters or numbers, it should be simple and effective.

Area: Size-Up

*COAL TWAS WE**A**LTHS*

The term "area" is defined as the square footage involved in fire as well as the square footage threatened by fire and its byproducts.

By definition, this size-up factor will assist in determining the maximum potential fire area that the incident commander might have to contend with. When responding to any type of building, it becomes extremely beneficial to have any information about the layout, configuration, and square footage of the area where you may be operating in. Getting a look prior to the building being full of smoke and heat will not only help quell some of the anxieties and fears about crawling into the unknown, it will also help the incident commander anticipate his or her strategic and tactical needs.

There are five areas of concern in this particular size-up factor that the incident commander must receive information about when available.

1. *Irregular shaped buildings.* Buildings that do not follow a normal square or rectangle shape, and initially cannot be identified from the street. An example is a building that might wrap around an attached building, or a building of irregular shape and design.
2. *Irregular shaped areas.* Areas that do not follow a normal square or rectangle shape. An example is an L-shaped cellar often found in a row of attached stores.
3. *Interconnected buildings.* Buildings that share occupancy with an attached building; they also could be known as walk-throughs.
4. *Hidden areas.* These examples are the unknown

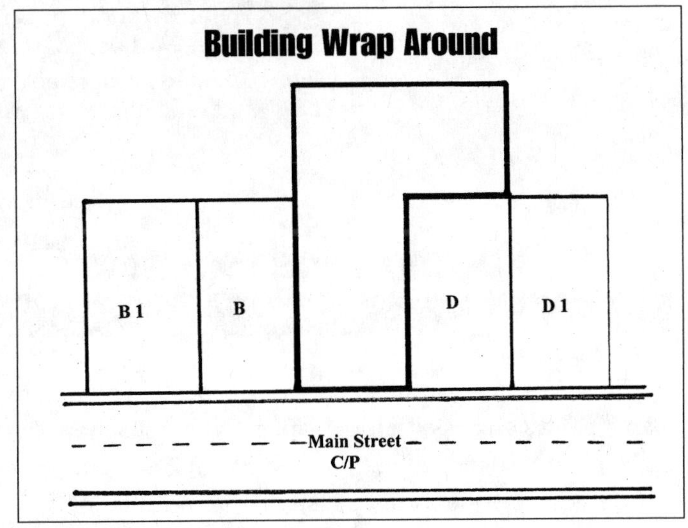

Fig. 1-35 *Area Size-Up Example*

rain roof, loft area, or building addition. Generally, any unknown or hidden area of significant size where fire could extend from or to.

5. *Overall square footage.* Information relating to the overall size.

Without the luxury of any pre-incident information, the incident commander must often rely upon what has often been termed "the eyes and ears of the fireground commander"–the ladder company. It is when this group of firefighters establish the roof position that (in addition to their assigned duties) they should also give a report based on their observation of the building layout, specifically reporting any irregular-shaped building, or a building addition.

Information about interior irregular-shaped areas, interconnected buildings, and located "hidden areas" will likely come from firefighters involved in either their search for fire or for life. Information that can be shared about the unknown will always help the fire officer rethink concerns specific to the life safety of the firefighter as well as to the stabilization of the incident.

As previously mentioned, the best way to gather information about the maximum potential fire area of a structure is through an aggressive "pre-incident information campaign". Having the ability to get into your districts and learn about the building's square footage, configuration, auxiliary appliances, or occupancy, are just a few of the critical elements of an effective, efficient, and safe size-up.

Fig. 1-36 *Unless you are able to access and view the rear of the building, this information should come from those assigned to the roof.*

Location and Extent of Fire: Size-Up

*COAL TWAS WEA**L**THS*

This size-up factor is defined as the location of the fire within the building, combined with the fire's extension probabilities and possibilities. The location and extent of the fire is the most influencing factor in the fire officer's size-up. It is from here that a fire officer will determine the severity of the life hazard, the initial resource needs for the incident, as well as the strategy and tactics needed to deploy them. As you view the location of a fire within the structure, your concerns can be categorized into four areas.

1. *Below grade fires* – basements, cellars, and sub-cellars.
2. *Lower-level fires* – areas immediately above grade.
3. *Top-floor, attic, and cockloft fires* – as listed.
4. *Upper-level fires* – those areas out of reach of fire department ladders.

Each of the four locations defined within this size-up factor will present specific thoughts that every firefighter and fire officer must be concerned with.

Below grade fires

The most dangerous and most difficult fire location in a building for the firefighter to attempt to work in, as well as for any occupant to be exposed to, is a fire that occurs in a below grade area. This association is put together from the fact that fire and smoke spread from a below grade area exposes everything and everyone above! Depending upon the class of construction and the building's features, this association can become more or less a concern, but undoubtedly it will be there.

Ventilation. Initially, it must be realized that any fire that is below grade will force life safety concerns to increase for any firefighters advancing and operating in this area. Just by knowing that the fire is below grade, we have to expect that we are going to be faced with a number of unique problems simply because of its location, the first being the lack of ventilation. Below-grade fires have limited, if any, ventilation options available to the firefighter. Limited or no ventilation of an involved below grade area will require advancing firefighters to literally crawl down into a chimney to seek the origin of the fire. This type of tactic is difficult and extremely exhausting even with the efforts of the best-coordinated ventilation options. Another concern specifically related to the lack of ventilation is the increased probability of carbon monoxide within the space. Areas that are below grade will prevent carbon monoxide from exiting the area. Dangerous levels of CO can accumulate and affect anyone without respiratory protection.

Fig. 1-37 *The most difficult and the most dangerous... the below grade fire.*

Fire loads. Once the members are operating below, they have to deal with the inherent concerns that are most often found in below grade areas. These areas are known to have increased fire loads from occupancy storage. Everything finds it way into the basement or cellar, filling the space from floor to ceiling. If the storage is as we described, the accumulation of combustibles in this space will also cause maze-like configurations that will increase concerns with firefighter disorientation. The large amounts of storage in these areas will also affect hose stream reach and penetration, possibly delaying early control and extinguishment. Below-grade areas of buildings are also known to contain the building's utilities, which when exposed and involved in fire will always create unique and difficult situations for operating forces.

Lower-level fires

The lower-level fire will for the most part come with many of the same concerns as the below grade fire. The concern of fire and smoke spread to everything and everyone above the fire floor still exists. These concerns will increase as the height of the building increases, as well as be influenced by the different classes of construction. We know that different classes of construction are more prone to interior fire and smoke spread than when compared to others. Avenues of fire travel can be numerous and inherent in a structure depending upon the construction methods used, the building design, building alterations, as well as by the building occupancy.

However, all lower floor fires in all classes of construction will come with significant concerns, regardless of how the building was constructed.

Top-floor, attic, and cockloft fires

The top-floor, attic, and cockloft fire will still present serious concerns to the firefighter, but somewhat less concern to the building's occupants. Obviously people trapped on the top floor, or in a possible attic apartment are definitely in harm's way. People still in the building, and below the fire area are, for the most part, out of direct danger and in many instances able to self-evacuate without the direct assistance of the fire department.

Concealed spaces. The firefighter, however, is never out of harm's way. Fires in attics and cocklofts not only present difficulties with access and control, but also encompass serious thoughts about fires in concealed spaces. It is after fire enters these voids, that the fire department reopens the space either ahead of or at the advancing fire to gain control and extinguishment. Dependent upon conditions in the space and the fire department's tactics, the possibility of a backdraft occurring in the space may become all too real. When conditions in a concealed space such as an attic or cockloft indicate

Fig. 1-38 *A top-floor fire example*

an advanced fire, it becomes the fire officer's responsibility to size-up these conditions and to direct the tasks safely. Introducing air into a concealed space from below may cause the ceiling to explode down, trapping unsuspecting firefighters in the middle of the room. Firefighters below must initially operate from refuge areas such as the public hallway or fire escape before moving into the room or area to expose the space. Firefighters from above must open up in an attempt to relieve pent-up gases. Walking into the middle of the room in an attempt to pull the ceiling from the light fixture because it provides easy access for your hook or pike pole, may find the ceiling between you and your exit.

As difficult as the fires will be from an access and control concern, fires in these spaces have injured and killed a number of firefighters from unsuspecting conditions. Just because you can't see it, doesn't mean it's not there.

Upper-level fires

Accessibility. Depending upon the height of the involved structure and our immediate means of access to these areas, the upper-level fire will continue to cause concern with overall fire ground operations. Fire department "Lead/Reflex times" (the time from when a resource is called to perform an objective to the time it starts to go to work on the objective) may be lengthy. A compliment of firefighters taking five, ten, or fifteen minutes to get to their assigned objective is not uncommon in tall structures.

There should be no doubt that upper-level fires will present the fire service not only with difficulties affecting the smoke spread, but their location will also greatly influence the time it takes to get a compliment of firefighters and equipment there to investigate it.

Fig. 1-39 *A typical high-rise*

Time: Size-Up

*COAL TWAS WEAL**T**HS*

Time refers to the specific time of the day the incident occurs, the day of the week the incident occurs, the time of the year the incident occurs, the burn time of the incident and all their influencing factors.

Time of the day and the day of the week

Both the time of the day and the day of the week are generally the first thoughts that we consider in this size-up category.

Occupants. The time of day is often referenced as the a.m. and p.m. hours of the day, and possibly further identified by some as the waking and sleeping hours of the day. The day of the week is identified by its reference to the calendar days, specifically Sunday through Saturday, as well as by the reference of the business day or business week, and the weekend.

There will be certain occupancies just by the nature of their business where we can factor increased or constant concerns with the numbers of occupants, regardless of the time of the day or the day of the week; (i.e., hospitals and nursing homes). Still there may be others where special events, social gatherings or additional working schedules may cause a building normally unoccupied during a particular time frame, to be filled with people.

The fire service must continue to realize that there is no patterned way of thinking for all buildings in this size-up factor. An attempt to fine-tune your decision-making using pre-incident information about the building in question will definitely lend some useful information; however, without any specific information, we have to expect the unexpected.

Traffic. Additional concerns with specific time intervals that may cause concern with fire department response and ultimately fire department operations will come from traffic patterns and traffic congestion. In many high commuter areas, firefighters and fire officers know that between the hours of 6 a.m. and 9 a.m., as well as between the hours of 4 p.m. and 7 p.m., certain streets and highways should be avoided due to "rush hour" traffic. In other more congested urban centers of the country, there seems to be no specific time frame for what always appears to be constant traffic. However, being able to at least consider, and possibly use, alternate routes to an incident may eliminate an excessive delay in getting where you're needed.

Time of the year

In addition to your time size-up concerns, you must also be aware of the time of year and how it may impact on your response and operation.

The time of the year and its related concerns can be influenced by the weather. As we know, weather specific to a certain time of the year, and certain regions of the country can play a key role in a fire department's ability to respond and operate at an incident scene, but there is at least one factor, regardless of the weather or region of the country, will be universal to every fire department across the country–the holiday season.

Fire/occupant loads. During the holiday season the fire service has to anticipate an increased fire load, an increased occupant load, as well as more of a concern with traffic than previously mentioned. In the commercial shopping areas retailers will stock their stores from floor to ceiling in an attempt to meet the demands of the holiday shopper. Increased stock will mean increased fire loads, possibility of blocked exits, as well as the probability of blocked sprinkler heads.

The number of holiday shoppers who could be in a store or mall at any given time can be staggering. This number will significantly increase as the holiday season approaches. Expect it to start in early November and end after the New Year, making all of the concerns identified for this particular time of year more demanding.

Fig. 1-40 *December 20th at 8 p.m. would see this store packed with shoppers.*

Burn time

Attempting to determine the burn time of a particular incident is going to be difficult. Not knowing how long the fire was burning prior to the receipt of the alarm is an initial concern that continues to plague fireground decision makers. Buildings with auxiliary protection will hopefully give us the first indication of a fire. Using this as a time reference can be questionable, especially if the fire originated in a void space, delaying an alarm transmission.

Once the alarm is transmitted, response times have to be considered. Response times can average from two to four minutes for some departments, with ten minutes or more for other departments depending upon distance and apparatus availability.

Attempting to put specific time frames on building collapse is dangerous and unrealistic today. There are too many variables and too many construction alterations that must force us away from this method of thinking. There is no rule of thumb that safely allows firefighters to operate in a building for a predetermined time. Our knowledge, education, and experience are going to be our best gauge of our operating time.

Height: Size-Up

*COAL TWAS WEALT**H**S*

This size-up factor refers to the height of the involved structure and its effect on the fire and fire department operations.

The height size-up factor accompanied with the square footage of the building are the two factors that the fireground commander uses to determine the maximum potential fire area of the building. As we give our initial radio report to dispatch, one of the first pieces of information that we relay to the fire dispatchers and others responding to the incident is the height of the structure. From here, we start to gather and compute information on how the height of this building will affect fire department operations.

Height accessibility

1. Are all areas of the building within reach of fire department ladders?
2. Are all areas of the building within reach of fire department hose streams?
3. Do we need to take the elevators? If so, can we use and control them?
4. Can we take the stairs? If so, how many are available, and what are their types and designs?

Some of the questions presented can be answered with size-up information gathered through your pre-incident survey. Others may be able to be referenced from preliminary radio reports of your assigned companies.

When height complicates your resource movement and operation, answers to some of the questions mentioned will definitely aid you in efficient and effective decision-making.

Fig. 1-41 *Initially, our thoughts focus on accessibility.*

Smoke behavior

The height of a structure could also affect fire and smoke behavior in a building, which in turn can impact fire department operations. Three unique and possible phenomena that are known to occur in tall structures are stack effect, reverse stack effect, and stratification.

Stack effect. Stack effect is known as the air and smoke movement in a building due to pressure differentials, which are caused by the temperature differences between the inside and outside of the building. When outside atmospheric air temperatures are cooler than air temperatures inside the building, you will encounter normal stack effect.

When we refer to stack effect, we find that it is not as unique a phenomenon as one might think. An everyday example of this would be how heat from a fire in a fireplace moves up through and out of a heated chimney. Similar in comparison with a building fire, this vertical movement of air and smoke is carried up through a stairwell or a shaft via the convection currents from the heat of the fire. The concern to the fire service with this phenomenon is that as the building and its inherent shaft(s) get taller, the force and movement of the smoke and heat can become more powerful. These rapidly rising products of combustion, if not channeled to the exterior, will mushroom down from the top floor or top of the shaft causing added concerns in those areas.

Reverse stack effect. Reverse stack effect is known as the downward movement of smoke or air due to the presence of cool, dense air within the structure. Here again, we must compare the inside air temperature of the building with the outside air temperature that surrounds the building. An example of this phenomenon would be on a hot summer day when the outside atmospheric air temperatures are greater (warmer) than the inside air temperatures within the structure. Cool air from the building's air conditioning system may pull products of combustion below, causing smoke conditions on floors below the actual fire floor. This type of condition forces firefighters to exit onto floors that may be well below the actual fire floor.

Stratification. Stratification is the cooling and ultimate layering of smoke and gases on floors above the fire. This becomes a concern in sealed buildings where the convected heat loses its buoyancy causing the smoke and gases from the fire to cool, thereby layering in a given area. This phenomena can occur from minor to moderate heat releases as well as from the convected heat's inability to be picked up by the building's normal stack effect.

Special Considerations: Size-Up

*COAL TWAS WEALTH**S***

This section will include any special thoughts, strategies or concerns at the end of each chapter that would be specific to the occupancy described. The material that will be presented in this section has been gathered from the size-up factors of the individual chapter, as well as from researched and tested strategies and tactics. It is from any of the listed thoughts, strategies, or concerns, where the fireground commander will be able to develop sensory cues for the direction and controlling of fireground activities.

Examples presented include a strategic and incident management plan for a fire in a high-rise building. Another includes a chief officer's general strategies for a fire in a row of frame buildings. Yet another will identify safety precautions for arriving and going to work at a vacant building fire. Whatever the occupancy, it is in this section where quick reference concerns, any specific information listed, as well as any thoughts that can be added by you, will assist with the overall outcome of the incident.

You're going to enjoy this section. It will put a significant portion of what is discussed in the chapters into a strategic and tactical review for the fireground commander.

Questions For Discussion

1. List the five classes of construction, and identify the concerns they present to the fire service.

2. Identify the different categories used within the Occupancy size-up.

3. What is "Resource Typing," and how is it beneficial for the responding fire officers?

4. Name the four Life Hazard groups.

5. Identify the categories within your Terrain size-up, and discuss how they can influence your decision-making on the fire ground.

6. Discuss the reasons why a sprinkler system might fail to control or extinguish a fire.

7. How can streets that are designated as one-way influence your thinking?

8. Identify and list the concerns within your Weather size-up.

9. What factors will influence your Exposure protection?

10. Of all the size-up factors listed within this chapter, which is referred to as the most influencing? Discuss the reasons why.

2 Private Dwellings

When individual members of the fire service refer to the term private dwelling, many different thoughts may come to mind depending upon the area they are assigned to protect. Private dwellings can mean manufactured housing like a mobile home, a small frame house referred to as the "Cape Cod," a 2½-story platform frame that dominate many areas of the country, a 3½-story Queen Anne or Victorian home that is in other areas of the country, or a massive 5000ft^2 mansion found in your more affluent areas.

Whatever the specific type of private dwelling, they represent the number one fire problem in the United States, with 70–75% of the civilian fire deaths occurring in this type of occupancy every year.

Construction: Private Dwellings

COAL TWAS WEALTHS

Just as we find different types of private dwellings in city neighborhoods, we also find hybrid construction features used in our private dwellings as well. When we think of private dwellings, most of us think of Class 5 wood-frame construction and the fire spread and collapse concerns associated with them. However, as the demand for more economical housing increases, with the design of larger, and more open spaces within the home, the fire service will increasingly encounter newer, lighter, and hybrid types of construction.

Hybrids

Buildings that were once completely constructed of wood now can have construction methods and materials that affect the building's ability to withstand fire. Lightweight wood trusses, steel beams, and pre-engineered lumber, among others, are finding their way into the private dwelling building industry. Each of these materials has their own inherent concerns for the firefighter. Examples of those concerns include gusset plate failure of the lightweight wood truss within a floor or roof system, the expanding and twisting of the unprotected steel that may support the building's first floor, and the de-lamination of a laminated timber within a structural span. The only way to prepare for these concerns is through identification during the construction or renovation phase in addition to continued education with the concerns of the specific materials. There is no doubt that the use of lighter and unprotected materials within any class of construction will affect our concerns with early failure and early withdrawal from the building. In private dwellings, identification becomes the key.

Fig. 2-1 *Sample hybrid construction*

Modular/prefab construction

As the title suggests, modular, or "prefab," construction is the practice of building individual sections of a structure at an assembly factory where, when finished, they are then numbered and labeled so they can be sent to a construction site and assembled together. The individual building sections can vary greatly in size from an 8′ X 10′ section of wall, to an entire floor space lifted and placed by an on-site crane. Within many of the assembled sections you may find lightweight constructional components, wooden I-beams, as well as thin gauge joist hangers. Each of these materials will present their own inherent danger to the fire service, but what we are increasingly noticing is the concerns created when the individual sections are assembled together. In many installations additional void spaces will be created. In others, assembled sections may be misaligned or unsupported. The reasons for the concern can stem from a number of areas, but are most notably from foundations not matching framing designs, improper on-site installation, or simply shoddy workmanship.

Their use throughout the building industry is increasing not only in the private dwelling sector, but quite extensively throughout the garden apartment and townhouse industry as well.

***Fig. 2-1a** Modular wall*

***Fig. 2-1b** Assembled sections*

***Fig. 2-1c** View from basement toward underside of the first floor. Note void space between floor sections, method used to fasten sections together, and the support of both sections.*

Balloon frame

As we proceed from one extreme to another, many of the wood frame private dwellings in existence today can date back 50 to 100 years. Earlier methods of framing, although no longer used today, still exist in many private dwellings.

Most notably, and at times most troublesome, is the balloon frame design. The balloon frame design found its way into the nation's building industry starting in the East as early as the 1800s. This design concept became very popular at that time due to the ease and speed at which structures could be built. Vertical timbers extending the entire height of the structure allowed builders to fasten floor beams or ledger boards at any level. This method of construction, although easy to design and build, presented an inherent flaw that continues to plague the fire service. A lack of specifically designed and placed fire stopping material within a vertical stud channel or wall bay allows fire to travel unchecked for that channel's entire height. This same lack of fire stopping is also a concern within any horizontal floor and ceiling bay, as well as where those same areas meet the building's exterior walls and interior partitions. To put it into perspective, if fire extends into a void within the building's structural framing, it could move vertically and horizontally throughout the building with little resistance.

The only built-in fire stopping that can be found within a balloon framed structure, (and it was not intentional) exists where a window or a door was installed within a wall space, or the possible presence of any diagonal bracing within the wall framing.

The width of the door and the window within the wall space, and the stud channels it spans, limit the vertical movement of fire within that area of the wall. This is a relatively easy concept to understand and identify.

Diagonal bracing, on the other hand, is not that easy to identify. Diagonal bracing is a method of shoring or supporting the exterior walls by erecting wooden members on an angle from the building's corners. Within this construction concept there seems to be no installation pattern. They may, or may not be present within a framing design. When present, they can differ greatly from the angle and the span found within one wall space, in comparison to another installation found within a wall space of the same building.

Fig. 2-1d *Unrestricted stud channel*

Fig. 2-1e *Diagonal bracing*

Queen Anne/Victorian homes

The problem of the balloon frame design was never more evident than in the Queen Anne type of structure. A Queen Anne or a Victorian style home could easily be considered the most difficult and task-demanding private dwelling fire you will every encounter.

Fig. 2-2 *Queen Anne architecture*

The Queen Anne architecture became very popular in the late 1800s to early 1900s. They can range in size from 2½ to 3½ stories in height, 25'–30' wide, to 30'–50' deep. They are built of Class 5 wood-frame construction with exteriors ranging from wood siding or asphalt siding, to brick or stucco veneer. The main reason they are considered so difficult is not only from their balloon frame design, but also from their many peaks and valleys, dormers, large overhanging eaves, cupolas, double-studded walls, knee walls, and numerous void spaces found throughout. Add all of this to the many variations in the size and number of the dormers and gables presented in just one building, and you will find a maze of areas into which fire can burrow. Any sizable fire in these types of structures is going to require additional resources, most notably ladder company personnel to aggressively open up the structure to expose any hidden fire.

Platform frame

As newer wood frames were constructed, code changes introduced the platform frame. In platform frame construction, each floor area is constructed as a separate section. Vertical wall studs within the building have a horizontal member or "plate" attached to the bottom and top of the stud. This framing design limits any fire travel in an exterior or interior stud channel beyond the floor of origin. Any vertical fire travel in a wall bay not compromised by poke-throughs would be limited to the height of that story. This was an obvious welcome for the fire service, and continues to be the method of framing used throughout the industry today.

Collapse concerns/private dwellings

Collapse concerns of the private dwelling will come with varying points of consideration for the fire officer. If the structure is of Class 5 construction (as it is in most cases) it must be remembered that the entire building is made of wood, most notably the exterior load-bearing walls. These "stick-built" homes will have nothing more than small dimensional lumber holding them up. The bearing walls in Class 5 construction typically are either 2" x 4" or 2" x 6" framing members. Remember in newer construction (post 1960) the sizes mentioned are nominal in number only, and are not the actual size of the lumber. The true dimension is somewhat less than the stated dimension, notably a half-inch less on both sides. Over the entire surface area of the wood, this becomes a significant loss of mass that would have greatly added to structural integrity of the building if it were present. This loss of mass is another building industry standard that directly affects the firefighter's ability to safely operate within the building. There is no such thing as a 2 x 4 or a 2 x 6 piece of structural lumber anymore; therefore we need to call it what it is, and identify the concerns.

Fig. 2-3 *Class 5 private dwelling fire*

When any of the hybrid construction features previously mentioned are known to be present in a private dwelling, collapse of the involved area must be considered even earlier. Wooden I-beams, pre-engineered lumber, or lightweight wood trusses were not designed to resist any type of fire involvement. The lack of mass of the lightweight materials, the integrity of the connection points used, or the structural member's ability to be affected by the fire, will undoubtedly promote early failure. When a hybrid is involved in fire, there will be no preliminary cracks, groans, moans, or leaning structures to give a telltale sign—just sudden, early, and without-warning, collapse of the affected area of the building. Their presence within a private dwelling will drastically change the rules of engagement.

Occupancy: Private Dwellings

COAL TWAS WEALTHS

Classification/occupant load

Private dwellings are classed as residential occupancies with the occupant load ranging from one to two families within a single dwelling unit. However, many fire service personnel are finding illegal conversions to multiple dwellings, where an original private dwelling may now have renovated apartments located in the basement and attic areas. This renovation may have been done by the homeowner to increase the rental income of the property, or simply to give a family member his or her own apartment. Unfortunately, this discovery isn't often apparent until during or after an incident. Occasionally during a district survey in the neighborhood, or even when arriving at the scene of a specific address, there might be clues to tip off the firefighter/fire officer that the building has an illegal conversion. Some of those clues may include:

Multiple dwelling cues (three or more apartments)

- Multiple doorbells
- Multiple electric and gas meters
- Multiple mailboxes
- Curtains and blinds on attic and basement/cellar windows
- Oversized attic windows
- Dormer attic areas
- Air conditioners at the attic level
- Skylights servicing the attic level

When you think you're arriving at a private dwelling and discover any of the above, expect an increased occupant life hazard. Resources must to be called and assigned to what is now a multiple dwelling fire.

Single room occupancies

In a similar instance, some private dwellings may have been converted to a legal multiple dwelling, but may not give the appearance from the building's exterior. Queen

Anne's and Victorian homes are commonly converted to Single Room Occupancies (SROs). SROs are rooms that are individually rented to tenants, with all the building's tenants sharing a common bathroom(s) at the end of the hall. These single room tenants may number in the double digits depending upon the number of rooms in the building. This type of conversion may be noted by the number of mailboxes near the buildings front entrance, as well as from the multiple locks on the individual bedroom doors as you begin work in the building.

Apparatus and Staffing: Private Dwellings

*CO**A**L TWAS WEALTHS*

Resource levels and resource types will vary from fire department to fire department. With many departments having limited numbers of firefighters responding and able to go to work at an emergency incident, the assignments and tasks that can be carried out will be minimal. It may be difficult for some to get a second hoseline stretched and operating, let alone to accomplish the tasks associated with ventilation and search. But at the very least, there are certain concerns that can be mentioned in the form of basic engine company and ladder company operations that will begin the process of regaining control of the building.

Engine Company Operations

Engine company operations at an incident will focus around the four following points of consideration:

1. Water supply
2. Hoseline selection
3. Hoseline stretch
4. Hoseline placement

Water supply

The engine company officer's decision to establish a water supply for the first arriving engine company, will be dictated by department procedure. We know that fires in private dwellings have a significant impact on the number of civilian injuries and deaths each year. Due to this crucial fact, the need for quick and sustained water supplies is a necessary factor to be considered.

There are numerous ways of establishing a water supply depending upon the department and the area served. Some departments will use the water-to-fire concept, where an engine company stretches a supply hoseline from a hydrant to the fire building. Others will use the fire-to-water concept, where an engine company stops at the fire building, drops off a compliment of hose, tools, and equipment and then proceeds to the water source. In others, engine companies will use their tank water and have their supplies augmented by a second arriving engine company.

Whatever method used has to meet two criteria for fighting a fire in an occupied building. First, the procedure or guideline must allow the quickest sustained supplies available, remembering that these types of buildings present our largest loss of civilian lives; speed and efficiency are necessary. Second, the procedure must work at every incident. This is the place where the problems can begin. Department procedures or guidelines should outline more than one way to establish a water supply for a building fire. As many would agree, each incident can be significantly different, responses may be altered, area geographics may vary, and incident priorities as well as the tactics used to accomplish them may change. Doing it one way all the time because this is the only way we do it is great, if you are guaranteed the results will give you the best outcome every time. However, we learn early on that there are no guarantees in this business. Being able to adapt with the right tools and procedures at an emergency incident will often give the best results.

What may aid you in your decision-making in determining water supply options is the information that can be gathered during the receipt of the alarm. It starts with simply knowing what specific companies are assigned to the incident, and where they originate.

As a former Captain of Engine Company 17, I knew that the surrounding engine companies on incidents in my district were Engine Companies 8, 22, and 9. When the alarm was transmitted, if I heard via radio or read from the alarm printout that Engine 8 or 22 were not announced as the second or third company due to the incident, I would most likely make the decision to bring in my own water supply or establish my own water from a nearby hydrant. This decision was simply made from the fact that the normally assigned Engine 8 or Engine 22 must be off-duty or assigned to another incident. Wherever they were didn't matter, the fact was they weren't coming. An engine company from a remote area as my second due engine would not only mean a delay in aiding my water supply, but it would also identify a delay in the deployment of a second hoseline to back us up. The information for my decision was there, it was just up to me to gather it.

Hoseline selection

Once your water supply is established, it must support the hoseline selection. Due to what is generally considered a light fire load with relatively small room sizes in the average private dwelling, the need for a quick, mobile, and flexible hoseline(s) becomes a necessity. Medium-size hoselines of 1¾" or 2" diameters will be the selection for most fires that involve private dwellings. With flows ranging from 180 to 225 gpm, and the possibility of a limited number of firefighters to move them, hoselines of these diameters should be easily maneuvered to handle a number of involved rooms.

Hoseline stretch and placement

The initial hoseline stretch and placement at a private dwelling fire is *always* determined by the location and extent of the fire in relation to any threatened life hazard. The initial hoseline stretched by the first arriving engine company, must use the quickest and shortest path to the fire's origin. The first deployed hoseline at an occupied building has a number of objectives to consider, but the most important is that it be placed and operated between the fire and any endangered occupants. Barring variations on accessibility, this hoseline is generally stretched through the front- or side-entrance door of the building, or by a means that allows the hoseline to proceed from the unburned portion of the building to the burning portion of the building.

In most cases members will quickly encounter a stairway with the front or side entranceway approach. Usually from the original building design we find the stairs for the second floor and possibly the basement in this same area. In a multi-floor building, note that the purpose of this hoseline is to protect and control the building's largest and most vulnerable artery while advancing to the seat of the fire. With control over the interior stair(s), fire forces will be allowed an aggressive approach to the fire encompassed with the hoseline's main objectives, its placement between the fire, searching firefighters, and any endangered occupants.

Ladder Company Operations

Ladder company operations at a private dwelling fire will consist of a multitude of tasks. The complexity and number of those tasks will depend upon the type and size of the private dwelling, the construction of the building, the location and extent of the fire, and the occupant life hazard within the building. When staffing allows, assignments should be broken down and delegated to inside and outside teams to ensure efficiency. Their duties will revolve around the five primary functions of these: *ventilation, forcible entry, search, laddering,* as well as *salvage and overhaul.*

Ventilation

Ventilation of private dwellings must be a controlled and coordinated effort with members advancing the initial attack line, as well as with those members involved in the search. Depending upon the location of the fire and its extension probabilities, options must focus around channeling the fire and its products away from the trapped occupants, as well as ahead of the advancing hoseline. This can follow with a number of avenues in and around the building, depending upon the information that can be gathered about the fire's location, and any building's features that we may need to consider and overcome.

Energy efficient windows. Today, the difficulties of identifying the fire's location may be complicated by the increased use of energy efficient windows in private dwellings.

Energy efficient windows will contain double- or triple-insulated glass within a wooden or aluminum sash. By design, their intent is to prevent air exchanges between the panes, limiting any heat loss through their openings. Energy efficient windows will make it difficult to determine the fire room, or even the actual floor when attempting to gather information from the building's exterior. Windows of this type are known to contain large quantities of smoke and heat produced by an advanced fire for a significant period of time before they fail, if they fail at all.

Fig. 2-4 *New energy efficient windows are often much smaller than the older ones they replace.*

This situation was evident at a recent response to a fire in a three-story frame building. The initial report transmitted by the first arriving engine company identified the fire as being on the third floor, when in actuality it was on the first. As firefighters made their approach through the front door of the dwelling, they were met with a wall of superheated smoke from the fire that was on the first floor in the rear. The heavy smoke that initially only showed from the third floor, came from windows that were left open. The windows on the fire floor were closed and still intact. This is reemphasized for many who operated at that incident, that when energy efficient windows are present, smoke doesn't always appear where the fire is.

When encountering this type of window, breaking all the glass, especially on a double- or triple-pane window will be difficult. Often what happens when a firefighter attempts to break a window of this type is referred to as "porthole ventilation". Without being able to view the window, the sound of breaking or shattering glass may give the impression the window has been vented, when it may not be vented at all. What often happens is that only one of the two or three panes will be broken, still leaving the window intact, with no ventilation taking place. In those situations where a firefighter breaks through all of the glass panes, the integrity of the window only allows a small opening or porthole, thereby limiting the ventilation effect of the opening. The porthole opening will remain unless the firefighter takes the time to clear all the glass and any remaining shards.

Education and training should not only include when and how to break a window of this design, but also how to remove it from its opening.

Concealed spaces and voids. Fire involvement of concealed spaces and voids in a private dwelling will continually make it difficult to determine the fire's location, its path, and where to open up to halt its spread. One tool that has greatly assisted the fire service with its objectives is the thermal imaging camera. This technology was originally developed back in the 1950s and 1960s for military purposes and has finally made it to the fire service. Thermal imaging used by trained members greatly assists the ladder company's ability to find the fire. Their ability to detect infrared radiation created by a heat source allows members to pinpoint the concerned areas depending upon the heat signature of the fire. By interpreting the screen, firefighters will be able to identify the fire location, fire's travel, the dangerous collection of heat at the ceiling level and the mushrooming effect of accumulated gases in a top floor or attic.

One building that continually causes difficulty with ventilation and fire control where thermal imaging is greatly appreciated is the balloon frame. When information dictates that fire is within the wall and stud channels of the building, vertical ventilation must be performed. Even if the fire is on a lower floor, roof ventilation would be justified to further halt the fire's spread.

These difficulties are never more pronounced than they are with the Queen Anne or Victorian type home. As we know, Queen Annes were originally constructed of balloon frame design. With the method of framing used, and the numerous concealed spaces and voids within the building, any significant fire involving this type of structure will require the difficult task of opening up, notably at the roof level.

Vertical ventilation of the main or center gable roof of a Queen Anne, will have significant strategic value, when and if it can be accomplished. Establishing a primary ventilation hole over the main gable will allow vertical ventilation of the attic hallway that serves the interior stair, the rear servant stair, the blind attic space, the knee walls, and any rooms directly below. This is obviously going to be a difficult task at best. In many areas where these buildings are located, they are known to sit well back from the curb, and are often on well-established, tree-lined streets. Strategic placing of the aerial device is a must if members are going to meet this objective quickly and safely.

Fig. 2-5 *Anticipate the difficulties.*

Forcible entry

Through the years forcible entry concerns for private dwellings were easily meet by conventional means. Not too many points around the private dwelling couldn't be handled by a flat head ax and a halligan bar. In many areas this is still the primary means of gaining entry. However, in other areas a more aggressive approach with more powerful tools may be required. Hydraulic forcible entry tools (HFTs) are not just coming off the apparatus for a fire in commercial or fire resistive building anymore. Door and jamb construction, as well as security-conscious civilians are forcing the fire service to be better prepared in gaining access to a private dwelling.

Search and laddering

Search operations in private dwellings take on the primary focus of getting to the bedroom areas, especially if the incident occurs during the late evening or early morning hours. The number of bedrooms and their location could be anywhere within the home. However, if we were going to prioritize and direct our energies, we can develop an association on bedroom locations from the building's original design.

In a single-family private dwelling of two stories, the usual location of the bedrooms will be on the second floor. This is not to say that bedrooms may not be located elsewhere, but chances are that most, if not all, will be on the second floor. If the building is a two-family occupancy with an apartment upstairs and an apartment downstairs, the normal bedroom layout will put the bedrooms at the rear of the building, on both the first and the second floors. In some cities researched, firefighters are finding other two-family occupancies with varied floor layouts that contain duplex apartments within a single building. This type of occupancy may use as a reference an "A" address or a "½" address for one of the apartments. Arriving at a two-story, frame building that indicates the address of 110 Main Street, when you are looking for 110A or 110½ Main Street, will warrant you to check for an alley or rear apartment door that leads to a separate unit. These types of units, although duplexes, will most commonly have their bedrooms located on the second floor, but their locations will be both the front and rear of the building.

The point that must be stressed here from our associations is that there will be varied layouts and entrances to a private dwelling. Thinking that room locations within the private dwelling will be the same for all buildings, is not efficient thinking. Different building and room association allows company and chief officers to initially assign and focus resources on protecting and gaining access to these vital areas.

Salvage and overhaul

Salvage and overhaul are two critical areas at any fire incident. Assigning the responsibilities of salvage work at a fire incident needs to be done as soon as staffing allows. Often with a fire in a dwelling (in this case a private dwelling) everything the occupants may own may be inside the building. Items, fixtures, and furnishings must be protected from the effects of the fire and fire department operations. Not only does this include protecting items with salvage covers and tarps, but also includes unnecessary or uncalculated damage. This concern often includes breaking glass and doors without additional or coordinated decision-making.

Overhaul at a private dwelling fire has to concentrate on the factors associated with building construction, and the method of framing used in consideration to the fire's location and extent. Various framing methods will dictate areas of concern, which will need to be examined early for hidden fire. Fires that have occurred in balloon frames will require different and more areas to examine as compared to that of the platform frame.

Lacking education or experience with these considerations has often resulted in a fire department returning to what they thought was an extinguished fire. A tool that will greatly aid the fire department's ability to successfully, as well as efficiently, overhaul a building is, again, the thermal imagining camera. Trained operators with a camera will not only eliminate unnecessary damage, they can also eliminate an unnecessary return to someone's home.

Life Hazard: Private Dwellings

*COA**L** TWAS WEALTHS*

Firefighter Life Hazards

Fires in private dwellings present an increasing life hazard concern to responding firefighters. Fires in these types of occupancies cannot be considered easy or simple as they are by some. Categorizing any fire incident as such can breed complacency and ignorance to situations or circumstances that could injure or kill a firefighter. From the brief list of reasons below, life hazard concerns for firefighters in these types of buildings must be constantly reviewed and revised in an attempt to best protect our people.

Combustibles/EEWs and R-values

In years past, firefighters responding and going to work in private dwellings encountered less heat producing combustibles as compared to the combustibles that they are experiencing today in the same homes. The number and type of combustibles represented in the average home are reacting and burning much hotter in addition to producing more deadly gases than years past.

Accompanying this concern, there should be no doubt in anyone's mind of the added difficulties that energy efficient windows are placing on the firefighter. Their ability to hold heat, as well as their ability to enhance or retard the fires growth is increasingly being observed and documented throughout the fire service.

Overall, energy efficiency in the home has not only increased from the energy efficient windows mentioned, but wall thickness of the newer private dwellings is increasing for added "R" or insulation space to save on the home heating bill.

Smoke detectors

Fig. 2-6 *A trapped firefighter*

With the added use of smoke detectors in homes in the late 1970s and early 1980s, a frequent and reoccurring set of events was being noticed. Because of the early notification of smoke detectors, firefighters were getting an early warning to home fires. With an earlier response firefighters were now finding themselves arriving and going to work just as the fire was progressing into the flashover stage. Undoubtedly an outstanding device to aid early notification and evacuation of building occupants, but now smoke detectors in the home are causing firefighters to exercise more caution when responding and going to work a fire in a private dwelling, specifically due to the timing of the firefighter's arrival and the fire's growth.

Protective clothing

All of the previously mentioned concerns forced the fire service to better protect its members. Initially, the response to these concerns came in the form of newer and lighter protective clothing, but for many it came without the awareness of the clothing's ability to insulate members from the fire environment. Undoubtedly it provides better protection to the firefighter from the heat being produced, but indirectly, it also allows the firefighter to penetrate much deeper into the fire building than ever before. For many in the business this is referred to as a plus. Being able to penetrate deeper and quicker into the building allows firefighters to find the fire and any trapped occupants more quickly.

However, all will agree that the aggressive nature allowed by the protective envelope must come with an added awareness of your surroundings. From this, many cities around the country began purchasing and training their members in the use of thermal imaging. In addition, heat detection devices are being purchased and carried by firefighters to signal when a rapid heat build-up takes place in a room or an area. These are definitely proactive and innovative steps toward safe guarding your people. However, many more departments researched are retraining, and continually educating their members in what is referred to as "back to the basics," with the emphasis on all the added concerns previously mentioned.

A class that is getting renewed interest is one referred to as "flashover and backdraft recognition signs". More so today than ever before, fire departments must continually educate their members on the signs of the two most dangerous phenomenon fire has to offer. The number of incidents where firefighters are being exposed to flashover and backdraft conditions is increasing, and it's becoming vividly apparent from our fire activity in the private dwelling.

Signs of flashover

1. *Rollover.* A telltale sign of a rapidly approaching flashover. Whether flames are easily visible rolling across the ceiling, or there are momentary dances of yellow and orange through the thick smoke at the ceiling level; plan ahead, it's coming!
2. *A rapid build-up of heat.* A difficult sign to foresee. With firefighters being totally encapsulated in the industry's most technologically advanced protective equipment, often when they feel this sign, they may be in the middle of it. When a sudden increase in the heat level forces you to the floor, it's here!
3. *Heavy smoke under pressure, exiting an opening.* During the latter part of the incipient phase, and the early part of the free burning phase there is a period of rapid development. This period of rapid growth will emit signs of the deteriorating conditions within.

Signs of backdraft

1. *Dense, black smoke turning a gray, yellowish color.* Combustion is a process of oxidation, which is a chemical reaction where oxygen combines with other elements given off by the burning combustibles. When enough oxygen is no longer available, large quantities of carbon are released during the combustion process indicating an oxygen-starved fire.
2. *Little or no visible flame.* Is another indication of an oxygen-starved fire. In order to have a visible, luminous flame, there must be a significant presence of oxygen within the atmosphere.
3. *Smoke puffing around door and windows.* When smoke puffs out of, or around a crack or opening, it's indicating that the room or area in question is under pressure and oxygen starved. It is between the intervals or puffs that the fire is trying to grab oxygen during the negative pressure change.
4. *Rapid movement of air inward, once an opening is made.* When a rapid movement of air is felt or seen going inward to the questioned area, this telltale sign indicates that the once oxygen-starved fire is getting what it needs. An oxygen-starved fire will create a negative pressure through the path of least resistance to pull in the air it requires.

The above concerns are not exclusive to fire activity in private dwellings. They are a concern to firefighters in all buildings. However, the reason why I take the time to state the concerns in this chapter is that the frequency at which these events are occurring is undoubtedly increasing, and we're seeing the increase where America's fire service responds the most; the private dwelling.

***Fig. 2-7** A typical private dwelling*

Occupant Life Hazards

The life hazard concerns of the occupants of a private dwelling are referred to as the most serious fire problem facing America's fire service. Through the years tremendous strides have been made in reducing the number of fire deaths in this area, but the number of civilians killed in home fires is still considered high. From educating civilians about fire prevention and exit drills in the home to the installation and use of smoke detectors, the number of home fire deaths has dropped significantly over the last few decades.

Research has shown that those who perish in private dwelling fires are caught while they sleep, attempting to escape the building, or by re-entering the building for a family member, pet, or a personal item. Even as members of the fire service stress the consequences of re-entering the building, emotions, anxiety, and confusion can override common sense. Awareness and education continue to be our best course of action.

With any fire in a dwelling, the speed and efficiency of your operations are going to be a critical factor. With most fires starting on the lower-level of a private dwelling, precious time may be lost awaiting the engine company to darken down the fire before assigning firefighters to the floors above. Aggressive, yet cautious attempts must be made to access areas of the building where occupants are expected to be. Incidents that occur during the late night or early morning hours will put immediate emphasis on the bedrooms areas, which can be anywhere within a building, but there is a pattern to bedroom location. Aggressive use and placement of ground ladders to the areas most known for their location may allow firefighters to enter and search a room, often with positive results.

The assignment of firefighters to a primary search on the floor above the fire is a difficult one. This task, although dangerous, must be attempted with your most educated and experienced people if the building's occupant(s) are to have an increased chance of survival. With this consideration may come many obstacles. Trees around the building may delay or even eliminate the placement of ladders to some windows. The window itself may be a significant obstacle.

Window size-up

Firefighters assigned to this task are reminded to perform window size-up noting a number of different items that could give information. The first consideration must be if the window is already open, closed, or broken as you quickly size it up. When I was first told this procedure many years ago by a veteran ladder company officer, I too said, "What?"

The logic behind the procedure was that if the window were open or broken prior to fire department involvement, your first step would be to search the shrubs and landscaping below for any civilians who may have exited that same window. The body of a small adult, child, or infant could be easily consumed by landscaping that surrounds the building. As unusual as this may sound to some, this search has been proven to be positive on more than one occasion. So take a look!

Window size, type, and features follow next. Many older homes, unless they have been renovated, will not have egress windows in the sleeping areas. Window size and design may present difficulties for a civilian, let alone a fully protected firefighter, to get into or out of. Window type and construction will also cause concern for firefighters attempting access. Window design, sash construction, locking devices, security gates, and window air-conditioners may delay your getting in as well as possibly getting out.

Firefighters should also look for unexpected or obvious living areas within the home. Dormer attic areas, curtains, drapes, blinds, screens, and air-conditioners in the windows are just some examples we have mentioned that may indicate living areas requiring your immediate attention. Child security or window guards on a particular window are another good clue to the presence of a child's bedroom that should not be overlooked. When time is of the essence, prioritize, and choose the windows that will hopefully give you the best results.

Through the years, attempts to give the fire service information on the elderly, the disabled, the young, and even information about pets in the home came in the form of stickers placed on the individual's bedroom window or on the front door of the occupancy. In the late 1970s and early 1980s a popular program called "Tot-Finder" promoted using stickers, which were placed on an infant's or small child's window. These alerted firefighters during their exterior size-up of any window that might directly access a tot's bedroom. This proved to be a valuable asset to on-scene firefighters where they would be able to prioritize a window in an attempt to gain access to a child's bedroom.

This type sticker, or a similar type sticker unfortunately also gave information to a person attempting to unlawfully enter a home as well. As these individuals did their own personal size-up, they quickly learned that they would receive little resistance if they attempted to enter through a particular window in a person's home. For this reason, many urban areas abandoned these information programs. Other departments who were not having a problem with unlawful entry also were forced to abandon the program over the years because it was now referred to as "outdated". These outdated stickers could not be relied upon anymore. Simply, children have now grown up, and are no longer considered a tot. It was also noted that people who may have originally placed the sticker might have moved leaving the stickers in place, furthering the inaccuracy of the decal. Others, who may now have small children, are unable to obtain a sticker; therefore there is no indication of a child's room other than possible presence of child window guards, or cartoon character curtains on the bedroom windows.

Fig. 2-8 *A former tell-tale sign of a child's bedroom.*

Many will agree that the program no longer gives accurate information. I also must agree, but maybe it is time for a new program that can be periodically updated, or revised with a change of ownership. Think about the reason why the program was implemented. If one sticker out of a million assisted in saving a child, then the program met its objective, and met it well!

As the number of civilian fire deaths in private dwellings continues to be our greatest problem, the fire service as a whole should not view or categorize these incidents as being simple. They should continually review, revise, and refine their activities to meet their greatest challenge.

Terrain: Private Dwellings

COAL TWAS WEALTHS

Setbacks

Buildings set back from the street or accessible area are more of a frequent occurrence in suburban and rural areas than in urban centers. This association is gathered from the comparison of an average building lot in a city setting, to that of the suburban or rural setting. With the demand for housing greater in the city, and the zoning of the building lots much different than that of the suburbs, buildings in many urban areas are actually spaced within a 25′ wide by 80′–100′ deep building lot. Not a whole lot of room, but as it relates to this size-up factor, one positive factor noted was that buildings constructed on this size lot have their front entrance door within a few feet of the street, thereby eliminating any setback concerns.

Private dwellings in a suburban or rural setting, present the most significant and most challenging setback concerns. Buildings can be placed on lots varying in size from 50′ by 100′, to a number of acres. Long winding driveways, landscaping, and large front yards are what some fire departments constantly find. Fire departments that are forced to deal with these types of situations must develop specific guidelines on bringing tools, ladders, and water to the front door of these buildings.

With structures that are built back from the street or other accessible areas, the fire officer's concerns begin with the increase in distance for any deployed hose lines. Longer hose stretches may force the engine officer's initial thought on hoseline selection to change not only due to the increased friction loss factors, but also the speed at which a second or back-up hoseline can be stretched and used. A fire which appeared to require the water flow of a single 1¾″hoseline may now require a 3″ hoseline with a gated wye to the front door of the building.

Longer hose stretches may also require the assistance from an additional company in order to get the first hoseline to its destination. Depending upon the number of people assigned to the first due company, a single engine company may not be able to get the first deployed hoseline in place without the direct assistance of the second or possibly the third due engine company. Even if the number of people assigned to that company seem adequate to handle the stretch, check and see if help is needed. There is no sense wasting resources starting a second hoseline if the first one is going to fall short.

Fig. 2-9 *A modified hose stretch.*

Landscaping

Landscaping terrain concerns can take on a number of different problems that can affect fire department operations at a private dwelling fire. We mentioned earlier the concerns of an occupant jumping or falling into the shrubbery. When the decorative landscaping is extensive, not only does it compound this effort, but it may also hamper and delay ground ladder placement to the building. Shrubs placed around the perimeter of the building will affect the placement of the heel of a ground ladder, the angle once it's placed, and possibly its reach to the required target height. In addition, any extensive or overgrown shrubs covering or obscuring a window can cause additional concerns from the blocking of a secondary means of egress, to playing a role in extending the fire. I can remember operating at a fire in a Queen-Anne home where fire vented itself out of a first floor window I had just "taken out" for ventilation. Fire roared right out into a dried-out evergreen tree near the window. This small forest fire allowed fire to extend into the large overhanging eaves of the structure. As you can imagine, this caused a little problem.

General Accessibility

Fencing is a common site at a private dwelling. When present it will generally line the property, and often have a narrow, gated entrance, which will access the front or side of the building. If ignored, it can present difficulties with the accessibility and movement of deployed hoselines, ground ladder movement and placement, and even firefighters trying to access the building. When time and people allow, remove a section or two of the fence in the heavily traveled areas. Having a couple of additional points of entry will make things a little easier, especially as more people and equipment go to work.

Here is an additional and personal note on fences. Countless times we have all operated at incidents where the fire has dropped the outside electrical service from the fire's direct impingement. Obviously the immediate concern is to watch where the wire has landed, but also take special note if it is lying over a chain link fence, or any other fence that could conduct electricity. If it is, there is high probability that the fence is electrically charged for its entire length, and this may include fences on adjoining properties if they are connected in any way.

This has happened to me on more than one occasion where my life hazard concerns had to extend for the entire distance of the fence.

Water Supply: Private Dwellings

COAL TWAS WEALTHS

Availability

Water supply with any type building will depend on the geographic area in which it resides. Rural areas of the country will, for the most part, rely on natural water sources, or sources exclusively provided by fire department apparatus. This will come from draft sites near lakes, rivers, and ponds, to tanker trucks carrying a large supply of water to the incident scene.

Firefighters in suburban and urban areas will have the advantage of a domestic or public water utility system, with some possibly having the additional ability to use natural water sources. Whatever the setting, the local fire service must be proficient in obtaining an adequate supply from whatever source is available.

Requirements and delivery

As we refer to water supply requirements and how water will be delivered to fight a fire in a private dwelling, we initially find that fire loads associated with these types of buildings are characterized as being low. If we reference fire flow formulas for these types of occupancies, we find a minimum required flow of 10gpm per 100ft^2 of structure. This would seem to be a fair amount of water for what is known as a relatively low fire load. However, situations and the conditions can vary. The building's contents can be amazingly different from one dwelling to another. People collect many objects for many different reasons. Rooms, attics, and garages can become makeshift warehouses that can require a larger and more concentrated flow of water to extinguish a fire.

Depending upon the combustibles being consumed, heat releases can be much greater than anticipated, also requiring larger flows to extinguish the fire. Another important factor overlooked with fire flow references and formulas, is the need to deliver water into the building's void spaces once these areas are involved in fire. These sizes can vary greatly according to the age and design of the building. Large frame structures as described in the Queen Anne or the Victorian style building will have numerous, as well as large void spaces, which in addition to the living area will require increased flows to extinguish the fire. This concern has to be factored into the officer's decision-making. Planning on establishing and flowing water for only a room and contents may find you needing much more as firefighters open up the void spaces.

The size hoseline that best meets the needs of a fire in a private dwelling will focus around a number of factors that must be considered by the fire officer. They are the

Fig. 2-10 *Aggressive efforts at a Class 5 fire.*

mobility requirements for that size and square footage of the building, the fire loading of the building, and the number of firefighters available to move the hoseline. Many would agree that the stretching of a hoseline larger than 2" in diameter into a private dwelling would seem counter productive. Due to the general reference of a low fire load and the need for speed and mobility of a deployed hoseline, the flow of a 1¾" or a 2" attack line seems to be the hoseline of choice. However, due to what can be vast differences in square footage, layout, and design, seldom do we stretch and operate one hoseline without stretching a second. Depending upon the size and type of building, fire conditions on arrival and your anticipation for its spread, fire officers should plan on a minimum flow of 500 to 600 gpm for a private dwelling. Using these numbers as a guideline allows for a three-hoseline stretch using medium-sized, manageable hoselines.

Auxiliary Appliances and Aides: Private Dwellings

COAL TWAS WEALTHS

Detection equipment

During the 1960s, 1970s, and 1980s, responding to multiple alarm fires with multiple fire deaths was not an uncommon occurrence in many of the country's urban areas. Shortly after this time period there was a large campaign on the installation and use of smoke detectors in the home. Any firefighter who worked during this time frame would agree when I say the fire activity dropped significantly after the code changes required the installation of smoke detectors. This rule without doubt has had a major impact on the number of civilian fire injuries and deaths in all residential buildings. A properly maintained smoke detector has become, and continues to be one of the occupant's best weapons for the survival of fire.

However, in recent years, fire fighters are starting to see a growing trend that is causing new concern. Improperly installed detectors, inoperative detectors, or detectors that have been totally removed are starting to become more of a common sight in the private dwelling. Generally after the second or third false alarm of a smoke detector in the home, homeowners or building tenants remove batteries or the detector altogether to eliminate the unbearable noise. With every intention of replacing the battery or detector after the smoke from an overcooked meal cleared, people would often forget for a period of time, or forget altogether. The complacency and misuse of this piece of equipment turned a once protected home now into an unprotected one. The point that should be gathered here is, when you see a concern, correct it.

Fire suppression equipment

Fire suppression equipment in older private dwellings is scarce. However, in some newer and modified construction we are encountering the installation of residential sprinklers. The residential sprinkler is a system that uses the domestic water supply for the home to deliver water through a series of pipes and sprinkler heads to specific areas of the building, most notably the attic, basement, and kitchen. Their installation has become a major asset to the protection of the building's occupants as well as for the protection of the building itself. Currently the enthusiasm for their installation has been slow to catch on in many areas. It is the hope for the future that the same enthusiasm of the smoke detector in the 1970s and 1980s will follow for the residential sprinkler.

Street Conditions: Private Dwellings

*COAL TWA**S** WEALTHS*

Unique or unusual street concerns for the fire officer will be for the most part area- or district-specific. It is obviously within these particular cities that firefighters and fire officers must be familiar with their surroundings. In many cities there may be no significant street concerns other than the occasional double-parked car. Streets in many areas will generally be wide and spacious, surfaces will be paved, and apparatus will be free of any obstructions or delays.

In many other cities, however, there may be more significant concerns presented from narrow streets, one-way traffic flows, to the constant concern of vehicles parked on corners that can limit apparatus from making turns. Wherever the town or city, firefighters and fire officers must be able to identify difficult street conditions and how they may affect their response and operations.

What doesn't help here in this size-up factor is the size of some of the apparatus. Some of these units are monsters; actually further limiting their ability to access certain areas they are expected to protect. While other designed and purchased apparatus are practical for the environment they are expected to work in, some make you ask the question, "What were they thinking about when they bought this?"

A significant amount of planning with a little common sense can go along way when designing and purchasing an apparatus. You know that hasn't happened when a new apparatus is delivered, and it can't fit into the firehouse.

Weather: Private Dwellings

*COAL TWAS **W**EALTHS*

Weather will affect all aspects of a fire department's operations. From slowed movement of apparatus and equipment during a heavy snow, to the early relief and rehab of firefighters necessary during hot weather periods, weather will affect us all in some form. The only difference according to geographic area is when and how much.

Wind

Wind, one of our greatest considerations within the weather size-up, will cause a number of concerns that may affect fire department operations. Depending upon the geographical makeup of your district, all private dwellings may not be built on one-acre building lots. For many, they can be attached or separated by a narrow alley. In these types of settings, the wind will play a significant role with fire extending to adjoining or nearby buildings, as well as carrying burning embers downwind of the main fire area. Winds of as little as 10 mph should concern a fire department; with buildings nearby, the concerns will become quickly evident.

Fig. 2-11 *Wind will easily extend fire to buildings nearby.*

Humidity

Humidity is a recognized concern where high levels can play havoc with visibility, ventilation efforts, and the fire officer's ability to accurately size up the building(s). In congested settings, firefighters on more than one occasion have had difficulty finding the actual building that was on fire as heavy, moisture-laden smoke filled the street. It is during these conditions where members, especially those assigned to the roof must exercise caution as they move about. If you can't see your feet, you definitely need to crawl.

Achieving an effective ventilation of a building may also become compounded when the humidity levels are high. As smoke cools and loses its buoyancy from the high moisture content, the smoke's ability to move as well as lift will become slowed. Conditions producing this type of situation lend themselves well to forced ventilation practices in an attempt to create a tenable and safe environment.

Another and major difficulty encountered with heavy smoke conditions in the street from the moisture content in the atmosphere, is that it can limit the fire officer's ability to accurately size-up the fire building and any threatened exposure buildings. Smoke hovering around ground level will affect every geographical assignment in and around the building. Attempting to gather reports from firefighters to aid in your size-up from the interior, the roof, or areas surrounding the fire building may be of limited assistance under these conditions.

One tool that again proves to be valuable to the fire service is the thermal imaging camera. Standing in the street and scanning the front of the building, its perimeter, as well as the building's roof deck may give significant information on the fire's location and extent. Making assignments blindly under these conditions will waste resources and possibly put firefighters in harm's way.

Exposures: Private Dwellings

COAL TWAS WEALTHS

Exterior exposure concerns of the private dwelling can take on different priorities depending upon a number of factors.

Proximity and construction

Exposure proximity is probably the most obvious concern. When the buildings are attached or separated by a narrow alley, fire officers need to be concerned about fire extension into adjacent buildings very early. We generally consider the closer exposure building to be the more troublesome, and at times it will be; however, proximity concerns can be further compounded by the fire building, as well as by the exposure building's exterior sheathing and its combustibility. Private dwellings can have every imaginable type of exterior sheathing available on the construction market. Wooden clap boards, plywood covered with vinyl or aluminum siding, stucco or brick veneer over wood, to asphalt shingles as the buildings exterior are some of the most common. Their ability to enhance the fire growth will play a direct role with the fire's ability to spread along the building's exterior, as well as to neighboring buildings.

Fig. 2-12 & Fig. 2-13 *Close proximity with asphalt shingle siding enhanced the fire spread in both illustrations.*

Asphalt siding. With those buildings that have asphalt shingles as the exterior finish, their presence will quickly be known by the heavy volume of fire they produce once involved. Asphalt, is a petroleum-based product that can become a highly combustible fuel when heated. Once ignited, tremendous amounts of radiant heat with large columns of superheated smoke will spread fire across alleyways or building lots to nearby structures with little difficulty. Their appearance can be very intimidating from the large volume of fire that may be present.

Where asphalt siding presents the most significant problem is the speed at which their energy can be transmitted to other areas of the fire building and exposure building. Fire transmitted to upper floors, building overhangs, cocklofts, and cornices of both the fire building and the exposure building will have a well-fueled heat source pushing fire into these areas if not quickly controlled. The conditions can be even further complicated when the exposure building and the fire building both have asphalt siding that face or expose each other.

With asphalt siding on both building exteriors radiation feedback is developed rapidly accelerating the fire growth. In the true sense of the definition, the heat that is generated by the involvement of one building exterior will quickly transmit its energy to adjacent building's exterior, igniting that building's sheathing. As the exposure building's exterior becomes involved, it duplicates the process and retransmits its energy back to the original building. In this type of setting, the energy-producing machine that has been created produces a large volume of fire involving the entire exterior of both buildings in a short period of time.

The volume and intensity of the fire described can be further accelerated by the buildings' proximity to each other. If there is considerable distance between the two buildings, the energy will take time to radiate across the opening. However, when the buildings are close to each other, the process is explosive, with the pressure and volume of the fire having nowhere to go but vertically up between both buildings, as well as horizontally through both buildings.

There is no doubt that these fires will appear intensive. Although a heavy fire condition may exist, a sweep with a large caliber hose stream generally darkens the building's exterior in a short period of time. However, I cannot stress enough that the main concern of an asphalt shingle fire is the amount of energy that the sheathing can quickly produce with its ability to push fire into adjoining or nearby buildings. Being drawn to the exterior and concentrating all of your resources on the large exterior fire will allow the extending fire to quickly take possession of the interior of both buildings.

Area: Private Dwellings

*COAL TWAS WE**A**LTHS*

When we discuss the area of a private dwelling, we find that they can vary greatly in size and square footage. In one town alone, variations in size can differ by thousands of square feet. With this comparison in mind, there also must come the considerations of compartmentalization within the building, and the square footage of those areas.

There is no doubt that larger buildings will obviously indicate a larger fire potential. However, in larger and more modern homes, buildings will notably contain larger square footage rooms with increased ceiling heights, which will initially allow for a more concentrated area of involvement. Smaller private dwellings on the other hand will present an overall smaller fire potential; however, smaller rooms with lower ceiling heights will quickly become involved spreading fire beyond the room of origin at a much faster rate.

Smaller buildings with smaller rooms and lower ceiling heights seem to compound conditions throughout the entire living space of the building much quicker than those in the larger more spacious private dwelling. This concern is quickly realized from our understanding of radiation feedback in a space and its contribution to flashover. This causes increased concerns with not only fire extension within the building, but more importantly with the occupant life hazard.

Smaller buildings don't necessarily mean fewer problems; in many situations they can present more.

Location and Extent of Fire: Private Dwellings

*COAL TWAS WEA**L**THS*

Anytime we mention the above size-up factor, it will be from the fire officer's ability in identifying the fire's location within the building that allows the anticipation of the fire's spread. In order to have that ability to act on anticipation, the officer must be able to access information that can be gathered from experience and education with that particular type of building.

Fires that originate in the private dwelling will present a number of opportunities for a fire to travel within the building.

Below grade fires

Below grade fires in a private dwelling come with a whole host of concerns for the responding firefighter and fire officer. The first and most obvious fact is that the fire is below grade. When we mention any fire that is below grade, we must take into consideration a list of common concerns for this specific fire location.

Accessibility. Depending upon the building layout, you will most likely find an interior entrance leading to the basement that is protected by a hollow-core type door or a panel type door at the top of a wooden stair. Exterior entrances into this same area of the private dwelling may be non-existent, but if present, they take the form of an exterior access door or opening that will allow an additional or an alternate route for the fire department if the interior stair is inaccessible. The commonly found, and often flimsy, door at the top of the interior basement stair is no match for a well-involved basement fire. For this reason this particular area becomes one of your primary concerns for placement of the first deployed hoseline at a basement fire in a private dwelling. Engine company members who have often found themselves in this position know that this limited accessibility into a below grade fire, will also come with a tremendous amount of heat. Hoseline advancement will literally be down into a chimney. But as difficult as this position may be, it must be attempted, or at the very least controlled if we want to prevent fire from accessing the living space.

Method of framing. The method of framing used within the building may cause additional or specific concerns with the fire's ability to extend beyond its below grade location. Platform framed homes, although resistant to fire spread via a stud or wall channel, have been known to allow fire spread from its below grade area around the building's chimney.

As we have stated, homes that are of balloon frame design will allow a fire in any unfinished basement to move vertically up throughout the building because of the design of building's exterior walls. Tactics often referred to with this type of fast spreading fire focus around the quickness and mobility of the initial hoselines, with hoseline placement concentrated on the inside of the structure at the exterior walls, and at the fire's ultimate collection point, the attic. We have all struggled with the removal of the wall area on the first floor of these occupancies in an attempt to direct water up the stud channels to slow the fire's spread up. (A difficult task at best, simply due to the fact that you are working on the inside of the structure under adverse conditions.) Add a number of small rooms cluttered with furniture, and the delay of opening up the walls to get a stream of water directed within the space is extended even further.

Alternate tactical options do exist if the members can access the structure's exterior sides. If the balloon frame private dwelling you are operating on is not attached or close to other structures, take advantage of the accessibility to the structure's exterior. Assign a ladder company to initially remove some of the structure's exterior sheathing at the first floor area in any wall space that does not have an inherent fire stop (windows and doors, and diagonal bracing are inherent fire stops within a balloon frame design building). By having the ability to remove vital areas of sheathing in a clear, uncultured atmosphere, a hoseline is allowed quicker and easier access to the void spaces in an attempt to slow the fire's vertical spread.

It must be noted, as successful as this tactic has been for many in the past, its use must be disciplined to ensure that it does not drive fire onto firefighters operating on the inside of the building. Coordination and communication are critical requirements.

Lower-level fires

Interior stairs. Lower-level fires in a private dwelling force us to focus on the most significant fire spread concern in the private dwelling–the unenclosed interior stair. This large, open, vertical artery will allow unrestricted movement of fire and its byproducts immediately to the building's upper floor. With the largest percentage of fires starting on the lower-level, namely the kitchen area, the importance of gaining access to the upper floors, as well as controlling the interior stair becomes quickly realized from the effects of this artery.

In older larger homes, notably the Queen Anne style, there may be multiple stairways within the building. The rear stairs in a Queen Anne were initially designed to lead from the first floor directly to the third floor, bypassing the second floor in the process. This stair was designed as the servant stair, where it would access from the rear kitchen area on the first floor, to their living quarters on the third floor. These stairs are known to be narrow with bends, sharp turns, and small landings most likely with wainscoting or paneling on its walls. Simply by their design, occupant removal and hoseline advancement will be slow and difficult, but fire spread will be very fast.

Fig. 2-14 *Anticipate rapid fire-spread to the floor above.*

Dumbwaiter shafts. It is also in the Queen Anne where firefighters may expect to find the presence of an abandoned dumbwaiter shaft. This shaft originally extended from the kitchen area to the upper areas of the building. Any fire that gains access to this space will allow fire to extend to all levels within the building.

Laundry chutes. In newer private dwellings there is a similar shaft—the laundry chute—that can, if improperly installed and maintained, allow fire to spread vertically as well. Both shafts, either the dumbwaiter or laundry chute can measure from $2ft^2$ to $3ft^2$ and run the entire height of the building. Depending upon design, age, or alteration, any fire that penetrates these openings will be allowed to move freely up or down.

Ductwork. In homes that contain central heating or central air conditioning, firefighters must expect fire and smoke penetration in and around duct openings. Openings for these systems will penetrate all levels of the home. Their presence can allow fire to spread into wall, floor, and ceiling spaces with little resistance.

Kitchens and baths. If the building is a two-family occupancy, fire spread concerns have extended to the presence of stacked kitchen and baths, or back-to-back kitchens and baths within the building design. Their installation, which originally served to meet the economics of placing all the utilities in a common area, will cause fire extension concerns to those attempting to confine a fire to one of these rooms.

Top floor and attic fires

Fire location and extent concerns at the top floor and attic level of the private dwelling will present varying concerns to the firefighter and fire officer notably by building design.

Fig. 2-15 *Involvement of attic areas will present accessibility concerns.*

For example any fire exiting a window on the top floor of a Queen Anne will cause a severe fire extension concern into the large overhang eave areas of this type of structure. The upper-level of a Queen Anne by design will have a significant number of hips, valleys, ridges and dormers where fire can burrow and take possession. When fire extends into the finished attic or dormer area of these buildings, firefighters must be prepared for a very difficult fight. The finished design of this floor space will incorporate a number of knee walls throughout the area that will compound the difficulties of opening up the void spaces, further allowing a fire to take possession of the entire top floor and attic.

Similar concerns can also be applied to the attic space of other private dwellings. Homeowners seeking additional storage areas in the attic will often install plywood over the ceiling joists to create a floor for storage. This will create extreme difficulty when attempting pull down or push down in any ceiling. Direct access to the attic space can be limited, and generally only attainable from a set of pull-down stairs mounted in the top floor ceiling, or through a small opening in a closet ceiling. If available, the pull-down stair will allow a hoseline the best access to sweep to area when the attic is involved in fire. It must be noted though, that opening this space before the attic space is vented will create difficult and possibly dangerous conditions on the building's top floor.

Time: Private Dwellings

COAL TWAS WEALTHS

Time of the day

Time of an incident and how it will affect our decision-making in a private dwelling will reflect on the occupancy and life hazard size-up. The occupant status at the specific time of the incident will definitely alter our concerns as we respond and go to work at a private dwelling. As a general concern we can say that the most significant time concern for a fire in a private dwelling will be during the late evening and early morning hours. It is during this time frame when most occupants are losing their lives in private dwelling fires due to the fact that most, if not all of the building's occupants will be asleep in their bedrooms. Any fire in a private dwelling during these critical times will require firefighters to aggressively direct their energies to accessing and protecting these areas. Search operations through the interior as well from the exterior are a high priority.

Time of the year

Time of the year plays another role in our time size-up concerns, specifically as it relates to the holiday season. From the most obvious holiday traffic and its association with delayed response, to the actual celebration of the season with the holiday trim and decorations, the difficulties seem to mount during this time of year.

Another and often overlooked fact that has to be taken into consideration during the holiday season is that more alcohol will be consumed during this time than any other. Alcohol will dull the senses of an individual affecting the ability to react. An individual under the influence will not only become careless, but once asleep may not awake at the sound of the smoke alarm.

It is also during a holiday period that we can expect an increased occupant load within the building. With the possibility of visiting family and friends, the number of people occupying a home during this time can be overwhelming. Just the fact that the guests will be unfamiliar with their surroundings will add to the concerns of the increased occupancy within the home.

Besides a normally congested setting, a number of cars in the driveway as well as in front of the home may be a tip that the building where you're about to fight a fire may be full of people.

Height: Private Dwellings

*COAL TWAS WEALT**H**S*

The height of the private dwelling can vary from one to three stories. Barring accessibility concerns in the form of overhead wires, trees, or terrain, most will be easily accessible by fire department ground and aerial ladders.

Peaked roofs

A significant concern with many private dwellings is the peaked roof design. The peaked roof causes operational considerations to those firefighters who may pursue roof ventilation if the fire involves the top floor, attic area, or if fire is extending vertically within the wall of a balloon frame building.

The pitch of the roof must be taken into consideration as firefighters attempt to gain height on a peaked roof building. A sloping roof is measured by its pitch. A pitch of "2 in 12" inches indicates that there are two inches of roof rise to every twelve inches of roof rafter span, subsequently a very low pitched roof. A "14 in 12" inch pitch indicates that there are fourteen inches of roof rise to every twelve inches of roof rafter span, an obviously very high pitched roof. The specific numbers associated with the roof pitch are relevant numbers for the building contractor who is constructing the roof.

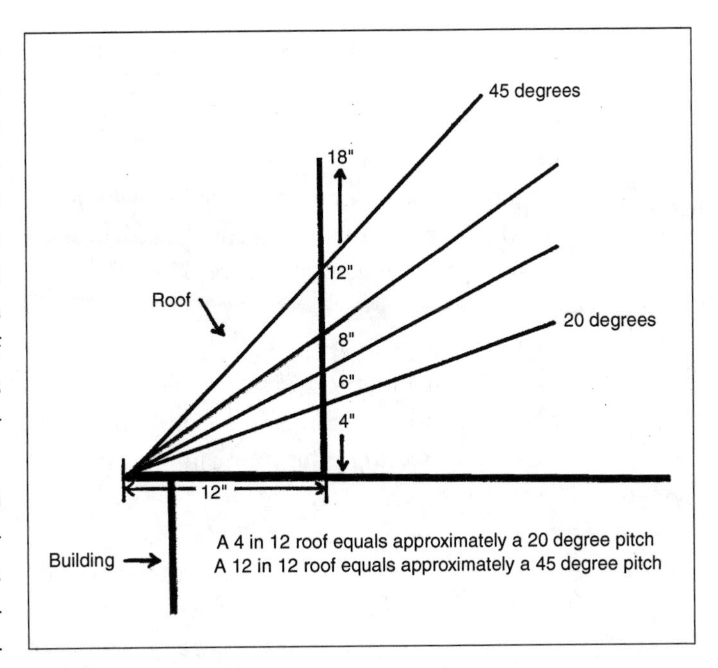

Fig. 2-16 *Roof pitch*

These numbers although good to know, may be difficult for fire fighters to apply. Besides pulling out your carpenter's framing square, firefighters new to the business can use a simple guide in determining pitch. We all know what 90 degrees from a horizontal plane looks like, as well as a 45-degree angle from the same plane. Use as you're size-up guide, 20 degrees or approximately half of the 45-degree pitch as a minimum number for use of a roof ladder to assist with member stability on a roof. Look up and take an estimated guess on the degree or pitch with the number 20 being your guide. If in doubt, simply pull the roof ladder out to enhance the safety of the work area.

It is important to note that this alone will not ensure a secure and stable roof surface to work on. Additional factors may warrant you to use a ladder even on roofs lower than 20 degrees, or eliminate the roof ladder altogether and work from the protection of an aerial ladder or platform. The type of roof deck and whether it has a wet, dry, or mildewed surface are just a few factors that may alter your decision-making. Others may include the type of roof support system.

The type of roof support system is a major factor that can eliminate any use of a roof ladder. Roofs that are supported by truss design will not contain a ridge beam that we so often associate with a framed roof. Generally found on newer and larger private dwellings, a lack of a ridge board negates the placing of a roof ladder over the ridge for support. If the roof deck fails near the ridge area, there will be no ridge board to support the hooked end of the ladder. The roof ladder will fall into the attic space bringing firefighters down with it.

Just as we gather information on obtaining height, we must also watch for items that may fall from heights at a private dwelling. Private dwellings of new construction are continually attempting to gain the aesthetic look of yesteryear. Slate roofs, which are often found on Queen Anne structures, are making a comeback into some newer constructed homes. Slate roofs whether new or old are very heavy. Not only do they increase the dead load of the roof deck (which in turn causes an increased collapse concern), but when these pieces start to fall they can literally become guillotines which could cut or shatter a firefighter's arm.

As we attempt to gain height at a private dwelling fire, firefighters must also be aware of two additional concerns that can cause serious injury to a member. They are the possible presence of "Yankee Gutters" and "Hollywood Fronts".

Yankee gutters

Yankee gutters (also referred to as wooden or box gutters) are still a common sight on many of our older private dwellings. A molded or fabricated piece of wood averaging from 6" to 12" in width, the gutter often spanned the entire width of the house as one solid piece. Designed to divert rainwater from dwelling roofs, it was constructed to form the classic "U" shape associated with a rain gutter. Gutters of this design can use a number of different methods in order to be fastened to the building's exterior. In many cases, they were often nailed directly to the roof rafters, eliminating the need to install a fascia board behind it. In other cases, copper straps attached the gutter to the roof deck surface, or a decorative wood support was fastened at the roofline, supporting the gutter from below. When initially installed (in some structures, this was more than a hundred years ago), this ornamental piece of wood was often painted, stained or wrapped in copper to match the decor of the building.

The threats the gutters can pose to firefighter safety arise from the fact that many of these gutters have not been properly maintained. At one time the gutter performed its function of catching and channeling rainwater from a building's roof deck to an awaiting leader (vertical pipe) that would eventually take the water to the ground. However, the continuous exposure of the wood to rainwater, other weather elements, and just simply the failure to maintain them has caused many of these gutters to dry out and rot. More often than not these once decorative features become nothing more than rotting, decaying pieces of hanging wood.

Fig. 2-17 *Yankee Gutter coated over the years with flashing cement*

The risk to firefighters will not be eliminated even on those structures where attempts were made to repair the gutter. Roofers will generally make two types of repairs to the Yankee gutter to save money for the homeowner. The first approach was to coat a leaky gutter with roofing (flashing) cement to prevent further rot, as well as allow some type of water runoff. If the gutter continued to leak, the roofer then applied a new layer roof material and covered the gutter altogether. In many cases this could make the gutter appear almost undetectable, and could actually give the false impression that there were 6" to 12" more of roof deck. This repair was often followed by adding a flimsy aluminum cornice that made the gutter even more undetectable from the street, possibly hiding from firefighters the inherent dangers the gutter presented.

Another major concern that was discovered with the Yankee Gutter was that building contractors were taking shortcuts when applying new vinyl or aluminum siding to the building's exterior. Unconsciously, many were removing the wooden supports below the gutter where the gutter's weight was being carried. Instead of boxing around the wooden supports,

Fig. 2-18 *Yankee Gutter covered by new roofing material*

Fig. 2-19 *Yankee Gutter hidden by aluminum cornice*

Fig. 2-20 *Wooden support for Yankee Gutter*

contractors would totally remove the supports so they could continue applying straight pieces of siding. With this type of installation, the gutter was now completely unsupported.

Hollywood fronts

Hollywood fronts are another concern for firefighters when gaining height at a private dwelling fire; when they are not identified they can become a danger. The concept of a Hollywood front first gathered its name from the motion picture industry of the west coast. In order to create the look of an actual building without totally erecting one, set designers would create just a facade. This facade was nothing more than the front or shell of a building with all the windows, trim, paint, curtains, and blinds in place to give the illusion of an actual structure. If someone were able to view the rear of this wall, all they would see is structural bracing holding a building facade in place.

This concept, although a little better constructed, is a common sight in many towns and cities with many upper-levels of our private dwellings today. On what was originally designed as a two- or three-story flat roof building, a false peak, or what it is more commonly referred to as a Hollywood Front, was erected to give the illusion from the street of a peaked roof building. This addition to the actual roofline was done for esthetics as well as to give the look of a much larger square footage building. If viewing the building from the side or corner, a firefighter would actually see a two-story flat roof building with a peaked front. This peaked front can extend above the roof deck surface from 8′ to 12′, and be as wide as 6′ to 8′ from the front of the building. The problem that this front creates occurs when the peak must be laddered from the street side at night or during difficult weather conditions. An unsuspecting firefighter may think he is climbing onto a peaked roof when in actuality a few steps may drop him 10 feet to the actual roof deck below.

Fig. 2-22 *An elevated view of a Hollywood Front from the "D" side.*

Fig. 2-21 *A Hollywood Front viewed from the "A" side.*

Another illusion created by a Hollywood Front that may add to the firefighters confusion is that some will actually have a window(s) installed in the peak, which from the street may give the appearance of an attic apartment. Although the width of a Hollywood front may vary, the space, if it is accessible at all, may be used as a storage/attic space. But never rule out the possibility of a room in this space, especially if the width of the Hollywood front is in excess of 8′. I have personally seen some creative designs with curtains and blinds installed which can further confuse those as they view the building from its front.

These types of designs further emphasize the need for firefighters to attempt a view of more than just the street side of the building as they arrive on the scene. At times when the building is setback and attached it may make it difficult to see the false peak. The width of the false front may exclude your ability to totally see it from any side. A visual cue from the street side that is often present with a Hollywood front is obtained by viewing the building's corner where the peaked slope meets the top floor. It is in this area where you may see a small section of the actual roofline, tipping off members to a flat, not a peaked-roof building.

One of the few positives with the Hollywood Front that might actually aid the fire department is the area that encompasses the windows or vent installed in the gable end. Depending upon when it was installed and how the area accessed the original construction, removal of the windows or vent may give direct access to the building's cockloft. By placing a tower ladder or aerial ladder up to the gable end, and then removing as well as enlarging the opening, accessibility into the space will aid in ventilation of the cockloft area, or depending upon conditions, the operation of a stream into the space.

Fig. 2-23 *When viewed from the corner you may see the actual roof line.*

Fig. 2-24 *Gaining access to the cockloft*

Special Considerations: Private Dwellings

*COAL TWAS WEALTH**S***

Any special thoughts, strategies or concerns that the fire officer must consider for the private dwelling will be, for the most part, address-specific. Simply, know your response district and the type of buildings represented within it. A fire officer knowing that the private dwellings in his/her district are all 3000ft^2 homes on 200′ x 200′ lots, compared to a fire officer who is responding to a row of closely spaced homes measuring 20′ x 40′ on 25′ x 80′ building lots, should definitely define each other's way of thinking.

It is within this size-up factor that any information that can be gathered from your previous experience, education, and anticipation about an incident should be mentally noted and referenced.

What follows are just a few considerations and suggested actions that a fireground commander can use for a fire in a private dwelling.

Private dwellings

1. *Occupant Life Hazard.* Prioritize areas within the building based on the fire's location and extent. Assign firefighters to the primary search of fire floor and floor above, with early consideration given to the bedroom areas by any means possible. With multi-floor homes, aggressive ground laddering is a must.
   A. If you arrive at a Queen Anne or Victorian home that is a converted multiple dwelling, or has SROs (single room occupancies), call for additional help; you will need it to replace those members consumed with the added forcible entry and search concerns.
   B. Place water between the fire and the building's occupants. This has to be the consideration of your first stretched hoseline. Depending upon the building design and layout, the hoselines approach may be from a number of different avenues, but it should use the quickest and most effective way possible.
2. *Fire Location and Extent.* As soon as this information can be identified, obtain information on the condition of the interior stair(s). If threatened, actions should protect and control this area.
   A. If the fire building is attached or closely spaced, assign companies to check exposures early. If there are any threatening conditions, assign firefighters and equipment to those areas in anticipation of the fire spread.
   B. Balloon Frames and Queen Annes are going be difficult and very dangerous fires. Never underestimate the fire's ability to eat away at the building's structural integrity as you continue to assign firefighters to these very old, large buildings. When fighting a fire in these type buildings take note of the following:
      1) Any fire on a lower floor will require examination and ventilation of the attic area.

2) If the fire is running the walls, pull walls inside the building, and pull sheathing outside of the building so you may direct water up stud channels.
3) If the fire has involved the knee wall area and/or the top floor, extensive and cautious efforts will be required to open up the involved areas.
4) Direct members to use extreme caution when opening up any concealed spaces and voids within this type of building. Unsuspecting conditions can overwhelm forces.
5) Use thermal imaging to determine the fire's location and avenues of spread. This is a tool that when available must be one of the first pieces of equipment off the apparatus. (If you don't have one, get one. Once you get one, you'll want more!)
6) Beware of radio reports to you that indicate the fire's increasing possession of the building's void spaces.

3. *Firefighter Safety.* Make sure your resources can support your assignments. Building fires are staffing intensive; don't hesitate to call for help. If you don't need them, you can always send them home.
   A. When present, advise members of the Yankee Gutters and Hollywood Fronts; notification can prevent a serious injury or death of a firefighter.
   B. Note the pitch of the roof. When greater than 20 degrees remind your people to take a few extra minutes and use a roof ladder.

Questions For Discussion

1. What is considered the most difficult and most task-demanding type of Private Dwelling fire that firefighters can encounter? Identify and list the reasons. Are there any of these types of buildings or similar types of buildings in your response district?

2. In the Life Hazard section of this chapter we took a look at the Tot-Finder program of years past. Discuss the pros and cons of the program, and what it would take to develop a new, as well as a continually updated program to better protect children in residential fires.

3. What factors determine the hoseline selection and stretch at a private dwelling fire?

4. Identify and discuss the life hazard concerns that a firefighter is presented with when arriving and going to work at a private dwelling fire.

5. What is an often-overlooked factor when determining fire flow formulas at a private dwelling fire?

6. What factors are considered when choosing a certain size hose line for a fire in a private dwelling? Do these factors differ in your own city? If so, why?

7. Identify and discuss how humidity can affect fireground operations. Once realized, what can we do about it?

8. What are the concerns with asphalt siding and how does it affect your decision-making?

9. Do you have any Yankee Gutters or Hollywood Fronts within your response districts? If so, how can they be identified? How can their hazards be avoided or planned for?

10. What considerations factor into how a firefighter accesses a peaked roof?

3 Multiple Dwellings

To gain a clear understanding of the fire concerns of the multiple dwelling, we need to categorize them for ease of reference. I found that the terms or building characteristic names differed slightly depending upon the city or area of the country visited. In this chapter I concentrated on the different types of multiple dwellings based on the more common terms and characteristics found. Nevertheless, if you have any of the types of buildings listed and described in this chapter, or if you have buildings that are similar in design and characteristics, it is important that you know as much as you can about them well before the receipt of the alarm. A fire in a multiple dwelling initially will be one of the most strategically, tactically, and task-demanding incidents a firefighter and fire officer will encounter.

A multiple dwelling is described as a building whose dwelling units are rented or leased, and is occupied by three or more families living independently of each other.

The types of multiple dwellings described in this chapter will include:

- The Tenement, old and new
- The Non-Fire Proof Apartment building, including the H-types
- The Housing Complex, including low-rise and high-rise

Note: Row Frames, Brownstones, and Residential High-Rise will be referred to in their own chapters due to their different and unique characteristics.

Construction: Multiple Dwellings

__C__OAL TWAS WEALTHS

Tenements

A building referred to as a tenement can be constructed from either a Class 5 wood frame or a Class 3 ordinary type of construction. Building heights associated with the frame tenement are generally from two to four stories with four to six stories average of the Class 3 constructed buildings. The widths of the structures can average from 20′ to 40′, with depths ranging from 40′ to 90′.

Fig. 3-1 *Old style tenement*

Old style tenements

In the older tenement, (those built in the late 1800s) firefighters will encounter a combustible interior within a wood frame building. The interior stairs in these tenements are not only narrow, but the treads, riser, and railings are all made of wood. The internal combustibility also extends to the stair that accesses the basement. This wooden set of stairs that leads from the first floor hallway to the basement will be found under the first floor staircase. With a wooden floor throughout the building and some wainscoting on the hallway walls, what you'll really discover in the older style tenement is a neatly arranged lumberyard in an occupied building.

Fig. 3-2 *New style tenement*

Newer style tenements

The newer tenements (those built in the early 1900s) made some improvements to the interior by eliminating some of the combustible design of the older version. Buildings built during this era were of Class 3 ordinary design. Basement/cellars in the newer tenements are (for the most part) going to be finished, protecting the building's structural members from any fire involvement below. Another significant improvement is that there will be no interior access to the basement/cellar entrance as is in the older style tenement. Accessing the basement of these types of buildings requires firefighters to exit to the outside and enter the basement through an exterior entrance. The interior staircase within the building also will be significantly different than that of the older tenement. Staircases have incorporated into their design, treads, risers, and landing platforms made of cut stone or marble placed within a steel frame design. Not only will these stairs be larger, but they also will be much stronger when compared to the wooden stairs of the older tenement.

Light and air shafts. A construction feature that is common to both the older and the newer tenements is the light and airshafts that serve the fire building as well as the exposure building. The original intent of the shafts was to do actually what its name implies—to allow light and some airflow into an apartment. Their presence creates a concern because once fire enters the shaft, it will expose all apartments served by the shaft in not only one, but also two buildings.

Railroad flats. Another construction design that is common to both types of tenements is the railroad apartment or railroad flat. Rooms in a railroad flat apartment are laid out from the front of the building to the rear of the building in a design similar to railroad boxcars on a single piece of track. Depending upon their layout within floor space, there are often two entrances into these apartments that may or may not be used by the fire department. The most commonly used entrance is generally right off the staircase landing. This doorway will lead directly into the apartment's kitchen. The other, if present, will be found at the remote end of the stair landing which opens into a bedroom. The later in most cases will be found locked, bolted, and blocked by furniture. This common association forces the fire depart-

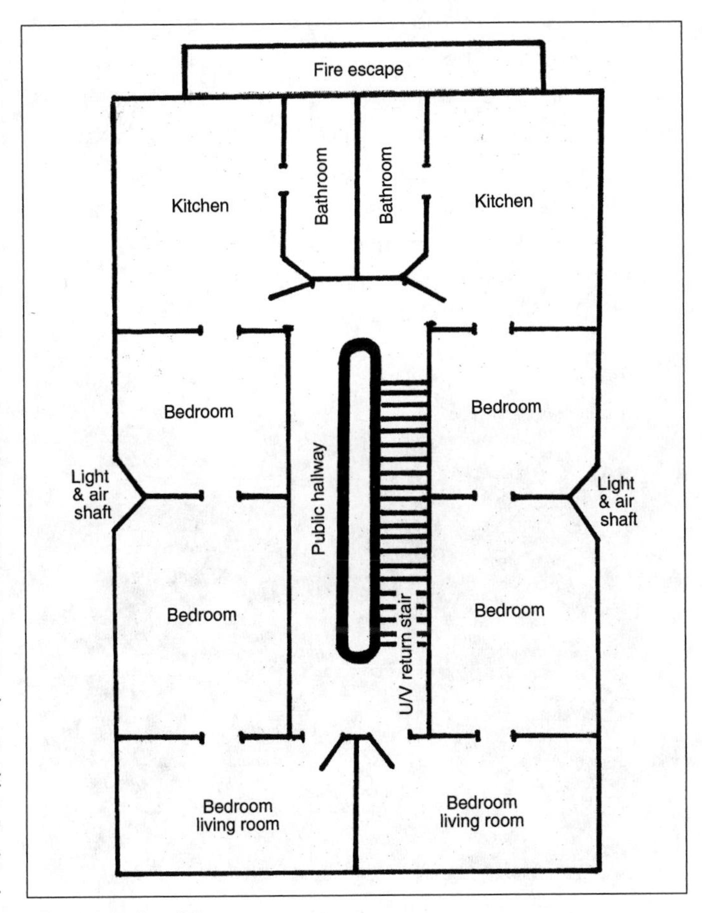

Fig. 3-3 *Railroad flat apartments*

ment to enter through the kitchen entrance, which, depending upon the fire's location, could create a difficult and long hose stretch. Remember, these can be deep buildings.

Kitchens and baths. Firefighters and fire officers must also be aware of the fact in their construction design that *stacked kitchens and baths* will exist in the building. Whether the tenement is considered young or old, the concern of stacked kitchens and baths will always be there. This is one construction feature that must be expected in all multiple dwellings.

H-Type and NFP Multiple Dwellings

The H-Type multiple dwelling, also referred to as a non-fireproof apartment building received its slang term from the unusual design and layout of the structure. When looking at a plot plan of the structure, these buildings can take on the shape of the letter H, U, O, T, C, E, or double E. The letter H shape seems to be the most common design used, therefore, the term H-type. Within the construction design of these Class 3 constructed buildings, the apartments will be located in the wings or vertical lines of the letter H, with elevators, stairs and a large entrance lobby located in the throat, that portion that connects the wings of the building. The original intent of the letter design of the building was to incorporate garden, playground, and sitting areas into and around the wings of the letter design of these large occupancy buildings. Basically what they were trying to create is a community within a community.

Fig. 3-4 *E-shaped apartment building*

Fig. 3-4a *Throat area*

Hybrid concern

To meet the needs of the expanding population of the urban areas of the early 1920s and 1930s, these large structures of five to seven stories introduced a new form of construction not directly recognized by one of the five classes. When initially viewed, the building took on the form of the Class 3 ordinary construction, where you would expect to find masonry exterior walls and wooden members that support the floors and roofs. In order to pick up the spans associated with the larger floor areas, as well as with the overall height of these structures, structural steel in the form of columns, beams, and girders were used to form a steel skeleton within the building. What causes the real concern for the responding and operating fire forces is not so much that it is not a specifically recognized class of construction, but the fact that the steel that is incorporated into the building will not be totally protected. In many cases the steel will be exposed in the basement as well as the void spaces it creates. The obvious concern with unprotected steel is that it is subject to expansion and elongation from the heat of the fire. Prolonged exposure may cause the steel to dump its load where it interconnects with the wood floor supports. But a more common concern will be from the open vertical channels it forms from the letter design of the I-beam. This open channel on either side of the I-beam, or channel rail, allows heat and fire to travel literally undetected in its opening until it breaks out into an open area like the cockloft, or involves its boxed out plaster and lath covering.

For the most part, the structural steel incorporated into the building design is going to be hidden from view. The structural steel supporting members may be found in any boxed out area within an apartment.

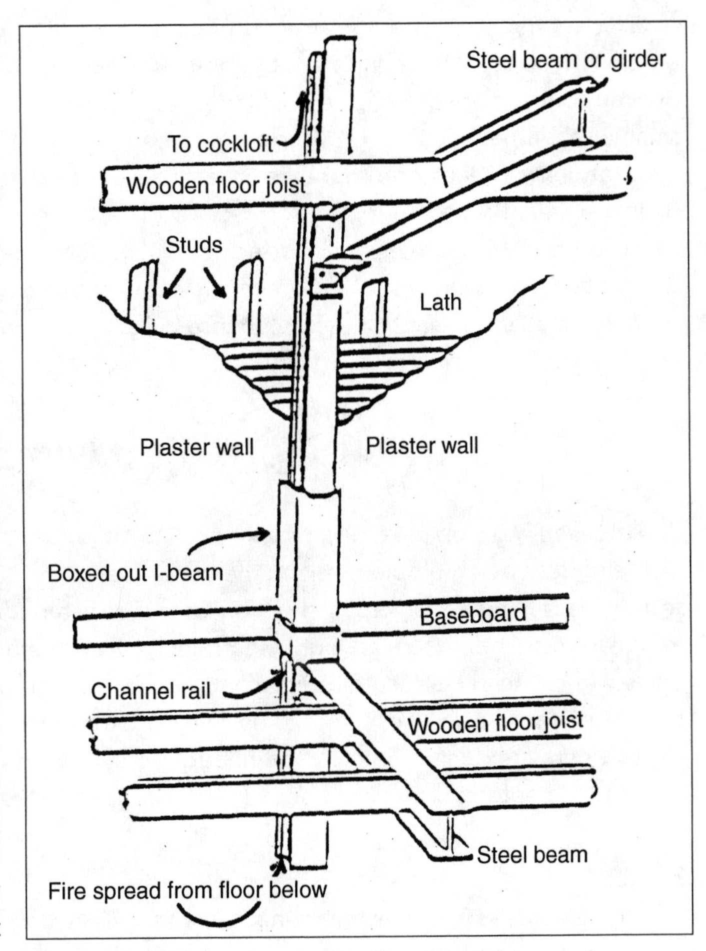

Fig. 3-5 *Steel I-beam construction/Channel rail fire spread*

Generally to eliminate their presence within the living space of the apartment, a boxed out area of plaster and lath within a corner of a room or within an apartment closet, will indicate their location. If fire is running the void spaces of a building of this design, these areas must be opened and checked for fire travel.

Cocklofts

The cocklofts in the H-type and NFP multiple dwelling encompass a very large and undivided area. These spaces average around 3′ to 4′ in height with spaces up to 6′ not uncommon. Often from the exterior you will see vents that serve these areas, not only tipping you off to their presence, but also indicating any possible involvement of the space. The fire loading of these areas will be substantial, with wooden roof beams, wooden roof decking, and a wood framework for the top floor ceiling with wood lath to hold the ceiling's plaster.

Roof design

Depending upon the age of the building, you will most likely find an inverted roof on the H-type and NFP apartment building. In a true flat roof, the roof decking is nailed directly to the roof rafters. But in order to incorporate a pitch to enhance water runoff, small lumber, namely 2″x 4″ wooden members, were nailed perpendicular to the top floor rafters with each one slightly smaller than the next as they progressed toward the drainage area. From here roof rafters were added to the vertical lumber with a roof deck nailed on top. Altogether, the inverted roof is another neatly arranged lumberyard.

It is from the wooden structural members throughout the building, combined with the use of unprotected steel, that many refer to these buildings as non-fireproof multiple dwellings.

Housing/project complex

The Housing Complex, or a project or public housing development, will be the best built and designed multiple dwelling the urban firefighter will face. This is specifically due to the Class 1 fire-resistive construction of the structure. The low-rise, which averages from two to six stories, to the high-rise project, which averages from 12 to 14 stories, will come with protected structural steel, concrete floors, walls and ceilings, all behind a steel framed door. Fires in these structures seldom extend beyond the apartment of origin. Cases where fire does extend to another apartment will most likely be from fire exiting a window and entering the floor above (auto-exposure or fire lapping).

Heat

In the housing complex, the benefit of a serious lack of fire spread will come with the added concern of high heat retention associated with these fortified structures. Simply due to their design, fire forces will not only find it more difficult to force apartment doors, but they must expect oven-like conditions from the heat of the fire.

In addition to the punishing effects from the fire, this type of building, namely the high-rise housing complex, will contain some building features that will give added concern to fire department difficulties.

***Fig. 3-6** High-rise housing complex*

High-rise housing

What you will find in the high-rise housing project that is a concern to fire department forces are compactors, incinerators, scissor stairs, and the odd/even elevator system. Neither one of the above mentioned will strike a blow to the fire department operations if prepared for.

Compactors and incinerator chutes. Compactor and incinerator fires will create some of the more difficult smoke problems encountered in these structure types. The fact that these shafts penetrate each floor area of these Class 1 structures, reminds us that the shaft can and will allow fire and its by-products to have access to each floor of the structure if given the opportunity. When present, each shaft will serve the public hallway of each floor of the building. The chute openings, if not properly closed, can emit smoke onto each floor. Often this is due to the hopper door of the chute not self-closing from general wear, vandalism, or the door being blocked open. The difference between the two different shafts is that one is designed to handle fire and its by-products, and the other is not.

Compactor chutes were not designed to handle any size fire at all. The shafts that serve the compactor are thin-walled, and in some installations, are required to have auxiliary appliances in the form of either a sprinkler system or heat detection system to help detect and suppress a fire within the shaft, however, don't bet on their reliability. A significant fire in a compactor room or in a compactor shaft will require examination for any possible spread along the shaft's route.

Incinerators, which are becoming more rare due to their environmental concerns, will resemble a large chimney. Residents on each floor take their garbage to the chute opening on the particular floor and drop their garbage down the shaft into the awaiting incinerator. Problems with the incinerator shaft will arise when the rubbish that is placed in shaft, is too large for the opening and gets stuck within. Two things can happen here: one, smoke will back down below the blockage through chute doors below the stuck rubbish; and two, heat within the shaft may ignite the stuck rubbish. In either case, the concern is that the fire must be extinguished and blockage must be cleared.

Primitive, but effective methods of clearing a blockage have included dropping a brick through a chute opening on a floor above the blockage. If you need more clearing power, a Belgian block or small cinder block dropped through the roof opening will generally do the trick.

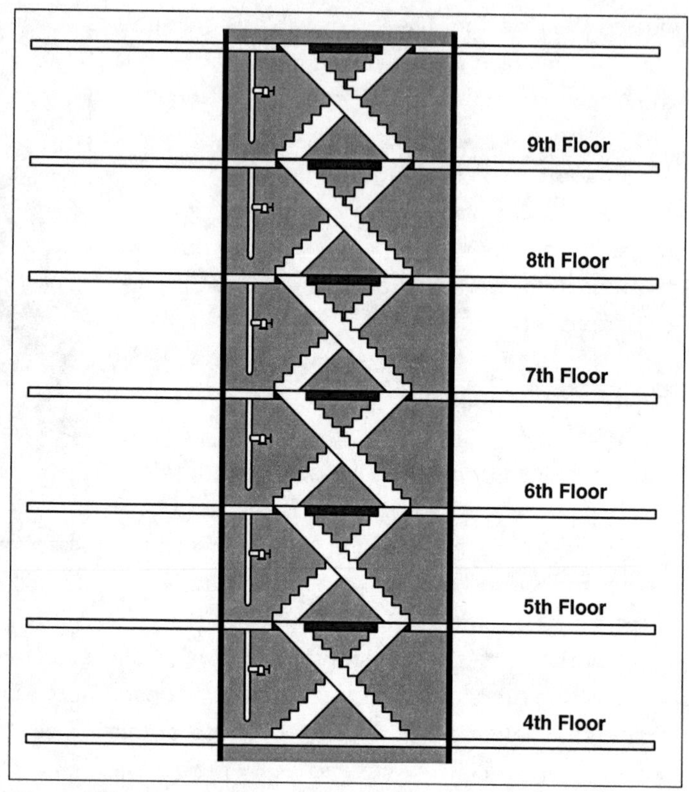

Fig. 3-7 *Scissor stairs*

Scissor stairs. Scissor stairs are a common site in the high-rise complex. When confronted with this stair design, you must remember it is important to determine which stairwell serves what area of the floor. In a typical scissor stair, each staircase serves alternate sides of the floor from a common stair shaft. Members exiting one staircase may find themselves on the opposite side of the stair core. This may cause an extended hose stretch, as well as a concern with firefighter disorientation.

Odd/even elevators. The odd/even elevator is commonplace at the high-rise housing project. With two elevators, one serving odd number floors and the other serving even number floors in a common shaft enclosure, firefighters knowing which one serves which numbered floor is necessary. Complacency may allow you to open the doors to the elevator on the fire floor, instead of on the anticipated floor(s) below.

Fig. 3-8 *Odd/even elevator system*

Low-rise housing

The low-rise housing project eliminates many of

the difficulties mentioned with the high-rise housing project. Due to their low height, you will find no elevators, no compactors, and no incinerators, within a building with a stair shaft that serves each floor area.

Where we found difficulties in the most recent years in many of these complexes researched was with their improvements, namely in the form of new courtyards, playgrounds, energy efficient windows, and rain roofs. I can remember as a young firefighter being able to drive the apparatus over the curb right into the open courtyard for quick access to any one of the buildings within the complex. Not today. Long hose stretches and long walks with ground ladders are a common practice.

Fig. 3-9 *Rain roof under construction. Note lack of ridge board, typical of the truss design.*

Another modernization concern is energy efficient windows. What we already consider a building known to produce a hot and difficult fire, has just gotten worse. Increased occurrences of flashovers and backdrafts are a constant concern for firefighters and fire officers who respond and operate in these types of buildings.

Rain roofs. Rain roofs are being added to many of the low-rise housing complexes to eliminate the constant roof repairs that plagued the original flat roofs of these structures. This addition was an obvious plus from a maintenance view, but a problem if fire extended into the lightweight wooden truss structure. In most cases, this caused a concern when a fire would exit a top floor window into the large soffit area of the roof overhang. Often the only barrier or protection you will find in the overhang is a thin sheet of plywood or particle board, if you were lucky, or nothing more than a piece of aluminum stapled over the framing.

Fig. 3-10 *Even when the construction is completed, a significant top floor fire will easily find its way into the truss loft.*

In what was once considered a fire that wouldn't extend beyond the apartment of origin, now may involve the building's entire roof structure and deck.

All multiple dwellings will present significant construction concerns to the fire service. Each will be unique in its own way simply based on age and design. It is critical to understand the construction design and features of each within your area of responsibility.

Occupancy: Multiple Dwellings

COAL TWAS WEALTHS

Tenements

To give general references on the occupant load of the tenement, we can state that on an average, there will be anywhere from two to four apartments per floor for what has been described as the older style tenement. To assist with this determination of the number of apartments that might be within the building, there may be a few clues that many in the fire service have used to give a quick reference.

First, is the number of mailboxes and doorbells in the vestibule of the first floor. This simple and often overlooked piece of information may at times be your most obvious. Another tip that may be less obvious because of obstructions is the size of the building. These structures are known to average the same street width, approximately 20′ to 40′; however, the depth of the buildings can vary as much as 40′ to 60′ with some reaching up to 90′ in length. Buildings in excess of 60′ will most likely have four apartments on each one of their floors.

Another possible indication of the number of apartments that are on each floor may come from the presence of a front fire escape in addition to a rear fire escape on the building. With only two apartments per floor, only one party balcony fire escape is provided for the occupants of each floor. This fire escape, that also serves the roof, will be in the rear of the building. If you find you have a front fire escape in addition to the rear, you can anticipate the presence of four apartments per floor.

In the newer version of the tenement, the structure was designed to be larger in order to house more families. As an average reference, you can expect to find four to six apartments per floor in these larger buildings. In this version of the tenement there are no clues to the exact number of apartments in the building other than the number of mailboxes and doorbells in the vestibule of the building. It is important to note that this observation will *not* give any information to the presence of any basement apartments or single room occupancies in either the old or new type tenements.

NFP/H-type multiple dwellings and housing complex

The non-fire proof apartment building; H-type multiple dwelling and housing project complexes will have a much larger number of occupants when compared to tenements. Size and layout of the structure may dictate the possible numbers of apartments, but on an average, you can expect to find anywhere from 12 to 16 apartments per floor.

Fig. 3-11 *First-floor commercial occupancy within a multiple dwelling.*

Depending upon wing design in the H-type, or the size of the housing project, the number of apartments per floor may vary based on accessibility. It is important to know how many, as well as which apartments will be accessible from a particular stair within these buildings. Complacency in choosing a stair may find you in another part of the structure.

Another consideration of the NFP and H-type multiple dwelling as well as the housing complex, is the possible presence of a commercial, health care, or educational property within the complex. In some of these complexes you may find a grocery store, a day care center, or even a health clinic on the first floor of one of the buildings. The intent of the added occupancy in these "small cities" was to make the complexes self-sufficient, eliminating the need for people to travel for these services.

Apparatus and Staffing: Multiple Dwellings

COAL TWAS WEALTHS

The complement of firefighters and equipment responding to these types of incidents will obviously vary from department to department. What must be remembered when responding to a fire in any multiple dwelling is that the main strategic factor at these incidents is always going to be the life hazard. There should be no doubt that these buildings can contain large numbers of people. It would be best, to have a sufficient number of troops responding to the incident with you.

If the response staffing is minimal, decisions and assignments must be prioritized until more resources arrive. It's one thing to be able to out-resource a fire, it's another to outthink it.

Although departments' strengths may vary on a first alarm assignment, there are some key strategic and tactical concerns of the engine and ladder company that can be shared for operations in these types of occupancies.

Tenements and NFP/H-Type Multiple Dwellings

Engine company operations

Each engine company operation will evolve around the four specific concerns of water supply, hoseline selection, stretch, and placement. Water supply concerns for these buildings will obviously be based on availability and the department's procedure on how to use it. What can be looked upon as general points of consideration for the engine company for a tenement or in another multiple dwelling fire is hoseline selection, stretch, and placement.

Hoseline selection. Hoseline selection for the tenement and the NFP/H-type multiple dwelling must consider small room sizes, light fire loads, the mobility to operate hoselines in these areas, and the speed required to place and operate them. To meet all of these needs, most fire departments will choose the 1¾" or the 2" attack line. This is not to say that a larger hoseline is not an option in a multiple dwelling. When forces are confronted with a large fire, a first floor commercial occupancy, or a significant wind affecting the fire's spread, the capabilities of a bigger stream with an extended reach is going to be required. However, for the large majority of incidents we respond to in these type buildings, speed, mobility, and efficiency are going to be the pre-requisites for choosing the medium-size hoseline.

Hoseline stretch. When identifying the hose stretch, the number of lengths of hose as well as their size, may be altered by the distance. As an engine company officer, if your size-up calls for a 1¾" hoseline as the initial hoseline, you'll need to determine the distance of the stretch first, so you can ensure an adequate flow of water once you get where you want to go. The tenement, for the most part, is easily accessible from the curb, with very little, if any, setback from the street. However, there may be an added concern with the total number of lengths for the tenement, especially when the fire is in the rear and on the upper floors. Remember some of these buildings can be deep. Any time your 1¾" hose stretch exceeds six lengths, you need to fill out any length beyond the sixth with 2½" hose. A disregard of the number of lengths in a long hose stretch, will quickly be realized as friction loss overrides the flow of the deployed hoseline.

Never has the concern on the hose distance been more realized than in the H-type multiple dwelling. Due to street setbacks, courtyards, playgrounds, and wing sections in what is already known to be a large building, you're guaranteed to be pulling a lot of hose. Before committing yourself, attempt to obtain the exact location of the fire as well as the best way to access it. Preliminary information gathered from the first due ladder company or a group of firefighters assigned to seek this information, can prevent a valuable waste of time, energy, and resources. Initially what may seem like a short and easy access from the street, may often be compounded by a wing or isolated stair, that would impede or even eliminate a hastily deployed hoseline.

Any time you are considering a long hose stretch, you definitely will need additional people to get the hoseline(s) to the intended location. It is not an uncommon practice to have the second, and even the third arriving engine company on the assignment assist with the first stretched hoseline. Whether it is a single hoseline stretch, or a more proactive approach with a large hoseline containing a gated wye that can be brought to the floor below, you're going to need help. If every company officer decided to stretch his or her own hoseline in these large area structures, chances are none of them would make it to where they were supposed to go. The most important hoseline stretched at a multiple dwelling is always going to be the first; assign the available resources to get it there.

Hoseline placement. The placement of the first stretched hoseline at a multiple dwelling will be between the fire and any endangered occupants. Depending upon the fire's location, this may be into the interior stairs of the involved building or into a store fire or similar occupancy on the first floor of the building. Another placement might be to a fire escape full of people surrounded by venting fire. Whatever the specific situation, the rule has always been, the first deployed hoseline in a multiple dwelling must be stretched, placed, and operated to protect as many occupants of the building as possible.

Ladder company operations

Specific areas of consideration for the ladder company for a multiple dwelling fire will obviously be dictated by the fire's location and extent. As a general reference for these occupancies, we can categorize the responsibilities of the ladder company with specific thoughts in each of the areas: apparatus placement, ventilation, forcible entry, and search.

Apparatus Placement. Placement of the apparatus must give consideration to best possible "scrub area" on the building. Scrub area is defined as that area on the fire building that can be reached by the placement of a ladder. Areas on the building where the apparatus ladder or bucket can reach, must be maximized when you arrive at a large building full of people. Often this includes the corners of the building. With the occupant life hazard as the main strategic factor, any occupants showing at the window on either the front or side of the building may need to be removed with an aerial ladder. With this thought in mind, the placement of an arriving ladder company will be critical. With only one ladder company assigned and responding, you will only get one shot at this, so make it a good one. In departments with a larger first alarm response your placement may be altered by the number of ladder companies assigned on the alarm. With two ladder companies assigned on an initial alarm, placing apparatus on the two front corners of the building, can give you reach and placement to possibly three sides of the fire building.

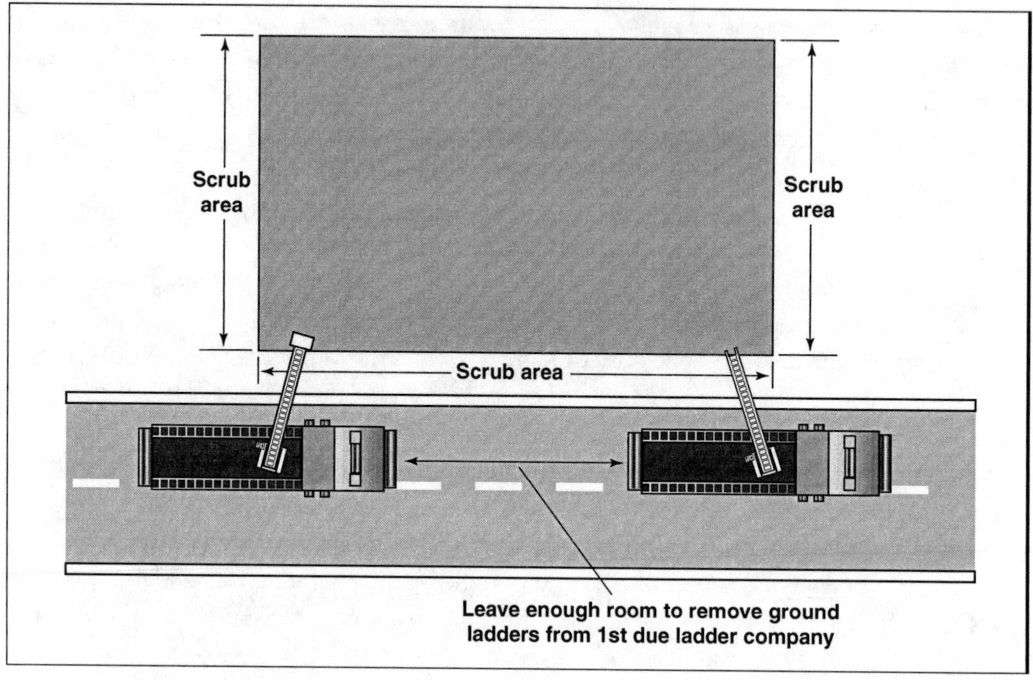

Fig. 3-12 *Scrub area/two ladder companies*

Undoubtedly, there will be a number of additional factors that can come into play here, but consider it when given the chance.

Ladder company placement must also anticipate fire spread. As an example, when fire takes possession of the top floor and cockloft of an H-type there maybe a key street position that will greatly aid in fire suppression efforts. Depending upon the building and your accessibility to it, a well-placed ladder company may be able to penetrate into the *wing* and *throat area* with a ladder pipe or elevated stream preventing fire from moving from one wing past the throat to another wing. If given the opportunity, a heavy caliber stream directed into the top floor and its ceiling will not only darken down a significant amount of fire on the floor, but with the stream concentrated on specific areas of the ceiling, the stream will be able to penetrate through the ceiling extinguishing any fire that is extending through the cockloft.

As we had mentioned in chapter one, knowing the capabilities of the ladder companies responding to the incident can assist in the above decision-making. As an example in Jersey City with two ladder companies responding to a reported building fire, a tower ladder would get preference to access the throat area in the previous example. Since the tower ladder is more versatile and better able to deliver a heavy concentration of water than the aerial ladder, it would take priority in this placement. Presented with the type of fire conditions described, and then knowing that this type of apparatus is responding to the incident, this information can allow a key piece of apparatus to be placed, only if it is factored into your decision-making.

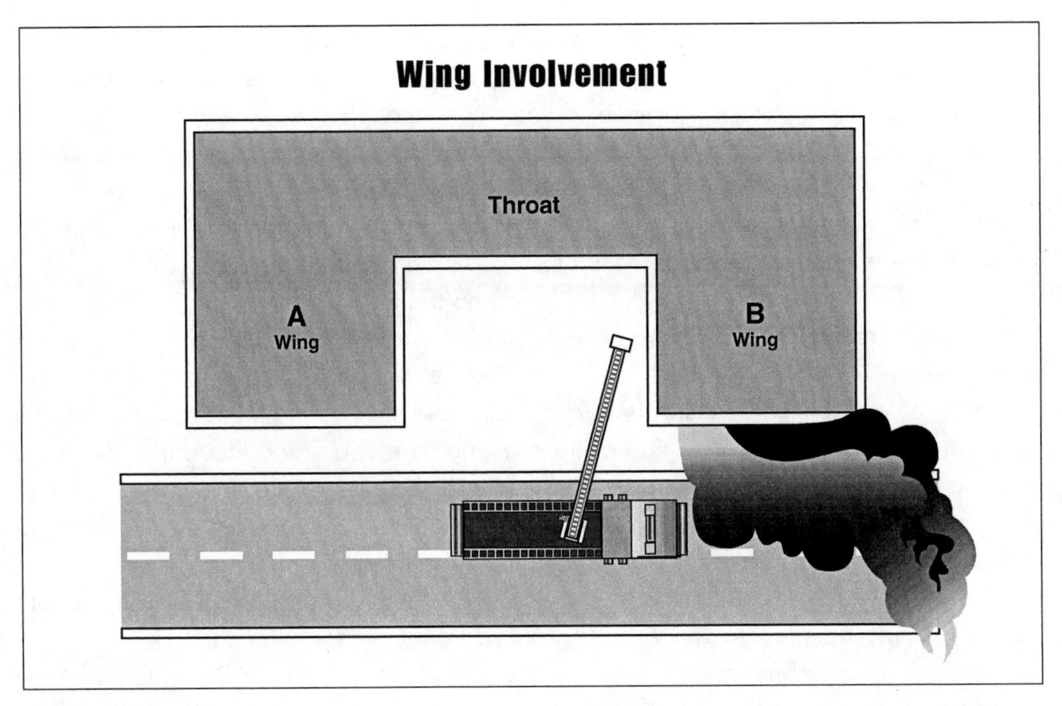

Fig. 3-13 *Throat placement*

Forcible entry. Forcible entry concerns for the tenement, NFP, and H-type multiple dwelling has generally in the past been handled with conventional means of forcible entry. Years ago there weren't too many different types of doors that a flat head and halligan bar with a couple of seasoned truckies couldn't get through. Today this thought has changed somewhat. From fortified doors with multiple locks, to steel doors in a steel jamb, the hydraulic forcible entry tool (HFT) is often the quickest and easiest way to get in. This tool was designed for the increased difficulties that the fire service was confronted with in these type buildings.

Fig. 3-14 *Establish ventilation operations as soon as possible.*

Ventilation. Ventilation operations can take on different tasks depending upon the fire's location and extent. What must be an "automatic" thought with a fire in a tenement or NFP/H-type apartment building is that ventilation at the top of the interior stairs must be established as soon as possible. When fire and its by-products enter this vertical artery it becomes crucial to keep this primary means of egress smoke free for any fleeing occupants, as well as to allow the engine company the ability to advance and gain control of the fire. Any delay of this specific task will increase the difficulty of conditions within the stairway, possibly allowing smoke and fire to enter apartments off the public hallway. This operation becomes crucial for tenement fire because of its smaller stairwell, which will become charged with fire and smoke much quicker than the stairwell of the H-type and non-fire proof apartment building.

Your techniques in this area will vary based on the task. Whether removing the stairwell skylight, removing a bulkhead door, or removing a bulkhead skylight, the objective is the same—take the lid off. An important note with skylights, especially in the tenement, is the possible presence of *draft curtains*. These are nothing more than thin pieces of cloth secured over the skylight ceiling opening to retard any drafts during the cold weather. They are usually there regardless of the time of the year. When venting a skylight over an interior stair, it is important to poke down and remove them. They could prevent a significant amount of smoke from venting through your opening.

Search. Search operations in the tenement, non-fire proof, and H-type multiple dwelling must be planned and coordinated. Priority areas in the building must be assigned. Firefighters assigned to the fire floor, floor above, and top floor may not only find occupants in the most difficult areas of the building, but also they themselves will be in the most dangerous areas as well.

Fires in tenements are fast moving events, with the floor above the fire often being referred to as more dangerous than the actual fire floor. When searching the fire floor, ladder company members will generally get a perspective of the fire's location and conditions due to their proximity. This, in addition to the close operation of the first deployed hoseline will afford members a better safety margin as compared to the search on the floor above. It is in this area, where you need to prepare your escape route before you start your search. Before entering the apartment directly above the fire, force the door of the apartment across the public hallway to give you a quick refuge not directly above the fire. If conditions deteriorate and the stairwell becomes blocked, the apartment across the hall will buy you some time or may allow access to a fire escape or a window if retreat from the floor is necessary.

If during your search of the tenement there exists any single room occupancies (SROs) within the building or individual apartments, an increased number of firefighters and equipment will be required for the tasks of enter and search. Each individual room is locked with either a key or padlock. It is critical that each room be searched even if the door is locked. Individuals when home, will lock themselves in for privacy or may leave the apartment and lock a child in the room. They all must be opened and searched.

One additional search tip that has proven valuable with any building that contains a bulkhead opening to the roof is to search inside it. Firefighters that are assigned to the roof will prioritize the opening of this area to remove pent-up heat and smoke from the building's interior stairway. Often the doors that lead to the roof will be locked, chained, or screwed closed to prevent any unlawful entry. Building occupants who are trapped above the fire may attempt to exit the building through this opening. If the opening is locked, they will become trapped in this area. Once the door has been removed for ventilation, a search inside the bulkhead opening is required in case someone is unconscious on the landing or staircase.

Housing Complexes

Engine company operations

The multiple dwelling referred to as the *housing project complex* will come with some variations for the engine and ladder company when compared to the tenement and the NFP/H-type multiple dwellings. This will primarily stem from the fire resistive class of construction associated with these buildings.

Water supply. Water supply concerns for the engine company officer will again be dictated by area water supply. The primary source of supply will be from the city's public water supply to the fire hydrant. What can occur, and must be anticipated with hydrants in some cities is vandalism. From this concern it has become a practice in many cities, even at routine incidents, that the engine company chauffeurs/engineers check nearby hydrants to ensure that they work prior to committing themselves. This procedure involves nothing more than operating the hydrant and flushing any unwanted debris. This practice of ensuring a water supply is further recognized prior to the supply of the standpipe siamese in the high-rise housing project. Many times cans, rocks, bottles, and debris will be found within the siamese housing. Supplying an already clogged siamese may delay or eliminate altogether any water that was intended to augment the system.

Hoseline selection and stretch. Hoseline selection for the engine company at a fire in a housing complex can vary from the moderate size 1¾", the 2" attack line, on up to the 2½" diameter hoseline.

The hose stretch in the housing project will vary greatly depending upon if you're going to be operating at a low-rise or a high-rise complex. In the low-rise complex, accessibility becomes your greatest concern. Many times the entranceways to the individual buildings will not be on the street sides, but on the inner courtyard areas. Low-rise housing complexes with an inner courtyard area will be littered with fences, playgrounds, and sitting areas which will eliminate the apparatus from getting any closer than the curb of the street. Once you know where you need to go, determine the distance, ensure the stretch is going to meet the required flow, and assign enough people to make the stretch from the street.

Fig. 3-15 *Fenced in courtyards and playgrounds will restrict apparatus accessibility.*

In the high-rise housing complex bringing enough hose and equipment as well as hooking to the standpipe outlet on the floor below is a standard procedure for the first deployed hoseline that many departments use. In some cities researched it is not an uncommon practice to hook up to the standpipe outlet in the same stairwell that serves the fire floor. Due to the fire resistive design of the building with its enclosed stairwells, a hoseline could be stretched right from the standpipe outlet on the same floor as the fire.

What members must be aware of with any standpipe operation is the number of lengths needed in order to make the stretch. Each engine company should bring in with them a minimum of three 50′ lengths of hose. With the appropriate fittings and adapters, three lengths is generally sufficient to reach the apartment from the stairwell if you didn't become disoriented with the scissor stair. Using the wrong stairwell may find you needing additional hose to stretch around the stair shaft to the fire apartment.

Ladder company operations

Ladder company operations at fires in a housing complex will always be governed by the fire's location and extent, with ventilation, forcible entry, and the search of the apartment as the initial assigned tasks.

First, you need to know where you are going. Large area complexes, scissor stairs, and the ever-reliable odd/even elevator will be awaiting you. Not only is the information on how to get there critical to the ladder company, but relaying any information to the engine company on accessibility is deemed crucial as well.

Forcible entry. Members assigned to forcible entry will need to expect difficulties in these types of buildings. Steel doors in steel door frames within concrete block walls is what you're guaranteed upon arrival.

Conventional forcible entry other than a through-the-lock procedure for the unattended cooking run should be discarded. Hydraulic forcible entry tools are designed for these doors when you need to get in.

Entry is further complicated on the exterior of these buildings by the inherent window design found in these complexes. From casement windows to metal screens and metal bars, any member attempting window entry or even ventilation will find it difficult at best. Even with the most innovative tools and techniques to remove some of these barriers, the fire is usually out before you're finished. Nevertheless, if your objective is ventilation, take out as much of the glass as you can first, before you attempt to remove these barriers. A sledgehammer seems to work well.

Search. Search operations in the housing projects are less intensive when compared to the tenement and the NFP/H-type multiple dwellings. This is specifically due to the class of construction of these buildings. Fire and its extension will for the most part be confined to the apartment of origin. Removal of the building's occupants is generally limited to the fire apartment and possibly the apartment directly above. Occupants in adjoining apartments do not necessarily need to be removed from behind the confines of their closed doors. The fire resistive construction will afford them this added protection.

Unusual situations may dictate otherwise, but this thought is generally referred to as the norm.

Life Hazard: Multiple Dwellings

*COA**L** TWAS WEALTHS*

The life hazard concerns of the multiple dwelling are considered severe for both the firefighter and the occupant. These concerns will come with some variations depending upon the specific type of multiple dwelling, but nevertheless they will be demanding for all types of buildings.

Firefighter Life Hazards

Tenements and NFP/H-types

Fire Spread

The old frame tenements will come with more of a concern to the firefighter than probably any of the other multiple dwellings, because of the age and construction of the building. An old building of wood frame construction with wood sheathing and an interior combustible stairway is a building that was inadvertently built to burn. Anyone operating in these structures, especially on the floors above the fire, is going to be at the greatest risk. Fires in the old tenement can spread at an alarming rate. Firefighters unaware of the features of this type of building will get into trouble early.

Staircases

Staircases in the older tenement are going to be made of wood, designed, and constructed within the building in a straight run. This means they are stacked over one another. As you walk down a flight of stairs in this type of building, look up, you'll notice that the backside of the stringer, tread, and riser are exposed and unprotected. This not only increases the stair surface area for fire involvement, but after both sides of the staircase have been subject to fire conditions, its integrity has to be questioned.

Stairs in the newer tenements will also be made of wood, but they will have their backside protected by a plaster and lath covering limiting the surface involvement, as well as extending their integrity.

When operating within a wooden stairwell, it's advisable to stay near the wall where the stringer is attached. Not only is this where the stair is its strongest, but also when the firefighters assigned to the roof remove the skylight that serves the stair, your location near the wall may eliminate some of the glass from falling directly on you.

The stairs in the non-fire proof and H-type apartment buildings are of a stronger design than those found in the tenement building. These stairs are referred to as a "V," or "U-return," due to their intermediate landing within the stair shaft. Within the stair shaft, the staircase will form the letter V. This requires a person ascending or descending the stair to change direction at a half or immediate landing midway between the middle of the V or U-return design.

The construction of this type staircase consists of a steel frame design, with the stringer, banister, and its railing all made of steel, or a combination of steel and cast iron. The tread and riser will consist of either marble or a slate piece of stone, which is placed within a 1½" to 2" steel pan lip of the staircase framework. Inherently these are very strong elements; however, because of the design of the staircase framework, the stone tread and landing will be directly exposed to the elements from a fire below. With this possibility also comes the heating and eventually cooling from a hose stream. This sudden heating and cooling of the stone may affect its integrity, possibly causing it to collapse under the weight of a firefighter.

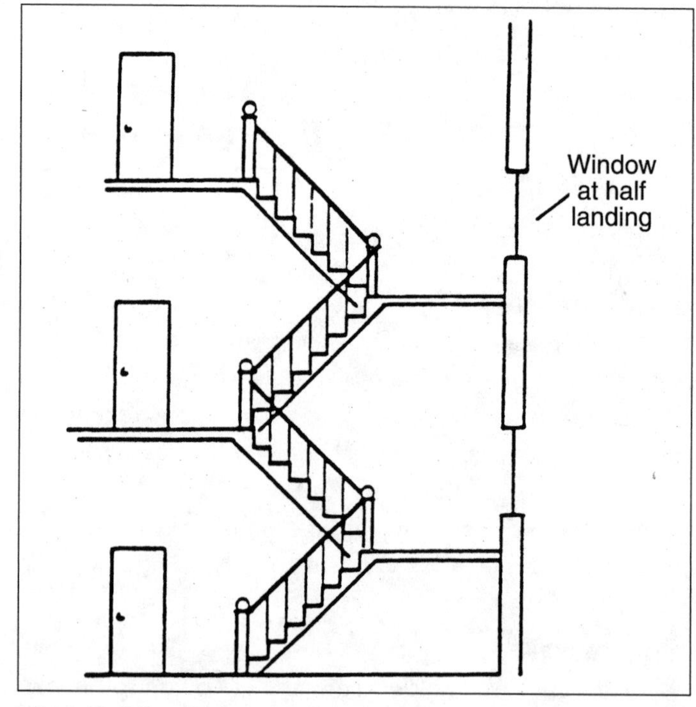

Fig. 3-16 *"V" or "U-return" stair*

While working in a building with this type stair design, take a look up and check the underside of the staircase. If the open pan type is present, alert other operating members to take caution while climbing, as well as operating under it.

Fire escapes

Still on the subject of stairs of the multiple dwelling is the fire escape. Due to neglect, age, and its constant exposure to the elements, this type of stair is your least reliable choice to move from floor to floor. Many firefighters have been severely injured and killed while operating on these steel traps. From rotted steps, railings, platforms, drop and gooseneck ladders, to a fire-damaged or rotted bearing wall where they are attached, many of the fire escapes firefighters use are not safe.

Parapet walls

While we're outside, let's discuss a few other building features that will cause concern to the firefighter's safety. The parapet wall and the attached cornice are two vulnerable areas of the Class 3 and 5 types of construction found in the tenement and the NFP/H-type multiple dwelling that will cause concern. When the top floor or cockloft of these type buildings are involved in fire, the parapet wall and the building cornice will be subject to early collapse. These construction features when directly involved in fire, or impacted by fire conditions, can come down with disastrous results. The parapet wall is a freestanding wall that extends above the building's roofline. This wall has been subject to the weather, which by itself will greatly affect its integrity. When a fire involves the top floor or cockloft of one of these buildings (primarily the Class 3) the slightest elongation or twisting of the windows' steel lintel could drop the parapet wall through the roof or outward to the street below.

In this same area, we must beware of fire department operations. An aerial ladder or tower ladder coming in contact with the wall, or a large caliber hose stream from an elevating platform that is being repositioned could strike the wall, bringing a section of the wall or cornice down. Beware of any activities in their collapse path. Firefighters operating on the sidewalk, advancing an additional hoseline up the

Fig. 3-17 *Street view of parapet wall shows its integrity is already questioned.*

fire escape or a too closely placed aerial or tower ladder bucket will be in harm's way. You need to get them out of there!

Housing complex

The housing complex will come with considerably less hazards to the firefighter. This again can be directly credited to the Class 1 fire-resistive construction associated with these types of buildings. However, the fire resistive design of the building will cause other concerns that the firefighter and the fire officer must be alert to.

Flashovers and backdrafts. Concrete floors, ceilings, masonry block walls with steel doors, and thermal pane windows with an apartment full of combustibles are the exact ingredients you need for a very hot, nasty fire. These components will allow a fire to progress more rapidly through its stages. Add the uncertainty about air flow through the room and the time the firefighter attempts entry into the area, and the possibilities of a flashover or backdraft at the apartment door become all too real. Don't underestimate the potential of these situations just because you're within a fire resistive building.

Occupant Life Hazard

All multiple dwellings

Occupant life hazard in the multiple dwelling must always be considered severe. This is primarily due to the large numbers of people associated with this type occupancy. It is a safe assumption to say that your occupant life hazard concerns will increase during the late evening and early morning hours at a multiple dwelling. But never underestimate the number of people you will find in these buildings during the day.

Depending upon the type of multiple dwelling and its class of construction, the occupant life hazard will be considered more or less severe. When we look at the fire resistive design of the housing complex, occupants other than the fire apartment and the apartment on the floor above are seldom directly threatened. Depending on the severity of the fire, you may be able to leave occupants in place in nearby apartments. Conditions are often safer and more tenable within the apartment.

Place the same fire scenario in a tenement or a non-fire proof apartment building, and the occupant life hazard concerns increase dramatically. Below-grade and lower-level fires in these buildings have often proven their deadliness to many people throughout the entire building in these differently constructed dwellings.

Due to the increased difficulties presented by this type of building, it becomes critical to assign your on-scene resources to handle the priority areas. Due to the severe occupant life hazard in the tenement and the apartment building, put your people where they will achieve the best results. Often you may find two, three, and even four companies of firefighters getting involved with removing people from fire escapes. If the people are not directly threatened, and they are not preparing to leap over the railing, a small compliment of firefighters can guide them down a fire escape or well-placed ground ladder. Resources may be best served someplace else—notably putting out the fire.

Fig. 3-18 *Creating additional egress points is critical.*

The occupant life hazard profile in the multiple dwelling fire must follow the assigning of priority areas based on the areas of greatest danger within the building. Those areas being the fire floor, the floor above, and the top floor. Following these areas will be any floors in between, as well as any floors below the fire. From what we learned as probationary firefighters, convection heat and smoke will move rapidly upward until they hit a barrier, where they will then mushroom out horizontally, and finally travel downward. Depending upon the fire severity, conditions on the interior can deteriorate rapidly. Getting the first deployed hoseline between the fire and the endangered occupants, opening up over the interior stair, and assigning members to the primary search of these critical areas mentioned are going to be the initial considerations.

Additional Life Hazard/Multiple Dwellings

Bystanders

Bystanders can be a problem at any fire if they are not confined and controlled. But in fires that involve dwellings, people will get more emotionally involved in an attempt to help their own or their neighbors. I have been at more than one incident where we fought our way into the building as occupants were forcibly removed.

It is important once you've gained control of these people that you maintain it. In Jersey City, the Mask Service Unit (a company that is assigned to every working fire to fill self-contained breathing apparatus) is assigned upon their arrival to establish the fireground perimeter with barrier tape. These simple pieces of tape seem to work well on keeping most, if not all, of the bystanders back.

Emergency service personnel

Another important group of people that also must be considered in our life hazard profile at a multiple dwelling fire is any other Emergency Service Workers that may attempt to get directly involved at these incidents. Members of the Police Department or Emergency Medical Services will often arrive at an incident before the arrival of the fire department. In their attempt to rescue the building's occupants, they themselves may become trapped. If upon your arrival you notice a vacant police car or ambulance near the reported fire, you now must consider them within your life hazard profile.

Terrain: Multiple Dwellings

COAL TWAS WEALTHS

Setbacks

Members of the engine company must be ready to deal with terrain obstructions in the form of parking lots, courtyards, and playgrounds that are often associated with many of these types of buildings.

Members of the ladder company will also be affected by many of the same setbacks as the engine company, with the added difficulties in many areas from any overhead obstructions. The ladder companies scrub area can be extremely limited if the apparatus can't get any closer than a parking lot will allow, or could be eliminated altogether due to overhead wires. Ladder company members will often take any steps necessary in order to gain access to a building with their apparatus. The cutting down of fences, driving the apparatus up onto sidewalks, and the setting in rear yards are just a few options used at serious fires in a multiple dwelling.

It is important not to get tunnel vision with any access problem. Larger diameter hose used to increase the flow of a hoseline stretched into the building to the expanded use ground ladders all around the building should be an automatic reflex, depending upon the objective and the location of the fire.

Fig. 3-19 *Ladders, ladders, and more ladders.*

Grade levels

Grade level differences around the involved structure can cause significant fireground management concerns for everyone operating. Not only does this affect the efficiency of your operations, but it could also prove to be disastrous in a May-Day situation when a firefighter calls for help from a specific floor. Structures that appear to be four stories from the street side of the building may be actually six stories from the rear. If you haven't experienced these concerns you can just imagine the confusion that could exist when a firefighter enters from a different level and requests help on what is believed to be the third floor.

As identified in chapter one, floors should be designated and assigned from the street address side. This is where your command post will be. It is from here that all assignments and orders will flow.

Water Supply: Multiple Dwellings

COAL TWAS WEALTHS

Water availability: All buildings

In the urban areas, water availability for the multiple dwelling will primarily come from a domestic/public supply to a street fire hydrant. Firefighters in the cities rely on hydrants, with an occasional opportunity to establish a supply from a nearby river, lake, or creek. In these same areas, many times water availability or its accessibility may be compromised. Availability problems can stem from the age and neglect of the system, as well as the constant problem of vandalism. It is not unusual to see hydrants with missing caps, damaged threads, rounded off operating stems, let alone being blocked by parked cars.

Fig. 3-20 *Typical solution to a parked car problem.*

Tenements

Water requirements

Water supply that is needed for a fire involving a tenement building will depend on a number of factors, those most notably being the class of construction of the fire building and its ability to retard or spread fire. If we were to compare all three types of multiple dwellings in relationship to the above factors, the tenement will receive most of your attention in the areas of quick, sufficient, and mobile water. From what we have already described as a highly combustible building, the frame tenement will burn quickly, spreading fire to their exposures, if they are not planned for. Speed, accompanied with a sufficient volume is the prerequisite here. Hose streams capable of delivering 180gpm to 200gpm and still being mobile enough to push in and throughout the building are what are needed for a fire in this type of building. The 2½" hoseline, for the most part, will not be practical for an offensive attack for a fire in a tenement. You will definitely gain control of the first floor stairwell with this size line, but that will be probably as far as you will get.

NFP/H-Type Multiple Dwellings and Low-Rise Housing Complexes

Water requirements

The non-fire proof and H-type multiple dwelling may vary your thinking on the amount of water needed. Due to the larger area concerns of these buildings, their potential for fire involvement, and the accessibility to them, fire officers must plan on multiple lines all being able to flow their designed capacities. A 1¾" and 2" hoselines will do well once inside an apartment, provided they are capable of delivering their intended flows. The key to your decision-making as a company officer must focus around being able to provide the necessary flows. As a reminder, some of these buildings will require long hose stretches.

Water delivery

Water delivery in the multiple dwellings listed will depend on the square footage of the building, the potential for fire involvement, as well as any terrain concerns in the form of courtyards, playgrounds, fences, etc. The two types of multiple dwellings that will initially cause concern in getting the required flows where they are suppose to go will be the NFP/H-type multiple dwelling and the low-rise housing project. All of these structures are known to contain fenced-in courtyards, playgrounds, and in some cases decorative landscaping. Any of the above listed obstructions will create a problem with the engine companies hose stretch. Fires even on the first floor may require a sufficient number of lengths from the street to go around and over the obstacles. Put this same fire up a few flights and your concerns in this area increase dramatically.

This problem can be further compounded with the H-type multiple dwelling. Many of these large and irregular shaped buildings do not have the luxury of a standpipe system within the building. Depending upon the fire's location and stairway accessibility, hose stretches can be very long and very difficult. Your first consideration with this concern is getting the hoseline to its assigned objective. If you don't have enough people from the first company, double, or even triple up with additional help. There should be no doubt in anyone's mind about the consequences of stretching a short hoseline, let alone an inadequately flowing one.

With this in mind, stretching more than 300' of 1¾" hose starts to seriously affect the hoselines flow. Stretching more than 350' of 2" hose starts to seriously affect that hoseline's flow. Stretches longer than these are not uncommon in an apartment building H-type multiple dwelling or even the low-rise housing project. Companies that routinely respond to fires in these buildings have their hosebeds sit for long stretches. Whether it's a 3" hose line with a gated wye to the floor below, or a manifold to the front door allowing multiple hoselines to be stretched, the objective is the same. When water delivery comes from a distance, plan on the friction loss, and ensure that the hoseline(s) is able to deliver the necessary water.

To further explore this thought, a larger 2½" hoseline should not be ruled out as the initial stretched hoseline in the large apartment building. This size hoseline has been used to gain control of a public hallway when fire from two apartments has exited beyond their doors. Without the reach and penetration of this big line, hallway conditions would have never become tenable enough to allow members to push down the hallway even with two 1¾" lines working side by side. With a significant amount of fire showing, it may be wiser to pool your resources to get a big hose line in place, instead of splitting them to stretch smaller ones. Big combustible buildings can overwhelm the smaller handlines. When in doubt, plan on the big water; you could always scale it down.

Fig. 3-21 *Be prepared for a long hose stretch.*

High-Rise Housing/Project Complex

Water supply

Buildings such as the high-rise housing project will come with their own unique set of concerns when discussing water supply. First, is the supplying of the standpipe system. Even though the standpipe system in these buildings is wet, policy and procedure should require that the fire department augment the system. In order to accomplish this you initially must know the location of the fire department connection. If companies respond there on a constant basis, as they most often do, their location quickly becomes known. If the incidents are few and far between, companies need to seek out their location before they need to go to work there.

Fig. 3-22 *A siamese connection plugged with refuse.*

Once found, many times the fire department connection in the housing complex can be damaged or clogged. Firefighters may find rocks, bottles, cans, and balls or the storage of drug paraphernalia inside the siamese housing. If you're going to make an attempt at supplying the siamese, never stick your hand in the housing to remove any debris. Even with a gloved hand, never place a hand into the opening. You'll not only risk cutting your hand, but also risk the chance of being stuck by a used needle. You should make this a rule regardless of what part of your city the siamese is located. If you still feel compelled to remove the debris and supply the system from the siamese, use a tool or a pair of pliers in an attempt to remove the debris.

As a young firefighter being assigned to an engine company that had its fair share of projects, I was quickly schooled on the best way to deal with damaged or clogged siamese problems. Simply, it was to avoid it altogether. Chances are you will never remove all of the garbage, and what you can remove, will find its way through the riser to the standpipe outlet only to clog a hoseline from receiving any significant water. A more viable way of getting water into the system that we constantly used involved elimination of the standpipe siamese on the exterior of the building altogether. By stretching a 3" supply line into the building and hooking it up to the standpipe outlet on the first floor, we would eliminate many, if not all of our problems.

After flushing the system and a minor adaptation with a double female coupling, the system was now supplied. One very important consideration with this procedure; if a pressure-reducing device is found on the standpipe outlet on the first floor, it should be removed. Its presence will restrict the pressure and flow of your supply hose line.

Auxiliary Appliances and Aides: Multiple Dwellings

COAL TWAS WEALTHS

Fire detection devices

Fire detection devices, namely in the form of smoke detection, started to become mandatory in multiple dwellings in the late 1970s and early 1980s, after the numerous deaths that had occurred in these occupancies through the years. The fire workload of many urban areas came to a screeching halt after the long and overdue installation of smoke detectors in the public hallways of these buildings.

Housing Complex

Fire suppression equipment

Fire suppression equipment in this particular type multiple dwelling is most notably found to be a sprinkler system in the compactor and incinerator areas of the high-rise housing complex, and the standpipe system within the same building. In some cities researched, local law also requires the sprinkler system to be extended into the public hallways as well.

The standpipe system within a high-rise housing project is generally a sufficient system when working and not vandalized. All too often these systems receive abuse from neglect and vandals. It is for this reason alone, that engine and ladder companies assigned to these areas must come to expect certain obstacles, and plan for them in order to meet their suppression needs.

The most obvious concern will be from missing equipment. When encountering the riser in the stairwell, realize that it may be the only thing you will find. The valve, the threaded adapter, and hose may be long gone. It is important that each standpipe tool kit carried by the engine company has all the necessary equipment to make the necessary hook ups with out any significant delays. Globe valves, wrenches, threaded adapters, a gated wye, hose straps, nozzles, and wooden door wedges are just some of the items you should find in a standpipe tool bag of the arriving engine companies.

Fig. 3-23 *Do you have the equipment to make this system work?*

Aides/Assistants

People or aides with the multiple dwelling that can generally be considered a help are the building superintendent within an apartment building, and the housing monitor or member of the City Housing Authority associated with the housing complex.

The building superintendent that generally occupies the same address of the apartment building, is sometimes helpful with such things as occupant and owner information, resetting of an alarm, notification to a fire alarm or sprinkler contractor, or the request for compliance with a code violation.

The housing monitor within a housing complex is a person that is assigned to a project complex with the responsibility to assist with alarm notification, alarm activation, as well as the eventual resetting of the alarm.

The City Housing Authority will generally take on the added responsibilities of building maintenance, elevator maintenance, and any other requests for repair or service.

With either of the above listed resources, reach out for these people because they can be an asset to your operations.

Street Conditions: Multiple Dwellings

*COAL TWA**S** WEALTHS*

Any time we refer to those conditions that may affect apparatus movement, placement and their eventual operation, we will obviously become very concerned. When this discussion involves the multiple dwelling, your concerns will increase. The fast moving fires associated with the tenement, and the large numbers of people associated with the apartment building will increase the concerns for apparatus placement, quick water, quick vents, and the placement of ladders.

Street width, traffic congestion, and double parked cars are going to be a constant problem around the multiple dwelling simply due to the congestive nature of these buildings. Whether it is a.m. or p.m. there seems to be little difference.

Street width/traffic flow

Narrow car lined streets and possibly only one-way traffic flows are a fact of life in these particular areas that the fire service must deal with. Due to the element of speed and efficiency required for a fire in a multiple dwelling, this size-up factor becomes critical for the first due engine and ladder company. Driving down a one-way street against the flow of traffic because you think it's quicker, can prove to be disastrous more times than not, especially if it is not thought out.

Let's think about it. First, civilian traffic will be moving one way. If you turn in to the block against them, where are they going to go? Backwards into a driveway? If so, you better hope it's only one car that needs to move, and there is a driveway somewhere on the block. Second, if the first due engine company has a clean path down the street against the traffic flow, the first due truck company who is responding from a different location may be proceeding with the traffic flow, and never get to the front of the building. In actuality they probably won't get anywhere close. There will most likely be a line of cars stuck between the first due engine company and the first due ladder company. The ideal procedure with one-way streets is to have the first due engine company and ladder proceed into the block with the flow of traffic. By following this established route you will push civilian traffic up ahead and out of the way, allowing for proper placement and positioning of the first due engine and ladder company. Once you gain control of the block, then additional equipment can move in any direction necessary.

Fig. 3-24 *Narrow one-way streets require a disciplined approach.*

This suggested procedure of the first due engine and ladder company comes with only one exception. If as the first due engine company, you are responding to a reported fire and the address of the fire building is near the street corner, or at the most a few doors down the block against the flow of traffic, stop and take a look. If you could see the building and have no vehicles between you and the fire building, proceed. Please note, that you must be able to see the building. With streets that have a low, high, then low grade again, you may head into a street against traffic and find as you climb the hill a line of vehicles approaching from the other side. Now you're in trouble. We must add that your decision could be further enhanced if the first due ladder company is responding with you and they are right behind you. This will allow proper placement for the first due engine and the first due ladder, working on the concept of quick water, quick vents, with quick searches. It is also extremely important with this exception that you immediately alert other responding companies to your move so they can make the necessary changes to ensure augmented water supplies, placement of the second due ladder company, etc. The exception to the rule has to be a calculated move. Think about it and plan for it before you act. If you don't have a procedure on streets with one-way traffic patterns, you should.

Weather Conditions: Multiple Dwellings

*COAL TWAS **W**EALTHS*

Weather conditions that affect fire department response and operations will cause concerns regardless of the type of occupancy. But it seems that this size-up factor, accompanied with a response to a reported fire in a multiple dwelling, can create a few more concerns for arriving forces to deal with.

Temperature

Extreme weather in the form of very cold, windy snowy days can take on added concerns with a fire in a multiple dwelling, specifically as it relates to occupant displacement. With a significant fire in these buildings, especially in the large apartment buildings, chances are you will be removing a large number of occupants. If the fire is on a lower floor it is a guarantee. On these difficult weather days shelter and eventual relocation for the burned out occupants will be a concern. Shelter in the form of a school auditorium, a church hall, or heated transit buses are possible considerations. Relocation agencies may vary in each city or state, but it seems like the Red Cross is always there to lend a hand when you call.

Fireground operations always will be affected by extreme weather. Whether it is from icy sidewalks or streets from water runoff during the cold weather, to heat exhaustion on a hot summer day, fireground conditions as well as firefighter rehabilitation are concerns that must be addressed early. Buckets of sand and salt, to coolers of cold water carried on each apparatus are just a start to combat these concerns.

Fig. 3-25 *Are you prepared for this situation?*

Wind

As we consider the weather and its effects on our operations, we must plan and prepare to deal with its eventual return. However, in my research, there seems to be one element of the weather that all too often plagues operating forces, and seems, for the most part, to be chocked up to a difficult fight without a plan for its eventual return. That is the wind. Winds that can take a hallway of an apartment building or a high-rise project and turn them into a blowtorch is the type we are most concerned about. Wind conditions that cause this concern are not always that obvious at the street level. Fires on floors as little as 90′ to 100′ above grade can be drastically different than at ground level. It is in situations similar to this, where street observations as well as the information shared by your ladder and engine companies becomes vital. The firefighters operating within and on top of the building will see and feel the effects of this first. Their observations, direction, and eventually operation, will be critical to the outcome of the incident.

Wind direction observations

1. From the Street Level:
   A. Movement, intensity, and direction of anything moveable above. (i.e., flags, clothes draped over balconies, smoke, etc.)
   B. Window Coning – refers to the flame tongues coning inward from the force of the wind against the face of the building.
2. From the Ladder Company:
   A. Wind movement, direction, and intensity as observed from the roof position.
   B. Conditions felt on the floor(s) above the fire as windows are opened or removed.
3. From the Engine Company:
   The force and direction of the heat and smoke from either the public hallway or upon entering through the apartment door.

Wind conditions and your actions

1. When possible, simulate ventilation conditions on the floor below the fire to assess the effects.
2. Stop any additional horizontal ventilation on the windward side of building.
3. Increase diameter and reach of hose stream.
4. Use two streams to push in—either in the form of two solid or straight streams or one narrowed fog stream to pressurize the hallway or apartment, and one solid or straight stream directed through the fog stream to push in. (Be careful with the use of fog in a superheated hallway or apartment)
5. Breach wall of adjoining apartment and attempt to knockdown the fire. In a Class 1 building, this will be extremely difficult.
6. Consider use of an outside stream to darken the fire down. This will require an evacuation of the fire area with a holding position by inside forces to prevent fire from traveling toward the interior stairwell.

7. Use the fog pattern of a Navy applicator and nozzle from the floor below into the window of the fire apartment above.
8. With alternate or additional stairwells as in a H-type, when possible, establish the untenable area and its direction to a vent stair, and use a new approach from an attack stair.
9. Positive Pressure Ventilation of attack stairwell and fire floor.

Difficulties associated with the wind are not as uncommon as one might think. Wind as little as 10mph can and will cause you havoc. Officers should develop and review contingency plans for this concern.

Exposures: Multiple Dwellings

COAL TWAS WEALTHS

Tenements

As we reference the exposure concern of the three multiple dwelling types discussed in this chapter, the tenement again, will present you with your greatest concern in this size-up factor. We state this primarily from the class of construction of the building, its inherent construction features, and the building's proximity to neighboring exposures. The tenement building, depending upon when it was built, could be either of Class 3 or Class 5 construction. The older tenement buildings of the Class 5 wood frame design not only have the obvious problem of a wooden exterior which will enhance the auto exposure concern, but two additional concerns will come from the building's proximity to others, and the shared light and air shafts of the fire building, and those buildings attached.

Light and air shafts

Light and airshafts come in two different designs, the open and the enclosed shaft. The difference between the two is that the open shaft, is surrounded by three sides of the building, with the one of the four sides open to the rear yard. The "enclosed shaft" is as it sounds. It is totally enclosed by both buildings and is not open to the rear yard. Although their intent was to provide light and air into a series of rooms that bordered the shaft, a fire officer must be aware of the fact that their design and placement in between buildings can greatly enhance the fire's growth and spread.

Any time fire enters a shaft there will be a great inflow of oxygen into the fire apartment opposite the shaft. This influx of oxygen will increase the release of heat and

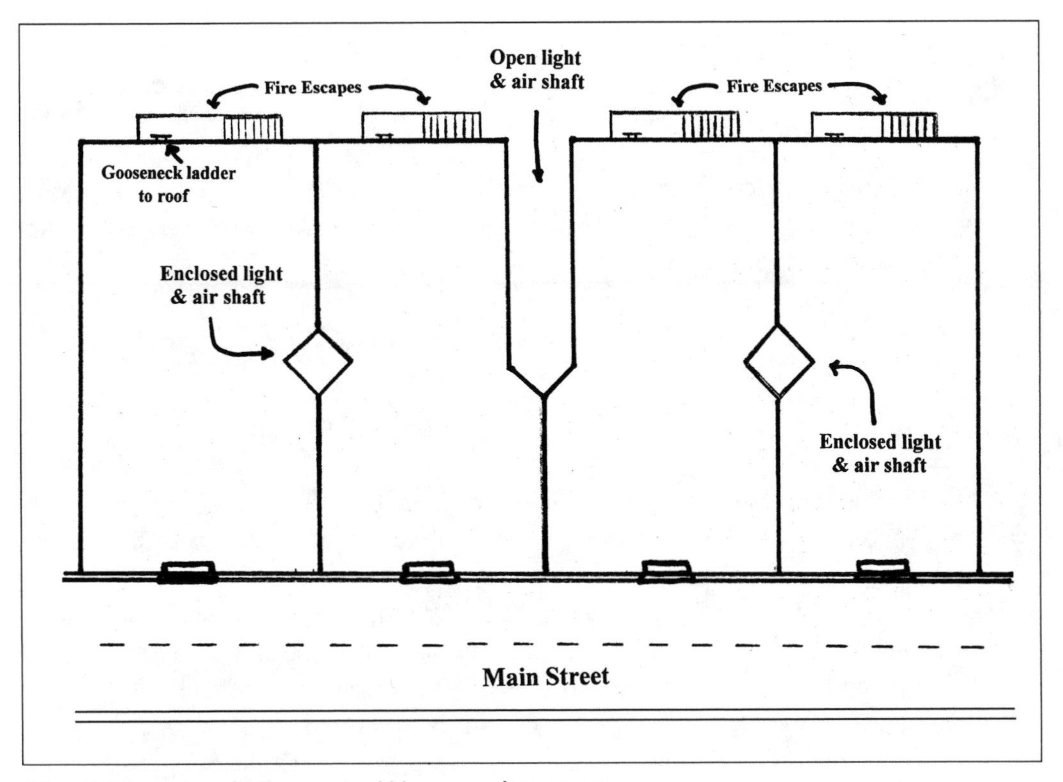

Fig. 3-26 *Typical shaft placement within a row of tenements.*

gases into the shaft. With the draft in the shaft always being out and upward, the highly heated gases will be moving and releasing energy in any direction that they can. The amount of energy being produced in the shaft will be further influenced by such factors as the sheathing that lines the shaft, and whether the shaft is open or closed. These factors accompanied with the window type and design, and whether those windows are open or not will determine the speed at which fire will spread into the adjoining building.

As fire enters the adjoining building, a door opened into the public hallway of the exposed building will accelerate the movement of heat and gases from the fire building into the exposed building. With this movement, fire will quickly take possession of the interior stairway of the exposed building requiring the placement of an early and well-supplied hoseline.

From all the reasons mentioned, it is a common practice by many chief officers that when fire has possession of a light and air shaft upon their arrival, barring any additional information, the second stretched hoseline is ordered into the interior stairwell of the exposure building.

If you placed all of your initial resources into the original fire building, you're going to be chasing this fire down the block.

Area: Multiple Dwellings

COAL TWAS WEALTHS

Square footage concerns, irregular-shaped buildings and areas, and the accessibility to them can be a tremendous problem if these are not reviewed and planned for well before the receipt of the alarm.

Tenements

The tenement housing of yesteryear never gave the responding firefighters any unusual concerns related to area size-up. Today in many areas throughout the country this has changed, and it's changed dramatically. In old neighborhoods that have been revitalized, rows of tenement buildings up to a city block long have undergone large-scale renovations. In many instances you may possibly find three or more of these buildings grouped and interconnected to create one large building, with a limited number of street entrances. In this type design, street entrances may be found at the end of the row of buildings, with long public hallways the only means to traverse the floors.

Confusing? With this type of renovation the builder will place a "new" brick veneer over the existing facade giving the false appearance of a new series of buildings, when in fact all the inherent "design" voids of the tenement are still in place, causing the same concerns as before.

In the downtown area of Jersey City these buildings exist in large numbers. Early one morning I responded as the officer with an engine company assigned on the second alarm to one of these complexes. Upon arrival we were ordered to stretch a hoseline into the top floor of Exposure D1, for what appeared to be a fire traveling the cockloft. After I was met with a false door on the first floor, it was radioed to

Fig. 3-27 *Renovated row of tenements. Entrances to all levels are at the ends of the buildings. Note false doors at the first floor level.*

me that the only entrances to the upper floors where I needed to go, were from two doorways at opposite ends, of what was originally seven buildings.

In actuality we were not stretching into Exposure D1, we were stretching into one large fire building.

H-Type Multiple Dwellings

Irregular shaped concerns for the H-type apartment building will literally come in all shapes and sizes. You may encounter any shape from the classic H to the U, O, T, C, E, and EE. Due to the irregular shapes of the buildings, not only must you contend with the larger square footages, but you must also expect long public hallways in some of the buildings, limited access hallways in others, with a guarantee of a large number of apartments and complex room layouts in all types.

Wings and stairs

In an attempt to eliminate the building's confusion, it is important to designate the geographics of the building. Fire departments will use varying methods of identifying the wings of an H-type design building. The two most common are the geographical references of east wing and west wing or the letter designation of the A wing and the B wing. The geographical follows the building's reference with our global association of direction. The letter designation is usually referenced from left to right of the fire scene, or from the front to the back of the building as observed from the street address side of the building

A lack of designation will surely cause confusion, a delay of resources, and some serious safety concerns to your people.

Obtaining access to the wings of an H-type multiple dwelling causes another series of concerns. The number and location of stairways in the building normally don't adhere to a common pattern. Planning for area access in these buildings is a must. Too many times, companies have stretched a hoseline up an interior stair only to find that the stair they committed themselves to, only served that wing, and not the entire floor of the building. Depending upon design, some stairways will be found in, or near the front entrance, while others may be found in the side or rear courtyard. Depending upon the type of stair, access to other wings may be from a totally separate stairway or may even have two stairways serving two isolated areas of the same wing. Sounds confusing? It can be if you don't plan!

H-type multiple dwelling stair types

- Wing stairs – You could have one or two of these type stairs located in each wing of the building. They can generally be found in the front and the rear of the wing area. This type stairway will "only" serve that wing.

- Transverse stairs –These stairs are located at points remote from each other. This type stair will allow an occupant or a firefighter to traverse from one stairway to another via the public hallway. These types of stairs will allow occupant evacuation and fire department access to all apartments on that floor from either stair.
- Isolated Stairs – This particular type of stair will only access the apartments that are served by that individual stair. There will be no access to other apartments in that wing beyond that stairway. Multiple stairs will serve the divided areas of that wing.

Fig. 3-28 *Well hole stretch*

Stairway construction in the H-type multiple dwelling is usually of the return type. Within this type design, some of the stairs may or may not have well holes—the space between the stair banisters at each rise and landing, which can range from a few inches to a few feet. Having access to a well hole will greatly affect the hoseline selection and stretch of the engine company. Well holes can eliminate lengths of hose in the stretch, expedite the movement of water as well as allow that third hoseline to be stretched into the stairway. If you have one, take advantage of it.

Some stairwells may even have windows at the half landing of the return that face a street or courtyard. The advantage of this window at the half landing is that, depending upon the fire's location and the wind direction, this could be an excellent place to ventilate. In addition, this window at the half landing on the floor below the fire can be used to haul up an additional hoseline if necessary.

Fig. 3-29 *Window at the half landing*

It is important with the different stair types in the H-type multiple dwelling to have the first due ladder company determine the type of stair, the fire location, and the best route for the first due engine company to take. With this information the engine company officer can plan for the proper hoseline selection and stretch.

Pre-plan and review will determine the types of stairs in the H-types in your district, without this information, have someone check before you stretch.

Non-Fire Proof Multiple Dwellings

Apartment layout and floor designation are two additional area size-up concerns for the multiple dwelling that can, and will cause operating forces extreme havoc if not planned for. It is not uncommon to find two apartments interconnected by inserting doorways in a common party wall to make one apartment. This type of alteration significantly increases the possible square footage of fire involvement, producing the probability of heavy fire involvement throughout both areas.

The number and layout of apartments in a wing may also vary greatly from one building to another. In one building you may find irregular shaped apartments in the form of a T design, and adjoining apartments of the same building wrapping around forming an L design.

Floor and apartment designation is another concern that may change from building to building. In most instances you will find the floor and apartment designation in a number, letter sequence. For example Apartment 5B, would mean the fifth floor, B apartment. However, it is important to note that this is not always the rule. I also have seen where the floors and the apartments were designated with a letter then number sequence instead of number then letter sequence. For example, apartment B5 would now mean, the B floor, fifth apartment. This can also be very confusing. By viewing the building vestibule to determine number and letter references, then proceeding to the floor below the reported fire floor to get the floor and apartment layout has proven beneficial.

Location and Extent of Fire: Multiple Dwellings

*COAL TWAS WEA**L**THS*

The location and extent factor, in association with the type of multiple dwelling will significantly affect your life hazard priorities when arriving at a fire in this type occupancy.

Tenements

When considering this size-up factor, again, the tenement will stand out as the most challenging for arriving forces. Due to the combustible design of these buildings, quick and decisive action is going to be a pre-requisite for the decision-makers.

Interior stairway

Any fire that is below grade, or at the lower-level, will quickly require water into the stairwell of this building. This vertical artery with its wooden stairs, banisters, railings and wainscoting will quickly add fuel to an already well-fed fire. Quickly placed water with top side ventilation over the stairs will slow down the lateral fire spread into apartments off the public hall. Gaining control of the building's lifeline is a priority with a fire on a lower-level.

Windows, doors, and more

Fires that originate or extend into a tenement apartment will also have numerous ways to extend up and throughout the building. As a young firefighter assigned to a ladder company, I can remember an old timer telling me after a good burnout in an apartment of one of these buildings, that it was important to pull all the door and window trim in the apartment due to the fire's ability to burrow into the furred out spaces around the door and window openings. The space between the rough opening and the window/door opening can vary from a ¼" of an inch to a few inches depending upon the ability of the worker who framed the opening at the time. This space, where the plaster and lath of the wall ends, and the door or the window begins, will only be protected by the trimmed covering. If the trim is compromised in any way, or if the trim covering has a significant charring, it must be removed. If fire or heat is allowed into this space, most notably to the backside of the lath covering, it can smolder for hours until you're called back.

While reminding of the above concern, he also spoke of two other overlooked trouble areas within any old building, those being the window ballast pockets and a sliding pocket door.

Window ballast pockets. Window ballast pockets are found in old double-hung wooden sash windows. A clue to their presence will come from the rope or chain lining the inside of the windows track. These pieces of rope or chain will be fastened to cast iron weights. This tubular shaped piece of cast iron is designed to slide up and down within pockets on both sides of the window to counter the weight of the windows slide. The cast iron weight which averages around 2" in diameter, 10" in length, and can weigh as much as five pounds, will ride this vertical channel for the entire height of the window.

Fig. 3-30 *Window ballast within its pocket*

In many buildings, renovations may have replaced the wooden window sashes with newer windows. Depending upon the renovation, replacement windows will be installed within the same opening as the older window. Often the newer window will be much smaller that the original framed opening. Many times with their replacement, the ballast pockets from the original window will still exist. By cutting the rope or chain, the ballast weight will fall and remain in its designed track, as the new window opening is re-trimmed to receive its replacement.

Sliding pocket doors. Sliding pocket doors were incorporated into an apartment to give privacy to a particular series of rooms. This recess area that allowed for the disappearance of an entire sliding door, could also allow for fire to hide as well as travel. If accessible, this opening could easily allow fire to gain access to the building's void spaces.

For the most part, many of the original pocket doors are still in place. Those that are no longer used may have had their openings trimmed over, but many still serve their intended design today. Take a look next time you are working in one of these type buildings, these may still exist.

Kitchens and baths. Anytime you have more than one, and they're stacked on top of one another, you must expect fire spread to and from them.

Light and air shafts. Their design and intent should be well understood by now. If the building is in excess of 60', plan on their presence and prepare for their involvement.

Cornices and cocklofts. Fires that reach the upper-level of the tenement will cause concern with the cornice and the cockloft. Cocklofts may be common between neighboring buildings. Those that were intended to prevent fire from traveling across may only have fire partitions that extend to the underside of the roof beams, allowing fire to spread in and around any gaps that remained to say nothing of any faulty work-

Fig. 3-31 *Sliding pocket doors*

Fig. 3-32 *Anticipate the fire involvement!*

manship of the wall itself. Nevertheless, anytime you encounter attached buildings of the same height, adjoining buildings must be checked for fire extension within this space.

Cornices, their design and attachment to the front of the building will present more of a collapse concern than a fire spread concern when they become involved. Caution must be used anytime a firefighter is above or below it. Often their support is questionable to non-existent. Their involvement will allow fire to spread to abutting cornices from attached buildings, but be careful operating anywhere near them.

NFP/H-Type Multiple Dwellings

Kitchens and baths

The NFP apartment building and the H-type multiple dwelling come with some of the same location and extent concerns that the tenement has. Most notably, the utility chases of the stacked kitchen and bath design. In any building that has a bath or a kitchen above one another that is involved in fire, you must anticipate fire spread, and check all *six* sides of the fire. With the apartment building it becomes more of a concern due the larger numbers of vertically and horizontally aligned kitchens and baths. This arrangement will compound the difficulties once fire enters one of the void spaces of either one of these rooms.

Channel rails

Next to the dumbwaiter shaft and the bathroom's soil pipe, the steel I-beam with its channel rails will create one of the largest concealed void spaces extending vertically within the building. As the square footage demands of these buildings grew, it became essential to incorporate steel within the design to support the large spans not allowed by wood dimension lumber. Structural steel in both the horizontal and vertical design, should be expected in your large Class 3 design apartment buildings. If fire has entered this vertical artery, you must check above, below, as well as around.

Garbage/dumbwaiter shaft

Another very large and significant void if it is penetrated, is the garbage/dumbwaiter shaft. Today these voids are sealed off and no longer accessible. However, they make great raceways for new electric and plumbing, and are often compromised for this reason. These shafts, which create a space of $2ft^2$ to $3ft^2$, will penetrate the entire height of the building allowing a large concentration of fire into the roof space. They are large in number and can be found in each apartment in the kitchen, or near the front door of the apartment. If fire penetrates this void space from any level, you need to check above for fire extension as well as below for fire drop down.

During extinguishment, be very cautious about placing any part of your body into the shaft. Obviously, placing your head in the shaft to take a look is out of the question, but at times when fire has extended into the shaft, firefighters will direct a hose stream up the shaft to halt any fire extension. Often, the actual dumbwaiter and its hardware are left intact in the shaft. Heat from a fire can bring all of this equipment down and possibly injure a hand if it is in the travel path.

Cocklofts

The largest concealed space in the apartment building and H-type multiple dwelling is the upper-level fire that involves the cockloft area. These areas are large with heights averaging around 3′ to 4′, covering the entire top floor area of the building. Height dimensions of the cockloft will fluctuate as the roof pitches toward or away from roof scuppers designed for water runoff. From this association it should also be expected that as fire enters the roof area near a scupper, it would accelerate away from this area into the rest of the roof space due to upward pitch of the deck.

To prevent fire from extending throughout the cockloft, firefighters will be required to create a sufficiently sized ventilation hole over the main fire area. The primary ventilation hole is an offensive tactic designed to limit the fire spread of the top floor and cockloft. Firefighters should attempt to cut a hole that will create an 8′ by 8′ opening over the main fire area. Creating an opening and enlarging it to this size will create a thermal updraft for the fire on the top floor or cockloft to flow too. There should be no doubt that creating an opening of this size is going to be task oriented, requiring the coordinated efforts of a number of people and equipment, so it needs to be well planned and staffed.

With an advanced fire on the top floor and cockloft, the officer assigned to the roof will also have to evaluate the need and placement of the trench cut or strip cut. Trenching a roof is a defensive operation that is performed to limit

Fig. 3-33 *Dumbwaiter shaft*

Fig. 3-34 *Dumbwaiter bulkhead*

the spread of fire in the cockloft. The trench is a procedure that creates a three-foot wide strip from wall to wall. Any failure to complete the trench from wall to wall, or from fire stop to fire stop, will allow fire to pass the trench and defeat the purpose of the procedure. The position of the trench should be at a location that will isolate the fire to a section of the buildings, as in the wing of an H-type, and still not pull the fire toward the opening when the decision has been made to pull the trench.

Roof trenching operations

1. The primary ventilation hole over the main body of fire is to be a minimum of an 8' x 8' opening. This opening is started out as a conventional roof cut and enlarged to this size. Its intent is to span a number of rooms within the involved area.

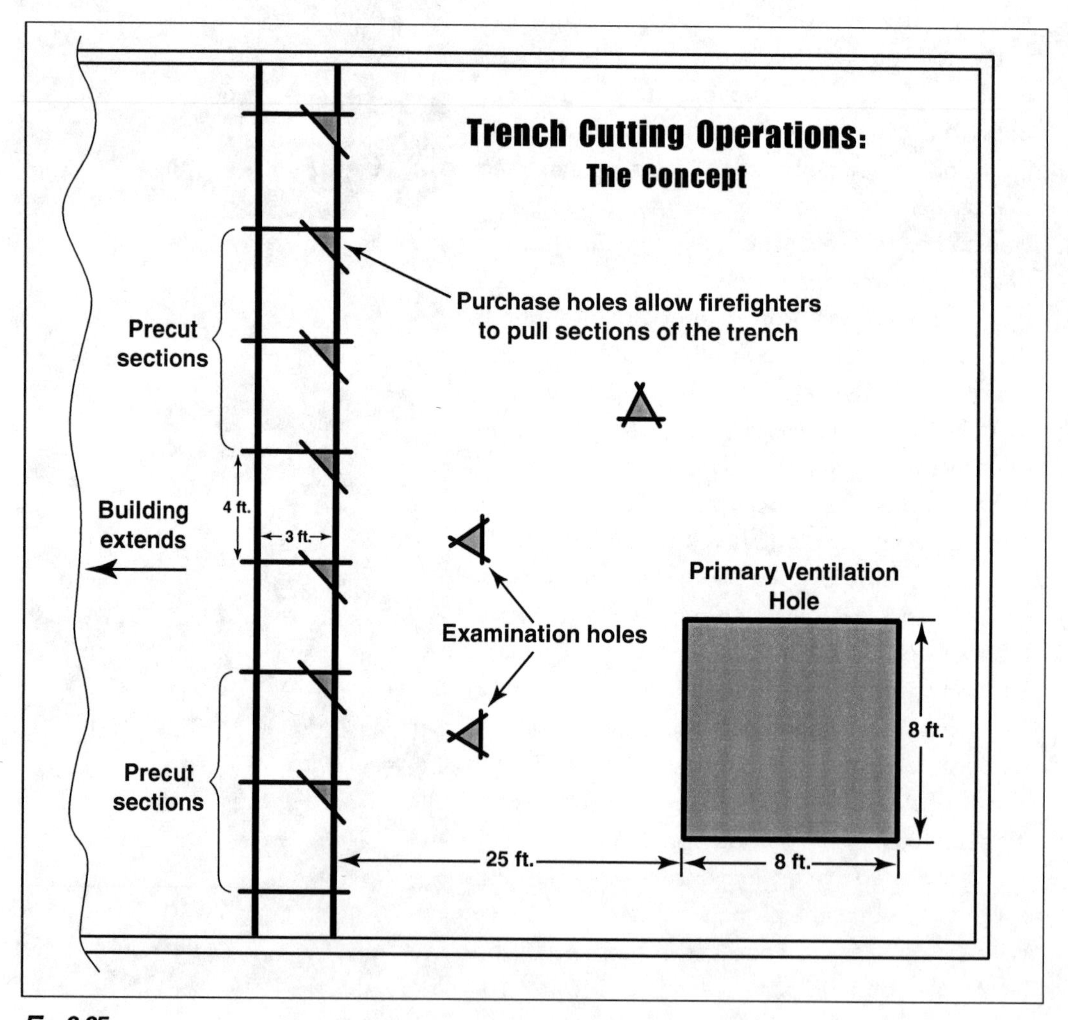

Fig. 3-35

2. The trench is a 3′ wide by a 4′ long precut from wall to wall or from fire stop to fire stop. The important point of consideration here is the precut concept. The entire trench is to be precut pending the completion or near completion of the primary ventilation hole and the trenches actual need. Premature pulling of a trench before the primary ventilation is completed will pull the fire toward the trench opening.
3. Inspection holes over the anticipated fire's spread, will help determine the trench's need.
4. The placement of the trench must be at least 25′ from the primary ventilation hole. This distance will allow you time to complete the cut.
5. If fire conditions in the cockloft are heavy, and during the trench cutting procedures it appears that fire is below and moving past the trench, stop and evaluate roof conditions. If allowed, drop back a greater distance and start a second trench.
6. Members must have at least two ways off the roof at all times.
7. A hoseline is to be stretched for member protection and limited operation into the trench if the fire attempts to move past the opening. Be careful with this one. The use of any hoseline into a roof space must be coordinated with the officer in charge of the top floor operation.

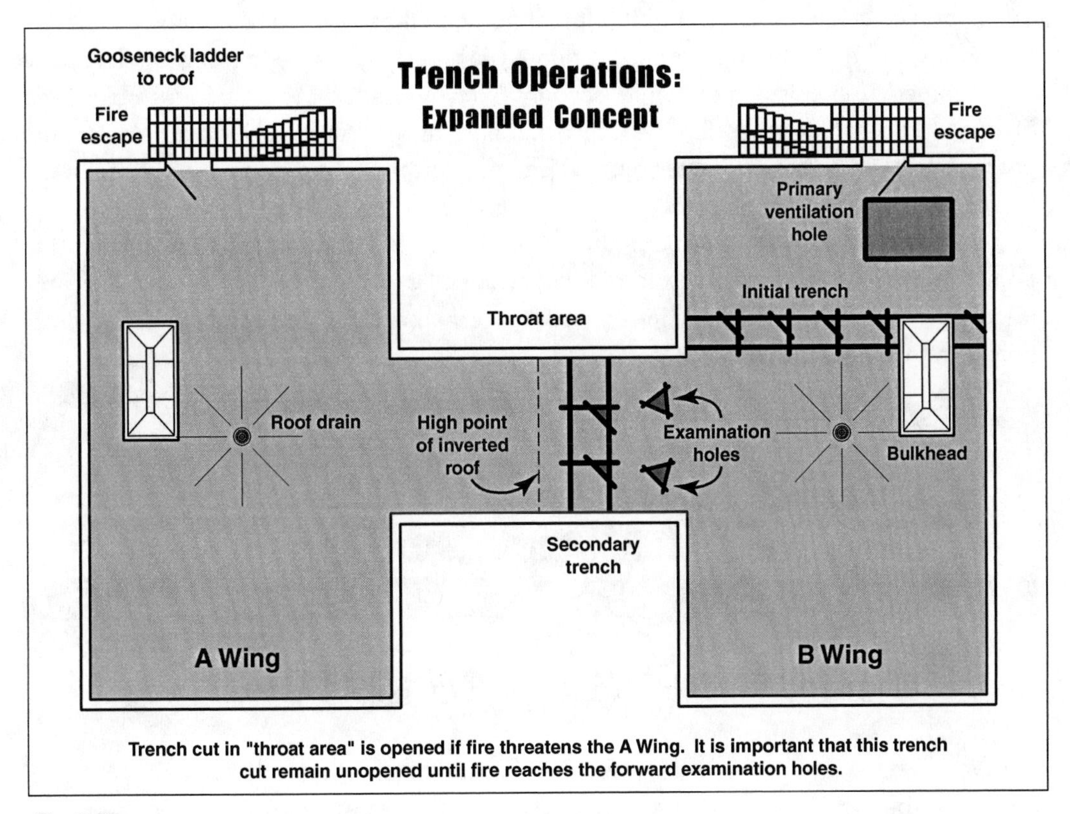

Trench cut in "throat area" is opened if fire threatens the A Wing. It is important that this trench cut remain unopened until fire reaches the forward examination holes.

Fig. 3-35a

Top floor/cockloft fires in these large area multiple dwellings will spread rapidly once this void space is compromised. Due to the layout configurations of the apartments and their wings, sufficient personnel is a must for hoseline advancement, pulling of top floor ceilings, and for roof cutting operations if we want to regain control of the building. Anything less than a coordinated step-by-step procedure will prove effortless at these fast moving top floor fires.

Time: Multiple Dwellings

*COAL TWAS WEAL**T**HS*

Time of day

Time of the incident plays a significant role at a fire in a multiple dwelling. Regardless of the type of multiple dwelling, any late evening or early morning fire must be considered severe. As we have stated, this is due to the fact that the majority of the building's occupants will be at home and asleep during the early morning hours. The hours of 12 a.m. and 7 a.m. have always proven to be the more difficult due the large occupancy population of these buildings and the fact that most, if not all, are asleep.

Adults and their children who have been awakened from a deep sleep are forced to make immediate decisions on family accountability and alternate exits in a smoke-filled building. This can at times force those occupants to act irrationally, doing unplanned, unrehearsed, or out of character acts which may further breed panic into surrounding family members to do the same.

Firefighters also are put through the extreme. Coming from a possible rest period to a situation which will require you to put just about all of your knowledge, skills, and abilities to the test within a two- to three-minute time frame, puts intense physical and emotional strain on each member.

The only plus of a late night or early morning fire for the firefighter will be the lack of traffic during the response.

Height: Multiple Dwellings

*COAL TWAS WEALT**H**S*

Tenements, NFP/H-Type Multiple Dwellings

When we review the height consideration in our multiple dwelling size-up, a number of items must be mentioned to ensure we handle one of the most important objectives quickly, efficiently, and safely once we get up there. Opening up over the interior stair because the interior stair in these occupancies is the lifeline and primary means of egress for the building's occupants. One of the firefighter's primary objectives is keeping this artery open to prevent smoke and heat from mushrooming out and down into other apartments, in addition to keeping the stairwell tenable for escaping occupants and fire department operations.

In this quest, there are two important requisites, speed and efficiency. Speed in the concept of how to get up there, and efficiency as it relates on what to do once you get there. There are a number of different avenues of approach on how to gain access to the roof of the tenement and apartment building that may warrant consideration.

Adjoining buildings

First is the adjoining building when available. Depending upon the area, you may find tenement buildings of equal height built in a row. Gaining access to the roof of the fire building from the adjoining is an early consideration, but not always a faster one. Many times building owners and even the occupants will nail or tar shut the scuttle openings leading to the roof in attempt to prevent unwanted entry from above. This may seriously delay members from getting to their assigned positions.

Aerial ladder

Another (and often used) consideration is the aerial ladder. This is the fastest means of getting to the roof, but is often plagued by obstacles, namely overhead wires. When not prioritized for occupant removals and able to access the roof, it becomes your best route.

Fire escape

A third consideration is the fire escape. This exterior means of accessibility is often the most dangerous. The fire escape comes with many areas of concern for the firefighter, namely stairs and railings. Checking each step with a tool, as well as the concentrated placement of your feet on the supported areas of the thread/stringer may eliminate a foot falling through a stair.

Don't forget to pull on the railings of the gooseneck ladder leading to the roof before attempting to climb up it. Originally these ladders were lag bolted into the roof deck, hopefully into a roof rafter. Over the years from shoddy roof repairs, age or rot, the lag bolts may have become loosened or removed. Not checking, may find you climbing a few steps and then hinging out over the rear yard as the top of the ladder comes free. This one will make you consider a reassignment with an engine company.

Ground ladders

A fourth consideration is the ground ladder. Depending upon the building's height and the compliment of ladders available at the time, this is an excellent means of obtaining access to the roof.

Remote stairs

A final consideration is use of a remote stair within the building. You must be careful with this option. The remote stair is a possibility with the H-type multiple dwelling and large apartment building. In a multi-staircase building, at least one, will lead to the roof. If the staircase you are considering is remote from the fire area, tenable, and leads to the roof, it may be used. But be prepared for a locked, chained, or bolted door.

Once you get up there, available natural vent openings over the interior staircase will exist in the form of a scuttle, skylight, or bulkhead opening.

Scuttle openings

With the scuttle opening, quick attempts at loosening three of the four sides may allow you to lift it and pry it up. If you attempt this, don't leave it hinged. I've seen a strong wind blow the hatch closed. It's best when working with the scuttle hatch to remove the cover entirely. If you accomplish this, place the removed cover on the roof upside down. The reason I suggest this is to draw attention to another firefighter walking along a smoky roof deck of an open scuttle hatch nearby. Lastly, if it can't be removed, cut it.

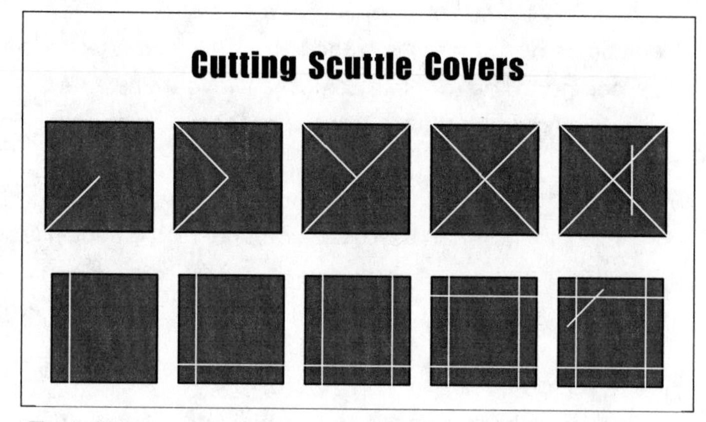

Fig. 3-36 *Two methods for cutting scuttle covers*

Skylights

With a significant fire, even on a lower floor, the skylight over the interior stair must be removed, and any obstructions below cleared. Unexpected shards of glass raining down a few flights of stairs will cause problems. If you're assigned to the roof, take out a few small panes of glass first to give a signal to members on the inside to stay toward the stairwell walls. This will hopefully eliminate any injuries when the rest of the glass from the skylight comes raining down.

Draft curtains

Another consideration with ventilation of skylight, notably in the tenement building, is the presence of a draft curtain. We mentioned them earlier within our ladder company section, but need to expand on their importance. As we stated, a draft curtain is a thin piece of cloth that has been stapled over the skylight opening on the top floor. Its intent is to limit any drafts that come from the opening on a cold windy day. Once a tenant or building owner has installed them, they are seldom removed. What a firefighter assigned to the skylight has to remember, is that once he or she removes the glass, they must also poke down to remove a possible draft curtain. As thin as they are, they could hold back a significant percentage of smoke in the stairwell.

Another important consideration with the draft curtain is the illusion or false picture they can paint. With the heat that could collect in this area from a fire below, the draft curtain could ignite. This quick observation from a vented skylight or firefighter looking up the interior stairwell and seeing fire, may give false reports that fire has possession of the top floor. When they do ignite, it is only for a short time before they burn away. Those that don't completely burn away have fallen on top of unsuspecting firefighters.

A draft curtain can be present in any type building with a skylight, but you will mostly find them in the tenement and row frame type buildings.

Bulkhead doors

The bulkhead door opening needs not only to have its door open, but also dislodged or removed. Again the wind can reseal a vital opening if left unattended. The bulkhead itself may also contain a skylight over its opening. Placing the fork end of a halligan bar into the roof deck to act as a small stepping stool will allow a member the added height to remove the glass of the bulkhead. A better option would be to take the removed bulkhead door and use it as an access ramp to the top of the bulkhead to clear the bulkhead glass.

Fig. 3-37 & Fig 3-38 *Removing the bulkhead skylight*

Additional height concerns

With the multiple dwelling, primarily those that are large and irregular shaped, there are a number of additional items that can be done from height that will aid fire department operations. Any relevant and useful information that can be given by radio to the incident commander, which could aid in the operation, should be looked for.

Roof radio information

1. The building configurations; H, E, U, etc.
2. If the building is accessible from more than the one side—another street, vacant lot, parking lot, alley, rear yard, etc.
3. Fire and smoke showing from any remote sides of the building.
4. Color and volume of smoke from any remote areas—for #3 and #4 include floor, wing, and number of windows. It is important to be specific here. State what is showing and where.
5. Fire extension probabilities and possibilities.
6. Location of any dividing walls.
7. Location of stairwells and fire escapes.
8. Location of any persons trapped.

Special Considerations: Multiple Dwellings

*COAL TWAS WEALTH**S***

Special thoughts, strategies, and concerns for the multiple dwelling are numerous. We say this primarily because of the different types, the different construction methods ,and the varying occupant loads. However, there are a few points that can be mentioned in this section that are universal in consideration for the arriving incident commander.

Tenements

1. Interior stairs. Get a report on conditions of the interior as soon as possible. Remember this is a small and highly combustible artery within the building that must be controlled and protected.
2. Location and extent of the fire. If the fire is in the basement, check on the integrity of interior stairs and all pipe chases in and around the area of involvement.
   A. If fire is on the top floor, check integrity of the cockloft area.
   B. If fire is in the air shaft(s), anticipate fire spread into exposures. Prioritize resources into these areas and call for additional help.

C. As soon as possible, get a report on conditions in the rear, especially if accessibility is going to be difficult.

NFP/H-types multiple dwellings

1. Accessibility. Get a report from the first due ladder company on stair type and fire location. Note with H-types, staircase accessibility may be limited. Depending upon the fire's location and accessibility to it, have additional engine companies assist with the initial hoseline stretch.
2. Location and extent of the fire. Floors above must be checked immediately for fire extension via pipe chases, I-beams/Channel rails, dumbwaiter shafts, etc.
   A. Continually monitor cockloft area.
   B. Use thermal imaging where and whenever possible.
   C. If cockloft is involved, assign an officer to the top floor, and one to the roof to coordinate what will be an extensive operation.

Housing complexes

1. Low-rise housing complex. Anticipate a long hose stretch; you may need to assign an additional engine company to assist.
2. High-rise housing complex. Check serviceability of standpipe system and be prepared to augment or bypass.

Questions for Discussion

1. List and describe the differences between old and new tenement, and how the differences may influence your decision-making.

2. What factors described give the classification of a multiple dwelling as being non-fire proof?

3. In what type of multiple dwelling will you find a scissor stair, and how their use could possibly cause a concern?

4. List and describe all factors that will affect your hoseline selection, stretch, and placement options for each of the three multiple dwellings described.

5. What is a draft curtain? Where are they located? What illusions can they create? What do we need to do about them?

6. List the life hazard concerns of the firefighter for a fire within a multiple dwelling. Your personal experiences will allow you to add to this list.

7. What is the big deal with one-way streets?

8. How many different types of stairs can be found within the H-type multiple dwelling, and how will they influence your actions?

9. Review the tactics listed for Roof Trenching Operations. Discuss each step, identifying the points and considerations listed.

10. What are the different avenues we can use in attempting to gain height at a multiple dwelling?

4 Taxpayers/Strip Malls and Stores

Fires that involve commercial buildings are more likely to cause death and injury to the firefighter than those in a residential building. Although the amount of fire activity in this type of building is far less than that of the residential building, more firefighters get into trouble here than in the private or multiple dwelling.

The uncertainty and infrequency of working in these type buildings combined with the different and unusual floor layouts, fire loading, as well as combined access and egress difficulties can create a difficult and dangerous fireground operation.

Taxpayers

"Taxpayer" is a term that originated as early as the 1920s from the practice of real-estate investors who, while holding land for speculation, constructed a building to offset their taxes until the land could possibly be sold or used for a more lucrative means. The building once erected was often divided into a row of stores

Fig. 4-1 *Taxpayer*

to increase the rental income for the property owner. These buildings are generally built of Class 3 ordinary construction. Heights can range from one to possibly two stories with a common cockloft and a common basement throughout for the entire row. Due to their age, combustibility, and inferior construction practices at that time, firefighters must expect a difficult firefighting operation.

Strip Malls

The strip mall is a newer version of the taxpayer. Reasons behind their completion are the same as the taxpayer; to create a number of leasable occupancies all on one parcel of land. What is primarily different between the two types of buildings is the method of construction used. The strip mall, which found its way into the urban and suburban areas in the 1960s, used newer and lighter building construction materials that became available at that time. Methods and materials associated with Class 2 limited combustible construction concepts started and continue in their design today. These buildings can range from one to two stories in height, with most constructed on a concrete slab.

Fig. 4-2 *Strip mall*

Stores

Within this chapter we have added fire department concerns with stores, generally the supermarket and retail type occupancy. The size-up concerns for these types of buildings will be similar to that of the strip mall. Where our concerns do increase will be with the fire loading and the building's square footage.

Construction: Taxpayers/Strip Malls and Stores

***C**OAL TWAS WEALTHS*

Taxpayers

The taxpayer is a building that is built of Class 3 ordinary construction. This method of construction consists of exterior masonry walls with wooden interior structural members. Originally, these constructed buildings contained plaster and lath walls to separate the stores, as well as to partition walls within the store. Depending upon when they were built, exterior walls may be brick and/or concrete block. Also depending upon when they were built, they may have steel incorporated into the design to carry structural loads over long spans. Any unprotected steel within the building that is subject to fire conditions can affect the integrity of the structures walls. Cracks developing in an exterior wall of the building are a reliable sign that the steel is expanding.

Tin ceilings

Within the building tin ceilings were common sights. These ornamental sheets were nailed directly to ceiling joists or applied directly to a plaster and lath ceiling. Many of these ceilings still exist today, but a dropped or suspended ceiling may hide their presence. The installation of a suspended ceiling will create a void for fire to travel. It is often in these void spaces where renovated utilities will be found.

Floor construction

Floor types for a taxpayer can also vary again depending upon when built. Original floors were made of either tongue and groove wood planking supported by wooden floor joists, which could all be held up by cast iron columns. Depending upon the occupancy type, you may also find terrazzo, marble, or poured concrete over the wooden floor joists and their wood surface. The purpose behind their installation was to eliminate the abuse and maintenance from a high volume of people that would frequent the store, at the same time providing some aesthetic value to its appearance. The fire service's main concern with a floor of this type is that

Fig. 4-3 *View of floor at store entrance way. Note multiple layers of flooring material.*

it can create a dangerous condition to firefighters when a fire is below. A fire below a floor of this design will weaken the wood supporting members while giving little to no indication of its severity to the members operating on it. The insulating effect of the material will hide the damage being done from the fire below as the floor suddenly fails.

Cocklofts

Cocklofts in the taxpayer are usually a common area extending over all the stores in the structure. This area can vary in height from 6" to 4'. Wooden roof boards, wood ceiling joists, and wooden lath for the top floor or store ceiling will be in this common void space that can be as long as a city block

Roof construction

Roof designs associated with the taxpayer can vary between two types. The built up roof was a design that came from the practice of nailing or stapling a layer of tar paper over wooden roof boards followed by a layer of liquid tar and asphalt roofing paper to give the protection and insulation it required. Also found within this design there may be a layer of tin, which was originally placed over the tarpaper prior to the layer of liquid tar and asphalt paper. The layer of tin was installed for added water protection. Some of the old-time roofers also insisted it had an insulating value as well.

The inverted or raised roof is a roof that is also referred to for all practical purposes to be a flat roof. Also found on the H-type and non-fireproof multiple dwellings, this type design was constructed so that one end of the roof is slightly raised to create a pitch for water runoff. The inherent problem with the inverted roof comes with its pitch. With all the stores' utilities in the rear, and the void spaces that normally accompany those utilities in the same area, a fire originating or extending into one of these spaces could extend into the cockloft only to be accelerated by the pitch of the roof deck.

Parapet walls

Parapet walls are a common association with the Class 3 constructed taxpayers. Parapet walls will normally be found extending above the building's roof deck on the front wall as well as possibly on the sidewalls of the building. Barring lateral support from the presence of any sidewall parapets, it will be unsupported for its entire length. With that concern, the weight of the wall is then dependent upon its placement and support within the building's front wall. Depending upon the date built, the construction methods used may have incorporated a steel beam acting as a plate at the first floor ceiling line. This steel beam is designed to carry the loads of the parapet wall for openings created in the front wall. With the steel spanning openings for large display windows or front entrance doors, any sizable fire venting through these openings could cause the steel to expand and twist, which may drop the parapet wall into the street or back through the roof. Within the design of the roof supporting members, additional steel may have been added to support additional wood joists. This additional steel is usually installed at right angles to the front and rear load-bearing walls. With even a remote fire within the store that exposes these cross-sectional steel members, their expansion from fire conditions may push the parapet walls supporting steel member, toppling the wall onto the sidewalk.

Any time there is a question about the integrity of a parapet wall the fireground commander must not only consider the area in front of or inside the store, but all the stores for the length of the wall. Depending upon how it was built, reinforcement rods or ties may have been installed throughout the parapet wall for its entire length. The slightest movement of any portion of the wall may pull it down for its entire length. In addition, many storefronts have advertising signs, canopies, or marquees that are supported by the building's front wall. Their installation may incorporate supporting fasteners, cables, or steel rods that pass through the wall and attach to the inside structural framing of the building. Any fire in the cockloft will weaken these supporting members and possibly drop the canopy or marquee, pulling down the parapet wall in the process.

Strip Malls and Stores

The modern taxpayer or strip mall is primarily built of Class 2 limited combustible construction. With this type building we find exterior walls of masonry block with a structural steel skeleton constructed on a concrete slab. Partition walls will be constructed of either wood or aluminum studs with a sheetrock finish, all covered by a roof supporting system of either lightweight wood trusses or steel bar joists. The roof supporting system and the type of roof deck construction is a major firefighting and operational consideration for the incident commander at this type of fire. Due to the required spans of the occupancy, the truss design must be expected in the roof.

Hybrids

In the lightweight wood truss design, the lumber's surface fasteners, gusset plates, or gang nailers will only penetrate the wood structural members on average of ¼" to ½". The connection points for the wood truss design are the weakest link in the design and fail

Fig. 4-4 *Sample hybrid truss construction*

with no warning other than their presence. Their failure can be within minutes of the arrival of the fire department.

The parallel chord steel bar joist is unprotected and subject to expansion, twisting, and bending from minor to moderate fire conditions. The unprotected steel truss supporting system in most cases will give visual indications of its failure from the roof level as it starts to sag.

This type of building has been associated with the built up roof system over the metal or Q-decking. Their fire and collapse concerns have been well documented in chapter one. In more recent times, this type of roof construction has been changing to the rubber or membrane roof design.

Membrane roofs are the latest concern for the fire service and can be composed of three different types. The *thermoset* type of roof is made from rubber polymers. The edges of the roof material are overlapped and sealed with an adhesive liquid at its seams. It generally looks like a giant bicycle inner tube that has been stretched and stapled to the roof deck. The *thermoplastic* membrane roof is made from plastic polymers including polyvinyl chloride. The seams of this type of roofing material are generally sealed with a torch and melted down. These roofs have been the subject of an increased number of fires due to their method of application. The last type of membrane roofing material is called *modified bitumen*. This type of material uses a number of new and old materials associated with roofing materials. From asphalt to fiberglass additives to polymers of plastic and rubber, this type of roof deck may give the initial appearance of a classic built-up roof deck of mopped tar and asphalt paper. What the fire service is experiencing with the different types of membrane roofs mentioned is an alarmingly fast spreading fire, that seems to be easily ignited from either surface ignition or from a below deck content's fire.

In older stores and supermarkets, the roof design incorporated large wooden timbers in the shape of a bowstring. Their construction could easily be seen as they formed the classic hump associated with this method of construction. The wooden members that formed the top and bottom chords as well as the web members could be of significant size depending upon the span.

Fig. 4-5 *Membrane roof*

As each web member contacted the bottom or top chord, an overlapping steel plate with steel pins would connect each section. In the large timber truss design, the connection point becomes the system's weakest link. The truss concept was an excellent one from an engineer's point of view. However, with each individual member of the truss acting on and relying on the other, any failure of the network will collapse the entire truss and any cross sectional members that it is tied to.

Occupancy: Taxpayers/Strip Malls and Stores

COAL TWAS WEALTHS

Within a row of stores, firefighters should expect to find every imaginable type of commercial occupancy. As an average there could be as many as six to twelve stores in the older style taxpayer, with twelve to fifteen different commercial occupancies, and the possibility of a large anchor store, movie theater, or restaurant within the strip mall complex. With the different types and numbers of occupancies that may become involved, firefighters must also expect multiple hazards that can be presented by each different type of occupancy. When available, pre-incident information is an extremely valuable resource not only to predict and anticipate the fire load concerns, but also to anticipate the content hazards to firefighters.

Contents

In most, if not all commercial occupancies, firefighters will encounter some type of hazardous material. When no specific incident information is available, arriving company and chief officers may need to gather information simply from the occupancy sign on the front of the building. Any on-scene-size-up information that is available must be considered and used when there is a concern for firefighters. Some examples of the hazardous associations with a particular type of occupancy are:

Occupancy hazards

Paint supply/home decorating centers:

- Numerous amounts of flammable and combustible containers from paints, varnishes and thinners.
- Varied amounts of building and roofing materials.
- Significant fire loads.

Clothing, rug, and carpet stores:

- Hydrogen cyanide gas from the buildings contents.

Supermarkets, meat markets, butcher shops:

- Phosgene gas from refrigerants.

- Corrosive and caustic materials.
- Projectile and BLEVE (Boiling Liquid Expanding Vapor Explosion) concerns from pressurized containers.

Drug stores:

- Hydrogen Chloride from the presence of plastics.
- Flammable and combustible materials.
- Uncertain activity when chemicals are mixed together.
- Uncertain activity when mixed with water.
- Pressurized Oxygen cylinders.

Pizza, bakeries, fast food establishments:

- Explosive potential and or rapid fire development from gas leaks from burned or ruptured gas lines.

Toy stores, general merchandise:

- Hydrogen chloride gas from general plastic contents.

Office supply stores:

- Nitrogen oxide from content involvement.

Pool and spa supply stores:

- Oxidizing agents, which involved in fire and/or mixed with water, can release chlorine gas.

With no material safety data sheets (MSDS) available, and no specific incident information from your pre-incident plan, viewing the type of occupancy may help in the decision-making.

Occupancy collapse concerns

Certain types of occupancies within the taxpayer may give added collapse concerns to responding and operating firefighters. In many of the older crime-ridden neighborhoods, property owners were forced to deal with a new type of unwanted entry, cutting the roof. These determined individuals would cut a hole in the roof deck, drop into the store, and take what ever they could get back up through the opening. After a number of these types of entries, store owners contracted roofers to install 4'x 8' roof sheets of ⅛" to ¼" steel over the existing roof deck. After the steel plating was lag bolted or screwed down into the roof deck, a new layer of waterproofing roofing material would be added over the steel plating. The steel plating with new roofing material gave the appearance of a conventional roof until firefighters assigned to the roof dropped their saw blade into the deck.

This new type of roof deck did meet its objective of eliminating any type of forced entry, but it also added a new dimension to the firefighter. The added weight of the steel created a new dead load that the roof supporting members were not designed to handle. Prior to the ladder company reporting their findings to the incident commander, members had very little information on their presence. In rare occasions there might be a small sign on the side of the building, "Steel Plating on Roof". Besides a sign, the only other tip that steel plating might exist was an association with a specific type of occupancy. In the inner city we found this mostly to be in high-value, high-need occupancies. Examples being jewelry stores, drug stores, toy stores, and camera shops.

Apparatus and Staffing: Taxpayers/Strip Malls and Stores

*CO**A**L TWAS WEALTHS*

You will generally get only one chance at controlling a fire in one of these buildings. Barring the slim possibility of a significant number of resources initially responding to a reported fire in one of these types of occupancies, first arriving fire officers will have to make decisions and assignments based on the priority areas and the available resources.

When a fire is reported in the evening after the close of business, firefighters will have to expect forcible entry difficulties, advanced fire conditions with an overall delay of getting water and significant ventilation in place. Add this to the fire extension possibilities and probabilities, and the tasks to be initially assigned, increase. With all of this in mind, a well-coordinated and managed attack must be implemented. Depending upon the location of the fire and its conditions, engine companies and ladder companies must have an orchestrated approach as they attempt to locate, confine, and extinguish the fire.

Engine Company Operations

Water supply

With the engine company operations initially focusing around establishing a water supply, operations must provide for continued, uninterrupted, and sufficient supplies for these buildings. Knowing water area availability is critical for a positive outcome. Hydrant location, accessibility, and main size will allow engine companies to take advantage of what is available.

Hoseline selection stretch and placement

The company officer's orders in the areas of hose line selection, stretch, and placement will depend upon the fire's location, extent, and volume. This is not going to be one of those situations where you can possibly play catch up. As we stated, you're only going to get one chance in controlling this fire. With moderate fire conditions, engine company officers must select and stretch their biggest size attack line. With heavy fire loading accompanied by large and deep floor areas, stream volume reach and penetration must be your deciding factors with hoseline selection.

The common request is 2½" lines with 1⅛" or 1¼" smooth bore tips. Whatever you may lack in maneuverability, you make up with the extinguishing capabilities of that size hoseline. If conditions allow, this size hoseline could always be gated down to a smaller diameter hoseline when needed. But what is important in this type of occupancy is that you're to prioritize the use of your water. If you select and stretch a small hoseline and then realize it's not doing the job, you have missed the opportunity to control the fire.

Fig. 4-6 *Commerical occupancies require large hoselines to be stretched.*

Hoseline placement will depend upon the location of the fire, its relationship to any life hazard, and the extension possibility. As a written rule, the first line must go between the fire and any endangered occupants. This has always been, and will always be a priority placement. With no occupant life hazard concerns, attacking the main body of fire and attempting to initially slow the volume of BTUs being generated will often be given priority consideration. With the heavy fire loads, the opportunities for fire extension, and the forcible entry delays of opening additional stores, giving consideration to the first deployed hoseline to darkening down and slowing the combustion being generated by the largest volume of fire may allow for a more timely and effective placement of protecting the exposure stores.

Additional hoseline selection stretch and placement of the second and third hoselines will be based on a number of factors that the incident commander must consider. The second line may be used to back up the initial line. It can be stretched into the most severe exposure. Depending upon the volume and extent of fire, hoselines may be ordered to skip the closest exposure store and operate in the next. Visual cues as well as reported conditions will determine their placement.

Ladder Company Operations

Ladder company operations will initially focus on apparatus placement, forcible entry difficulties, areas to be searched, and ventilation options associated with these buildings.

Placement

Apparatus placement for the ladder company should focus around maximizing the ladder's scrub area in association with the anticipated travel of the fire. In many cities researched, the front of the building belongs to the ladder company. With only one ladder company responding initially to the incident, the area in front of the building must be given prime consideration. We find this a critical position even when dealing with a one-story building. When all else fails and fire takes possession of the building, the tower ladder becomes the most versatile tool available if it is allowed this position. The ability of this type of apparatus to deliver and maneuver large quantities of water is often the incident commander's best weapon in delivering a significant blow to any sizable fire.

The same consideration must also be given to the aerial ladder. Fires that involve the two-story taxpayer will require a sufficient number of ground and aerial ladders for roof ventilation, window ventilation, window rescues, and hoseline advancement in the front, but let's not forget the rear of the building. With this in mind, fire departments that have two ladder companies responding to a fire incident, should give obvious consideration to placing a ladder company in the front of the building, and when possible, a ladder company in the rear of the building. It is this area where forcible entry will often be most difficult, and where you may very well find a number of occupants trapped on the second floor, awaiting rescue.

Search

Areas within and around the building that have to be searched can be numerous and difficult. Officers and their members must be alert to the possible occupant life hazards in cellars, rear store areas, second floor occupancies as well as exposed occupancies. In some cases their presence may not be that obvious. From the unexpected basement apartment to the obvious second floor occupancy, all areas must be searched.

Second floor occupancies commonly found in the taxpayer may house people all hours of the day. Many of the occupancies on the second floor of a taxpayer may have no (or very limited) secondary means of egress. Often the occupant's only means of exiting the area will be from a common public hallway that traverses the building. Depending upon the occupancy, many people could be trapped above a rapidly advancing fire. Aggressive laddering and searching of the second floor will become a priority to members assigned to the incident.

Forcible entry

One of the biggest problems and concerns for the fireground commander is early accessibility to the stores currently involved as well as accessibility to those stores threatened. This concern is not generally a problem during the business day when all stores are open, and access is not a problem. However, after the close of business, when store merchants and their workers have gone home, quick and early access to these stores becomes a concern. As any ladder company member will tell you, forcible entry in these types of buildings can be a slow and tedious process. It is for this reason, accompanied with the anticipation of early fire spread, that fireground commanders should consider assigning a number of firefighters to force all the doors in the front and rear within the row that are questionable. Forcing the openings of only one or two stores in the row and then later realizing that there is a possibility of fire spread into adjoining stores beyond where companies initially forced, will only allow fire to gain substantial headway and take possession of a number of stores as members try to play catch-up.

***Fig. 4-7** Forcible entry will be difficult.*

Ventilation

Vertical ventilation, when available, is often the key to a successful outcome at a difficult fire in a store or taxpayer. We state "when available," from the risk analysis of the roof supporting system. When the building involved has a lightweight wood truss support system, their involvement should negate operations on top of, or below this type of system. The failure of a wood truss system has been documented on numerous incidents within minutes of the fire department's arrival. If there is a serious enough fire below this type roof deck that requires a ventilation hole to be established, many agree that you shouldn't be up there.

When information lends itself to a roof deck supported by a solid lumber system, being able to create a substantial thermal updraft will slow down the fire's horizontal progress and allow companies to regain possession of the store of fire origin and any adjoining exposure stores. This tactic does not come without an extensive and well-managed effort by a complement of firefighters. Initial consideration must be given to opening up any natural ventilation openings, that in most cases should come with minimal effort. However, when observations present more advanced conditions, well-planned and coordinated cutting of the primary ventilation hole is a critical objective in our ventilation options. This hole must be planned and pulled in order to support the largest possible hole over the main fire area. A number of small holes will not create the same thermal updraft as a large hole. Efforts should be made to obtain a hole as large as 8′ by 8′.

In a further attempt to halt the fire's extension, a defensive *trench cut* or also *strip cut* may be established to eliminate the fire spread throughout the row of stores. This is a decision that should be made by a fire officer assigned to supervise roof operations. Many departments will attempt to assign a Chief officer to the roof due to the extensive and coordinated efforts required of this type of building.

There should be no doubt that the cutting of a large primary ventilation hole and the cutting and pulling of a trench will require a small army of firefighters and tools that more and more departments are finding it difficult to muster. With a limited number of people and equipment to initially work with, energies should be concentrated on a large hole over the main fire area.

Roof observations

As a final note with our ladder company operations, it has often been said that the ladder company is the eyes and ears of the chief officer. This in many cases is a true statement, and is primarily mentioned because of their ability to see and report specific observations that will aid the incident commander in the ability to make efficient, effective, and safe decisions.

With the rapid fire development possibilities of the taxpayer/strip mall and store, members assigned to the roof must report specific observations to the incident commander. The taxpayer/strip mall observation report should include these:

1. Size and shape of the building and its attached exposures.
2. Smoke volume and intensity around the structure–be specific.
3. Observations and indications of fire present in the cockloft.
4. Location of parapet/dividing walls separating stores.
5. Location of any light and air shafts within the complex.
6. Roof support and deck construction.
7. Heavy dead loads on the roof deck. (i.e., air conditioning, ventilation units, etc.)
8. Heavy live loads on the roof deck. (large accumulation of ice, snow and water)
9. Differences in levels of the roof deck surface, which may indicate a structural add-on.
10. Location of structural add-ons.
11. Life hazard concerns in the rear.
12. Accessibility to the rear.
13. Other useful observations.

Life Hazard: Taxpayers/Strip Malls and Stores

*COA**L** TWAS WEALTHS*

Firefighter Life Hazards

According to statistics that are compiled on an annual basis from the National Fire Protection Association (NFPA), firefighters are two and half times more likely to die in a fire in a commercial building than they are in a residential building. This comes to the fire service as an alarming figure especially due to the fact that the majority of the fires that we respond to are in residential buildings. As members of the fire service community started to investigate the life hazard concerns and reasons for these statistics, the list became extensive.

Familiarity

The first association with the difficulties focuses around the number of fires in these types of structures and the uncertainty that comes with every individual building. As the majority of our work load has been and continues to be with residential buildings, the fire service like any other industry gets better at what they do the most. But as the statistics indicate, we must constantly review and educate ourselves about the dangers that await us in the commercial building fire.

Advanced conditions

As we start to look further at factors that cause concern to the firefighter, it is important to start with the conditions and difficulties presented to us on arrival. On arrival concerns become greater to the firefighter after business hours when most taxpayer/stores are closed. With no or limited early warning devices associated with many of the type occupancies described, by the time the fire department receives the call and then arrives, the fire conditions may be well-advanced. Add the delay from forcible entry difficulties commonly associated with these occupancies from roll steel doors, iron gating, and fortified rear exteriors, the fire's ability to take possession of the building and its adjoining properties becomes obvious.

Contents and square footage

Further considerations must be given to the fire loading and large floor areas associated with this type of building. As we discussed in the occupancy/content section of this chapter, the fire loading concerns can vary greatly based on the specific store. Ordinary combustibles of significant quantity, to untold numbers and types of hazardous materials cause the firefighter to encounter this uncertainty.

Floor areas associated with each individual store will add to the content concerns. Floor areas for the taxpayer can average from 25′ in width to 75′ to 100′ in depth. Store sizes can encompass the width and length of an entire city block; the bigger the building, the bigger the fire load.

Disorientation

The large floor areas are not only going to cause concern for sufficient stream reach and penetration, but also for the advancing firefighters to navigate the maze-like configurations and high-piled stock. The uncertainty of the large floor areas and their layout can create confusion and disorientation among firefighters. High-piled stock has often been known to fall on top of firefighters, or bury hose lines preventing their means of egress. The same maze and high-piled stock concerns will also allow fire to move around the sides, over, and possibly behind, advancing firefighters. Consider all the above with the possibility of high ceilings or multiple ceilings within the building, and firefighters at floor level may receive no indications of developing or hidden fire conditions.

Collapse

Collapse concerns for the taxpayer/store occupancy will primarily focus around the class of construction, the building's age and any known structural alteration or addition. From the design defects of the parapet walls, the uncertainty of hanging marquees and canopies, multiple ceilings levels, terrazzo floors, and truss roofs to the steel plating on the roof deck, the taxpayer/strip mall and store will present numerous potentials for collapse.

Occupant Life Hazards

Occupant life hazard concerns can be considered quite a challenge to arriving firefighters depending upon a number of factors. During the daytime hours, places of business can be fully occupied causing concerns with occupant search and evacuation. The general benefit with the daytime hours is that even though the store is full, most occupants will vacate the store at the earliest signs of fire. The difficulties most often lie with unsuspecting occupants of a rapidly developing fire in the two-story taxpayer. With public gathering establishments, offices, or residences often found on the second floor of these type buildings, people may become trapped from a fire in a store below. With limited access from these second floor areas, panic will quickly set in causing numerous and difficult rescues to be made above a quickly spreading fire. Members will be hard pressed to get ground ladders up to the second floor to create as many added means of egress as possible for escaping occupants as well as for advancing firefighters.

Other unique concerns to the occupants will be unsuspecting occupants sleeping, living, or working in the rear of the store or in the basement. Although illegal and uncommon, these areas can surprise the unsuspecting firefighters. Individuals may be found in the store for a number of reasons. They can be stocking shelves, cleaning the store, or staying overnight as a watchman.

Fig. 4-8 *A two-story taxpayer with residential occupancies on floor 2.*

Terrain: Taxpayers/Strip Malls and Stores

*COAL **T**WAS WEALTHS*

As we review the three categories within our terrain concerns of setbacks, grade, and accessibility, the one that seems to cause the most common concern with the above type occupancies is the accessibility to the rear of the buildings. This seems to be more of a concern with the older taxpayers when compared to the newer strip malls or larger stores.

Taxpayers

The older taxpayers were erected on any plot of land available at the time. It didn't matter if they were erected between two apartment buildings, had buildings built around them after their completion, or they themselves consumed an entire city block. What became evident was that their congestion and accessibility to the rear received little attention during this process. In older areas of many cities all that is often found in the rear of some buildings is a narrow alleyway just wide enough for a person to traverse. In others, a single-lane driveway may allow access to the rear. In many areas, these small passageways are all that separates the rear of the stores from occupied apartment buildings.

Regardless of the type of accessibility, it is essential when the fire department responds to a fire in a row of stores placed in this type setting that early attention be given to the rear not only to gain access, but also to obtain information on any threatened exposures and trapped occupants.

Strip malls

In the more modern version of the taxpayer/strip mall, the rear accessibility concerns changed somewhat. It became apparent, as storeowners became more money conscious, that in order to allow for an uninterrupted flow of shoppers to the front of the stores, deliveries had to be made from the rear. This allowed for a larger design in the rear areas behind the stores for the unloading of delivery trucks. This consideration inadvertently allowed for greater access for fire department vehicles and equipment to the rear of the buildings. This area however, even with its added space, does not come without some additional accessibility concerns to the fire department. The rear area of a row of stores becomes the collection point for garbage and its storage. From dumpsters, accumulated piles of trash, to the buildings utilities, this area cannot only be the source of fires, but can also cause apparatus placement and maneuverability to become difficult and dangerous when we need to get people and equipment there.

Fig. 4-9 *Expect limited access and forcible entry difficulties in the rear of the building.*

Just as this space allowed for easier access for delivery people and the fire department, it also allowed unviewed accessibility for anyone attempting unlawful entry into a store. As the frequency of store burglaries increased, so did the number and types of locks on the rear of the individual stores. For this reason, the rear of the stores became fortified creating a more difficult and time-consuming operation for the fire department.

Water Supply: Taxpayers/Strip Malls and Stores

COAL TWAS WEALTHS

Water availability for a fire in a taxpayer/strip mall or store will depend upon the area in which the occupancy resides. In many instances this type of occupancy was often built in sections of the city or town well after residential housing was established. In other areas their construction was somewhat more planned with a larger emphasis placed on ensuring an adequate water supply.

Taxpayers/Strip Malls

Requirement and delivery

The water supply needs for the taxpayer and strip mall will quickly become evident soon after row after row of stores are consumed by what firefighters continually refer to as fast moving fires. Quick, large, and sustained water flows are necessary to mount an effective and efficient attack on a fire that involves this type of commercial occupancy. Engine companies with large supply and attack hose lines, supported by sustained supplies, will be the rule rather than the exception. Creating a number of operating or attack pumpers each supplied by a supply pumper is the policy many departments use to ensure adequate water volume.

Fig. 4-10 *A 2000 GPM Engine Company is assigned as a Supply Pumper at this incident.*

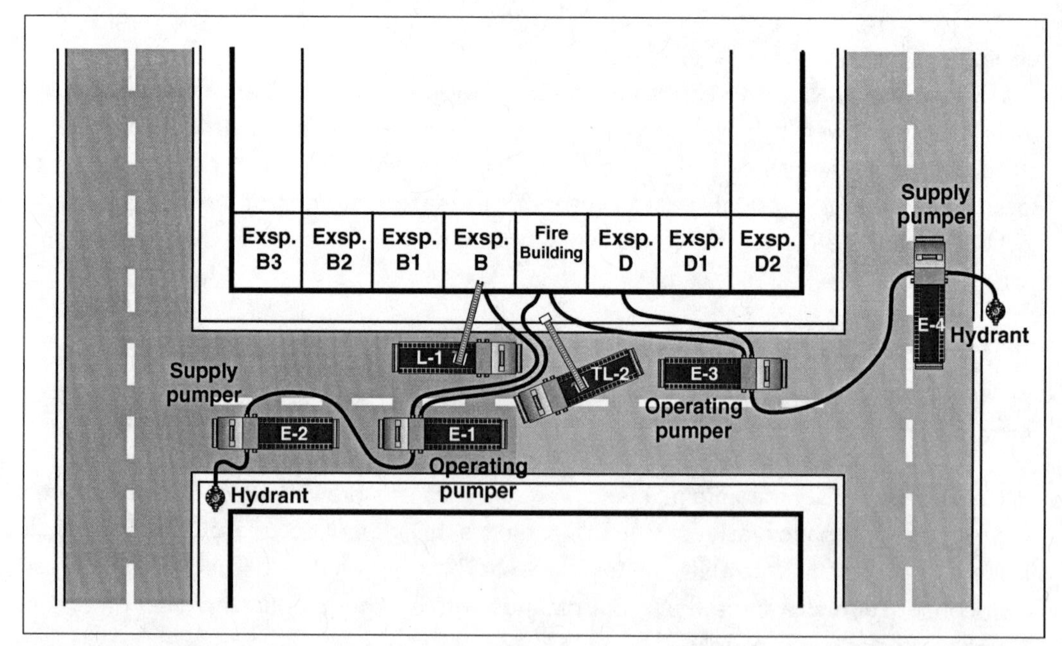

Fig. 4-10a *Operating pumper and supply pumper concept*

Departments who strictly rely on in-line pumping with minimal size supply hose lines, will quickly be overtaken by what initially seems like a small to moderate size fire. When past history or research shows the need for large water supplies to be delivered to multiple hoselines, a series of operating and supply pumpers is often your best grouping of your engine company compliment. Placing an operating or attack pumper in the tactical zone and using a second pumper on a hydrant supplying the operating pumper will allow your resources to move large volumes of water to the front of the building, with significant hose available to be stretched for handlines and master streams. As stated earlier, this is not the type of building where you can play catch up.

Auxiliary Appliances and Aides: Taxpayers/Strip Malls and Stores

COAL TWAS WEALTHS

Fire prevention codes will vary from jurisdiction to jurisdiction. What is seen in one jurisdiction may differ from what someone may see in another. Even those areas with the most stringent codes will only be as reliable as they are enforced. The reliability of this enforcement is essential not only during the construction or alteration phase, but also more importantly, during the occupancy use phase.

Taxpayers

Research shows that the most you will find in the older taxpayers will be a local alarm system that will sound only on the premises when a fire starts. It is with this type of system that you hope someone calls the fire department as soon as they hear the alarm. With the taxpayer primarily being in the downtown areas, most residents in these areas are so accustomed to hearing noises especially from car alarms, burglar alarms, etc., that when a fire alarm sounds, no one pays any attention until they smell or see smoke.

In addition, in some older style taxpayers we found, on rare occasion, a sprinkler system that served the basement area. This finding is generally considered to be the exception rather than the norm with the taxpayer.

Strip Malls

Strip malls, in addition to a local and/or monitored alarm system, may also contain a sprinkler system that serves the store area. Their presence will depend upon code requirements within the jurisdiction. The advantage of a sprinkler system anywhere in this type of building is an obvious luxury, as long as the water gets on the fire. Anytime a sprinkler head is activated and hits the fire, it is preventing firefighters from having to deal with an uncontrolled and advancing fire. Unfortunately this doesn't always happen.

System concerns

The concern with a suppression system can prove to be detrimental if not considered. Initially we have to be aware of their presence in older, partially renovated buildings. Depending upon the amount of renovation and the strictness of the codes at the time, many of the building's original deficiencies as well as inherent voids will still exist. If a fire starts in a void space or is given quick access to one, the fire's ability to spread is unchanged.

Poor housekeeping of the occupancy will often impact the sprinkler systems effectiveness. With storage rooms piled high with stock as we often see during the holiday season, the ability of an activated head to discharge its designed spray could be seriously reduced.

Reliability of the system is another concern that must be considered especially when present in the older occupancies. Uncertainty about when the system was installed, inspected, or tested may find firefighters putting faith in a system that has been inoperable or inadequate for a number of years.

There is no doubt that a well-designed and functioning sprinkler system is the firefighter's best weapon in controlling a fire. But when the reliability of the system is in question, firefighters and fire officers should continue to plan for the worst.

Street Conditions: Taxpayers/Strip Malls and Stores

*COAL TWA**S** WEALTHS*

Fire service concerns with street conditions and their relationship to the taxpayer/strip mall and store, will generally focus on whether the store is old or new, and the type of area in which they were built.

Taxpayers

The older type taxpayers can be found anywhere within the congestion of a town or city. Their location will most likely be on streets in residential neighborhoods that in many cases offer no off-street parking. When landowners gave consideration to erecting a row of stores to offset their property taxes, allocating some of that property for the use of parking rarely entered their minds. Why waste rented land when they can park on the street? Early on this concept wasn't a problem for anyone. But as communities became more crowded, parking a car near a person's residence, let alone parking a car near a market or store, became impossible. Streets in these areas were not originally designed to handle the vehicle load of today. Today we find cars parked on both sides of the street often with double-parked cars, double-parked delivery trucks, and blocked fire hydrants just to name a few. What this all adds up to for the fire department is a delay in response and possible elimination of effective apparatus placement.

Strip malls and stores

Times changed and more thought was given to how to get the most for your money. Property owners and building developers soon realized that if you give them parking, not only will they will come, they will stay for a while. Newer versions of the strip mall and store will generally have off-street parking lots to allow more people to visit their developments and to stay and shop. Not having to worry about change in a parking meter or double parking gives an individual the opportunity to do what the storeowner wants—spend money.

Fig. 4-11 *Off-street parking does not guarantee the FD total accessibility. In many cases it can be worse.*

This is not to the say you won't find cars parked illegally in fire zones in front of the stores, but it generally eliminates our inability to get there at all.

Weather: Taxpayers/Strip Malls and Stores

COAL TWAS WEALTHS

Weather can and will continue to play significant roles at a fire regardless of where it is located. Wind, temperature, humidity, and precipitation, when present, will always add concern to the decision making of the firefighter and fire officer. Each of the four comes with its own specific set of concerns.

Humidity

Certain elements of the weather can mask fire conditions, ultimately compounding the officer's ability to make effective decisions. Anytime you are dealing with large square footage or maze-like configurations (often associated with the taxpayer/store fires) finding the actual seat of the fire can sometimes prove to be very difficult. What can further compound this problem within our weather category is the humidity. I'm sure many of us have been at incidents on hot, humid summer nights when as you enter the block of the reported fire, finding the actual building that is on fire is time consuming. In incidents as described, the humidity is often the contributing factor. With high humidity, the atmospheric air becomes moisture-laden and quickly cools the smoke affecting its buoyancy. Low-hanging, cooled smoke will not only compound the concerns of finding the main body of fire, with the added probability of ten to twelve stores all with locked roll-down steel doors in the row, determining the fire spread is made more difficult. Picking, prioritizing, and going to work on the right roll-down door in these types of conditions will be complex.

Combating humidity. One of the most advanced and best-used tools by the fire service is thermal imagining. This should be one of the first pieces of equipment off the apparatus when it is available. It seems that initially this piece of equipment was only finding its way into the fire building to chase down hot spots after the fire was under control. As training and education with its use continued, the camera was finding its way into the front door of many fire buildings with the first deployed hoselines. With any difficult-to-find fire, thermal imagining is the firefighter's best friend. Even in the case of ten to twelve roll-down steel doors in a block-long row of stores, a scan with a thermal imaging camera may give you indications on where to prioritize forcible entry, as well as where to stretch and place hoselines. Under heavy smoke conditions this should be one of the most widely used tools in your department's arsenal.

Exposures: Taxpayers/Strip Malls and Stores

*COAL TWAS W**E**ALTHS*

Taxpayer/Strip Malls

The probability of fire extending into adjoining exposure stores is very high, and must be considered very early. In this section we will concern ourselves with early recognition of the exposures, accessibility, and their identification. As you can already imagine, the types of construction, the anticipated alterations, and the high fire loads should all lead the incident commander to anticipate early and rapid fire spread into adjoining stores. As we stated in earlier sections within this chapter, this type of building can encompass an entire city block. Many will be sandwiched between larger and higher multiple dwellings within the block, or they can be designated their own section within a downtown area. In either of the cases, their numbers could exceed 12 to 15 stores in a row.

Incident management

In anticipating and dealing with fire spread, incident commanders must be able to assign companies into any store within the row to check for fire extension, as well as also be able to account for them as they manage the incident. Management and accountability of the forces can be a difficult task considering the number of stores members may have to work in. Exposure store identification becomes critical especially with companies assigned and operating from the rear. From the street or command side, the fireground commander will obviously designate the fire store and reference the exposures by letter and number, in addition to having the luxury of furthering that reference by the store name displayed (if it becomes necessary). From the rear, storeowners may be reluctant to display the name of their store in fear of attracting a certain element of people, especially if the store is a high-value occupancy like a jewelry or camera store. For this reason early assignment of a person in charge of the rear of the complex with the information of the designated fire store will allow early and quick reference as companies go to work in adjoining properties.

In New Jersey, the state uses a standardized statewide Incident Command System, whereby companies from as far away as Atlantic City could operate at an incident in Jersey City and not be confused on geographical assignments and responsibilities within that incident. Exposure designations for the State of New Jersey use the letter reference, followed by a number as exposures increase on the lettered side. At times this has proven somewhat difficult especially with the similar sounding letters of B and D. We learned early on that it was best to reference exposures by stating the words *Bravo* and *Delta* instead of the letters to eliminate any confusion. This exposure designation accompanied with the assignment of a division formally referred to as a sector, allowed for

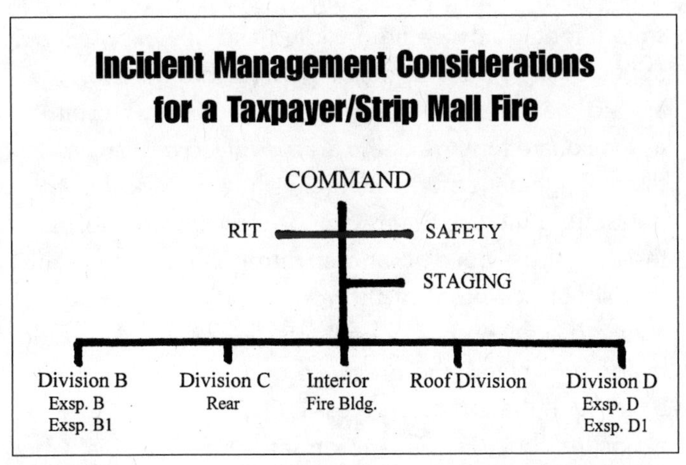

Fig. 4-12 *Sample of organizational chart for Incident Management considerations*

quicker and more efficient management of the exposures within the overall incident. Assigning responsibility early on to the specific areas, as well as to the specific stores, proved to be valuable for the management and accountability of the incident.

Area: Taxpayers/Strip Malls and Stores

*COAL TWAS WE**A**LTHS*

The square footage of the fire building, its layout, configuration, and the members' ability to operate safely and effectively are the considerations in this size-up factor. A key concern with our efforts in this area is that no two commercial occupancies are the same. Not only will they differ in square footage and configuration, but also their contents, fire loading, and accessibility to vital areas will vary.

Initial considerations for incident commanders as they arrive at any incident are always going to focus around getting the answers to the questions "what do we have?" and "where is it going?" These structures don't lend as much information as we would often like, especially as it relates to the answers to these two questions.

Taxpayers/Strip Malls

Shapes and sizes

As Chief officers attempt to manage these types of buildings, it is difficult (at times impossible) to be able to stand in front of a row of stores and be able to predict the square footage of the fire building and any exposed buildings. Pre-incident information about these particular buildings and information received from vital positions in and around the fire building is necessary. Information about the layout, configuration, as well as the square footage of the area involved and any areas threatened are not only vital for the management of the incident, but also critical to the firefighter's safety.

Barring the use of any pre-incident information, obtaining initial information about the size of the building and any irregular shapes associated with the building in question should come from the roof position. A quick walk around the perimeter of the involved area by one member will lend valuable information about the size of the building involved, the presence of any extensions, whether they wrap around another building, if there is rear accessibility, if there are any trapped occupants, as well as any exposure concerns. This information paints a valuable picture to a fireground commander who often only has the opportunity to see nothing more than a block long series of storefronts.

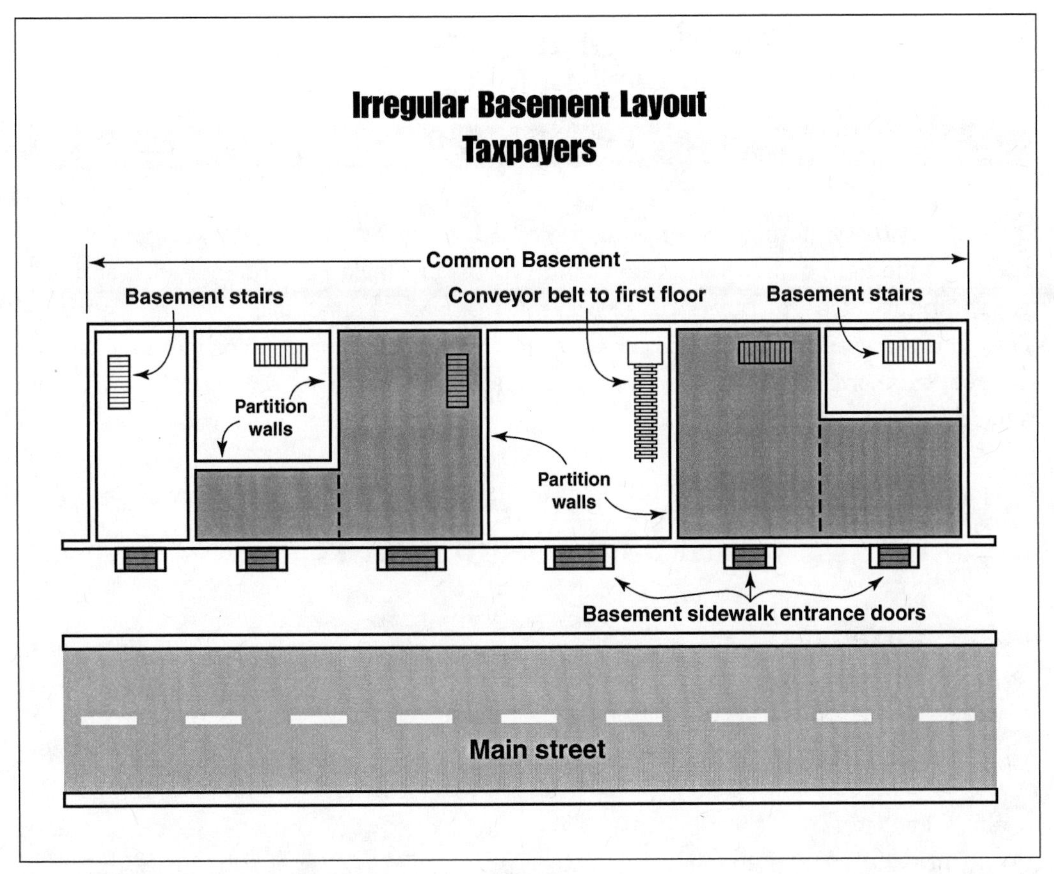

Fig. 4-13 *Irregular basement layout found in some taxpayers*

Just knowing the uncertainty that accompanies these types of buildings, fire officers must be aware that the interior layout, configuration, and square footage may also be drastically different on the inside of the building as compared to the outside of the building. These buildings are notorious for cellar layouts that do not conform to the layout of the first floor. Irregularly shaped cellars (often L-shaped) are a common occurrence in the taxpayer. If the storeowner needed more storage area and was willing to pay for it, simply moving a partition wall in an already common cellar could give him the square footage he desires.

Interconnected/walkthroughs

Another, although less frequent, occurrence in a taxpayer is the possibility of interconnected buildings or walkthroughs. Owners who owned two or more stores adjacent to each other may create openings in the division walls for their own, as well as their patrons', passage. This possibility automatically doubles or even triples the square footage of the fire building.

Location and Extent of Fire: Taxpayers/Strip Malls and Stores

*COAL TWAS WEA**L**THS*

As previously stated, the importance of a fire officer's ability to anticipate the fire spread in the type building they are going to work in. Having prior experience and education about a type of building will hopefully provide a successful and safe outcome. From this association you could gather by now, that when a fire begins in either the cellar area, store area, or cockloft area it will cause serious concerns with early and rapid fire spread to adjoining stores.

Taxpayers And Stores

Below grade fires

Without any doubt, fires that originate and extend from a below grade area of a taxpayer/store are going to be extremely punishing, difficult, and dangerous to fight. Officers must weigh the risks of sending firefighters into a possible no-win situation. With maze-like configurations, heavy fire loading, limited access points, inferior and most likely altered construction, the difficulties will mount. When decisions are made to advance down into the seat of the fire and cut off its extension, members must know how the enemy might behave, especially how it might move from this below grade area.

Fig. 4-14 *Sidewalk access to cellar*

In the taxpayer, a common cellar can exist for the entire length of the building. Within the common cellar, areas may be divided with flimsy partitions sometimes no more than wood-frame studded sections covered by chicken wire. Partition walls, even of significant construction material (i.e., studded 5/8" Sheetrock™) can still allow fire spread concerns to the floor above as well as into adjoining cellar areas. Partition walls in many of the installations will have their top plate attached to the underside of the first floor joists at right angles. Depending on whether the cellar ceiling is finished or not (in most cases it never is) a fire originating in the cellar has a high probability of entering into the ceiling bays between the joists and traveling the length of the bay. Not only does this allow fire to directly attack the first-floor joists, but also once fire is introduced into the bay, it can travel throughout (even if adjoining cellars have finished ceilings).

Fire can extend from the cellar area vertically up to the first floor store area from numerous openings in the floor. From *cellar entrance doors, conveyor belt chute openings* that serve the store, *boxed out display windows* near the buildings front entrance, to the common *utility chases* found throughout, there is no doubt the fire will have plenty of avenues to choose.

Some of the openings mentioned may prove to be advantage points in creating a thermal updraft, allowing companies to advance below. Chute openings, although often located in the rear of the store, will provide a quick opening that generally will allow significant fire to exit up through the opening. Once this situation is recognized members could assist with the updraft by opening the roof directly over the first floor chute attempting to slow the horizontal spread of fire.

The boxed-out area below the first floor display windows, although a common area for fire to extend to, also makes an excellent area to open in attempting to allow the fire to vent as companies advance in from either a rear or side cellar entrance. One point worth mentioning here is that as the fire is allowed to vent out these openings, any structural members and the loads they may be supporting should be observed. Some of these load-

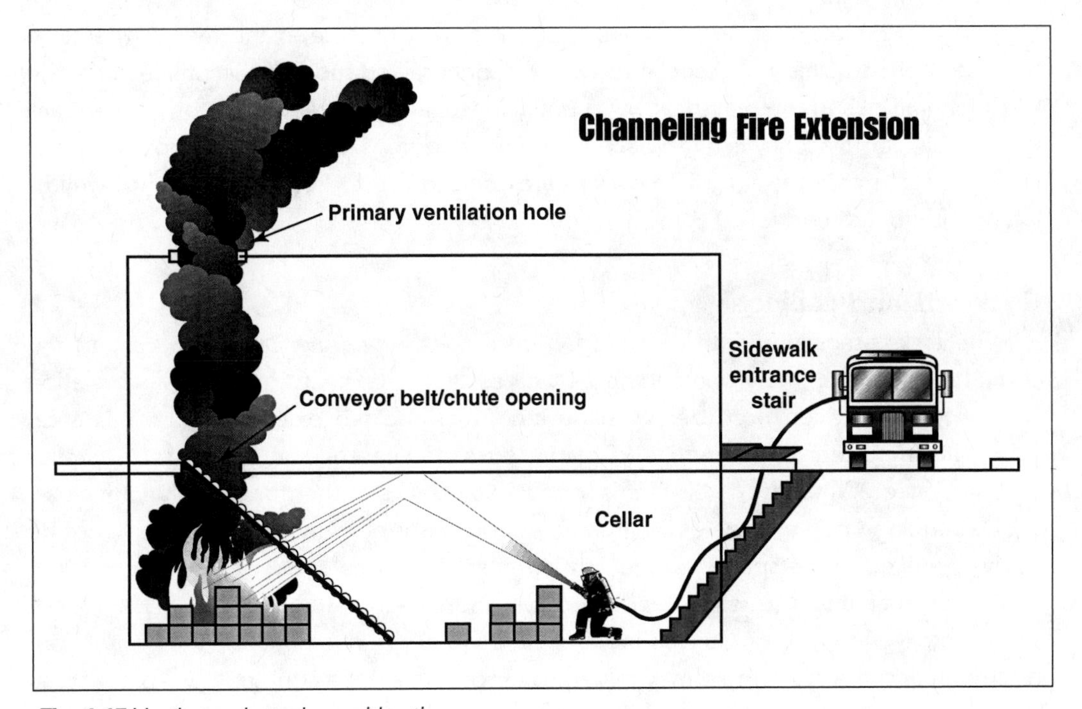

Fig. 4-15 *Venting and attack considerations*

bearing members that are now being attacked by fire may be supporting overhangs, marquees, or ornamental facades. In one such incident, engine companies were making a successful push into the basement from a side entrance when a ladder company member narrowly escaped being injured by a large display sign that was supported over the front display windows. One action will always cause a reaction.

Fig. 4-16 *Boxed out display window*

Taxpayers/Strip Malls and Stores

Store-level fires

Statistics dictate that a large percentage of fires originate in the rear of the store level. Reasons for this are associated with the general location of the utilities and the accumulation of storage materials. With poor housekeeping practices a common hazard in this area, complacency often breeds ideal fire conditions. It should also be noted that from this same area the fire will most likely extend into the dropped ceiling, as well as into the building's cockloft.

Cocklofts and ceilings

The tasks associated with regaining control of the hanging ceiling space and the cockloft will be extremely demanding as well as dangerous. Critical vertical ventilation with water application from below is the key in halting fire extension from this area. Again, it's important to be reminded of the possible presence of the roof's truss supporting system. When fire involves this space, early failure of the truss must be anticipated. Depending upon the size of the truss loft, indications of its involvement might not be easily identified. If the space is large or triangular, the space could be used for storage (furthering the fire and structural load concerns) and the large area will contain a sufficient supply of oxygen producing a well-fed, free-burning fire with minimal smoke production. Upon entering the store there may a light to moderate smoke condition when in actuality there is heavy fire overhead. Conflicting reports from the roof area and the interior of the building, as well a scan from a thermal imagining camera, will hopefully give you clues to aid in your decision-making.

When conditions allow us to bring the fight inside to check and halt fire extension in the cockloft, the pulling of ceilings (especially multiple ceilings) will be difficult, time consuming, and exhaustive. The tasks may be further compounded by high ceilings and/or originally installed tin ceilings. These are going to be extremely difficult ceilings to pull. Sheets of varying sizes nailed directly to ceiling joists or a plaster lath surface will not lend themselves well to ceiling removal. What is often done once a hole is started is to make a ripping or "can opener" type of opening in the tin, producing minimal access. Not only will this be exhausting, it will be frustrating.

Attempting to pull from a tin seam may allow larger or entire sheets to be removed without ripping the tin. However, this is not as easy as it may sound. Accumulated layers of paint will make it impossible to find (let alone grab at) a seam. A corner or fixture opening may provide a more efficient starting point. Other tactics suggest taking the blunt end of the tool and pushing the tin sheet upward, in an attempt to pop a seam. These are all great options, provided smoke at that level is minimal and you can see the ceiling.

Occupancies that contain meat markets, butcher shops, and related areas within a supermarket will have insulated ceilings over their frozen storage areas. In some occupancies this storage may consume large areas. These ceilings will contain plywood that is nailed or screwed to the ceiling joists followed by sheets of insulating material, all covered by a water and rust proof finish in the form of tile, plastic, or marlite™. When the fire gets above these spaces, nobody is pulling this ceiling. In order to get access into the cockloft to achieve extinguishment, unconventional means may be necessary by directing water through roof openings over these unusual areas. When this type of occupancy is present, Chief Officers must recognize that altered tactics may be necessary to extinguish the fire.

When occupancies other than the above are present, the task of accessing the cockloft area will still be demanding. Officers must recognize the problems and attempt to supply and replenish people to handle this task. Company officers and firefighters must be able to use the tools and techniques not only successfully, but also safely. Pulling ceilings from refuge areas and working your way into the room is a must. Members who walk into the middle of a room and attempt to pull the ceiling may find themselves buried below a mountain of material. At the same time, officers must understand what conditions they are dealing with prior to introducing air into the concealed space. Introducing air into a highly heated concealed space may backdraft the ceiling down. When these conditions present themselves, allow topside ventilation to relieve the build-up.

Water applied from below is, at best, a difficult task. Depending upon conditions and the current operational mode, this will come from either an intense fight from interior hoselines or large caliber master streams directed from the exterior. Earlier in the chapter we mentioned the versatility of a well-placed tower ladder company. When operations go defensive because of a well-advanced fire in the cockloft, few ceilings can withstand the power of a tower ladder stream as it is directed from the sidewalk. By directing a solid stream toward the ceiling space of the involved stores, the pressure of the stream will penetrate through the ceiling(s) allowing large volumes of water into the space. A word of caution with this operation; although an extremely effective tactic, members must be alert to the condition of the parapet wall when the fire attacks either the front display window area or is heavily involved in the cockloft. As we have stated, any movement of the unprotected steel within the building can drop the parapet wall outward.

Time: Taxpayers/Strip Malls and Stores

*COAL TWAS WEAL**T**HS*

The consideration of time at a taxpayer/store fire can take on a number of different concerns.

Time of the alarm

Time of the alarm can indicate to the responding firefighters the probable fire conditions and associated difficulties. If an alarm is transmitted in the early morning hours after the occupancy has been closed, arriving forces must make a careful size-up and assessment of conditions before determining their strategic and tactical approach. In order to understand the meaning behind this statement, we need to reflect over some of the previous size-up factors discussed.

Factor	Problem	Result
Construction	Roll-down steel doors and windows within a Class 2 or 3 constructed building	Delay in forcible entry, ventilation, and water application
Occupancy/Content	Commercial	Significant fire loading
Auxiliary Appliances	Minimal to none	Delayed discovery of fire, advanced fire conditions
Time	Late night or early morning; business has been closed for an extended period of time	Advanced conditions

The fortified design of these buildings coupled with the time of the alarm, are significant concerns for firefighters and fire officers who first arrive. Late night and early morning fires in these occupancies are known to produce backdrafts. This possibility must be anticipated. It is from the above associations, that many in the fire service know the taxpayer and strip mall as *backdraft factories.*

Time of the year

Time of the year may also affect a fire officer's decision-making when dealing with a reported fire. During the holiday season, merchants will often stock up to meet the demands of the shopping season. This means increased fire loads within the shopping space, as well as increased fire loading within the storage space. The amount of combustibles stored within a given area in a store during this time of year can be overwhelming and not obvious to the average shopper or store employee. During the holiday season it is critical for members of the fire department to get out and visit this type of establishment to correct any violations concerning blocked or obstructed exits and sprinkler heads, as well as general housekeeping practices within the store during this time of the year. It is also important to be realistic with your inspections. If you think once you corrected a violation it is going to stay corrected throughout the entire holiday season, you're mistaken. Frequent visits may help eliminate the same concern from reoccurring, as well correct any new violations as they may appear. In a number of towns and cities researched, shopping malls are visited on a daily basis throughout the holiday season to eliminate the concerns.

Height: Taxpayers/Strip Malls and Stores

*COAL TWAS WEALT**HS***

Taxpayer/strip malls and stores are limited to two stories in height, generally making them easily accessible for the fire department.

Taxpayers and strip malls

Regardless of their limitations vertically, ladder companies must still provide as many alternate means of egress to and from this area. With the possibility of the fire extending into the cockloft, members can find the fire between themselves and their initial means of egress if not planned for. As a standard, two ladders should be placed to the front of the building, and when practical, two additional ladders should access the roof from the rear.

It is also a good practice when placing those ladders, especially those from the front, that they be placed to indicate the boundaries of the fire building, or the locations of division walls in the row (as long as their intended placement does not interfere with sidewalk operations).

A ground ladder that has three to five rungs above the roofline is a nice visual tip on where to drop inspection holes to determine if the fire is extending into an adjoining store.

Special Considerations: Taxpayer/Strip Malls and Stores

*COAL TWAS WEALTH**S***

A fire in a taxpayer/strip mall will initially tax the resources of any department. Upon arrival, fire officers must obtain answers to key concerns from their own initial size-up as well as from reports of companies operating within, above, and around the incident. Observations and information will aid in the decision-making.

Taxpayers/Strip Malls and Stores

Considerations for the Incident Commander

1. *Time of the alarm.* Late night or early morning fires in this type of occupancy are known to produce backdrafts.
2. *Location of the fire.* Obtain information as soon as possible on where the fire is located, especially where it originated. A fire that has started in the basement of an old style taxpayer may have already weakened the basement ceiling/floor joists prior to you putting companies to work. The type of occupancy and its association with a concrete, terrazzo, or marble floor can further enhance this concern (i.e., laundromats, drug stores, dry cleaners, etc.).
3. *Occupancy.* Obtain information about the second story occupancy types when dealing with a fire in a two-story taxpayer. They may contain residential properties, which will cause additional life hazard concerns regardless of the time of the day.
4. *Area.* This information is a critical concern when dealing with these types of occupancies. Size and configuration pose concerns; obtain information as soon as possible about these:
   - The size and layout of the complex. Some taxpayers that border two or more street fronts may wrap around the block forming a large L- or U-shaped complex. Late information here may find you chasing the fire around the block.
   - If there are any parapet/dividing walls within the roof layout. Large ornamental signs or high front parapet walls may hide the presence of a true separation between the stores from your street view.

   - Accessibility to the rear. Assign at least one engine company and one ladder company to the rear as soon as possible.
5. *Incident management.* Divide the incident into manageable areas early in order to enhance the effectiveness of the operation as well to account for your people. Pay particular attention when assigning companies to the front and rear of the stores. Coordination of assignments is a must. For example, if the fire is discovered in the basement, the engine company assigned to the rear may have the shortest and safest hose stretch to the basement entrance, which is often located in the rear of the building. With this order given, companies in the front would then be directed to vent the first floor, as well as the boxed out areas below the front display windows to assist with the engine company's movement into the basement.
6. *Fire extension.* If fire extends into the cockloft of a taxpayer, roof cutting will be critical as well as resource extensive. Assign a Chief officer to coordinate and manage the operation. If fire breaks through the roof, give early consideration and evacuation to any higher buildings attached within the row.
7. *Unconventional.* Fires that involve cockloft areas of meat markets, butcher shops, supermarkets, etc. will be difficult to access. Occupancies similar to these may require unconventional means of getting water into the cockloft area. If this tactic is considered, it must be in coordination with interior efforts.
8. *Auxiliary appliances.* Are there any sprinkler systems within the complex? Are they operating? Can they be augmented by the fire department? If so, where is the location of the Fire Department Connection (FDC)?
9. *Firefighter safety.*
   - Have members exercise caution when operating below dropped or suspended ceilings.
   - The presence of any large equipment on the roof should be radioed to the incident commander.
   - The accumulation of large amounts of ice, snow, and water on the roof should be radioed to the incident commander.
   - If any truss construction is involved, this should be immediately identified.
   - Beware of conflicting radio reports from the roof and the interior of the building. This may indicate an advanced fire in the truss loft or cockloft. For example, "Light smoke, no heat inside" from the interior teams; "Heavy smoke and heat venting from roof openings" from the roof teams.
   - When available, put thermal imaging to work early. Obtain information about the troubled areas of these buildings early. These include basements, multiple/suspended ceiling areas, cocklofts, and truss lofts.

Many factors will be included and considered in a chief officer's size-up. The above is meant to aid in the decision-making, as well as to further the information gathering process when a fire involves a taxpayer/strip mall.

Questions For Discussion

1. List the occupancies that may contain steel plating on their roofs.

2. What obstacles and hazards exist when steel plating is present?

3. Describe the benefits of a ladder company's ability to create a significant thermal updraft for a fire in a taxpayer/strip mall.

4. List and describe the life hazard concerns to the firefighter at a taxpayer/strip mall incident.

5. Identify some uses for thermal imaging when a fire involves a taxpayer/strip mall.

6. Where do most fires start in the taxpayer/strip mall?

7. What type of occupancies may require unconventional means to extinguish a fire?

8. What factors contribute to the possibilities of backdrafts in the taxpayer/strip mall?

9. How does the time of the year add to the difficulties when an incident occurs?

10. Discuss the pros and cons of marking division walls with ground ladders.

5 Garden Apartments and Townhouses

Garden apartments and townhouses began to appear in the building industry a number of years ago from the demands of an ever expanding population and the lack of affordable housing at that time. Many people attempting to own a piece of the "American dream" during that time soon realized that owning their own private home became a challenge for monetary reasons or a lack of available housing. For those seeking a temporary place before they could afford their own home, the garden apartment and townhouse became the answer. As real-estate developers soon realized, the demand for this type of housing could barely keep pace with the sale or rent of the properties. The high demand for affordable housing often came at the expense of code enforcement. Early construction methods where often met with little resistance lending to some devastating fire incidents in these types of occupancies.

Garden Apartments

The term "garden apartment" applies to a grouping of apartments within a building that can range in height from two to four stories of varying types of construction. Often many of these buildings will be grouped together to form a complex. Within the complex, buildings will often be setback and surrounded by grass, landscaping, and courtyards. Each of which will cause accessibility and operational concerns.

Another significant consideration for responding firefighters will be the interior layouts and entrances that serve individual buildings. Individual buildings within the complex can have several entrances not interconnected within the building, while others may contain only one entrance that serves all apartments in the building. Obtaining this information can be difficult and time consuming if not known in advance.

What is referred to as modern day multiple dwelling is quickly becoming the fire services newest nightmare.

Townhouses

The townhouse concept quickly gained in popularity due to its relatively high degree of privacy when compared to the garden apartment, and the affordable prices they presented when compared to a detached private home. The townhouse is a one- to three-story dwelling (again of varying types of construction) that is attached to several other dwelling units within a complex. Depending upon where and when they were built, these units may or may not be separated by firewalls.

Townhouse dwelling units are normally grouped into complexes of various size and design that could be as small as a six- to twelve-unit complex, or they could be a gated community housing hundreds of families.

What is referred to as the "modern day row house" is becoming the fire service's latest challenge in many areas of the country.

Construction: Garden Apartments and Townhouses

COAL TWAS WEALTHS

Garden Apartments

During research of the construction uses and types in garden apartment complexes, I found many cases involving different classes of construction with hybrid construction components mixed throughout. In an attempt to classify or give direction to what should be some of the fire service concerns with the building construction of these types of occupancies, I have categorized three different types of garden apartments.

Garden apartment type 1

In this type of construction, buildings will resemble Class 2 non-combustible construction. Here you will find exterior walls constructed of brick and masonry. Some units contain balconies that are accessible from sliding glass doors. The building's floors, ceilings, and roofs will also be constructed of concrete. Within the interior walls you will find metal studs covered by fire-rated gypsum board.

The Type 1 garden apartment's height will range from three to four stories, which can vary depending upon the terrain. Floor areas can contain two to four apartments per floor all served by an open interior stair. Although considered to be the most desirable construction type of those listed, it is the least common due to the cost associated with the building materials and their installation.

Fig. 5-1 *Garden apartment type 1, note below grade terrace apartments*

Garden apartment type 2

We categorize this type of construction as Class 3 ordinary construction with hybrid materials found in the areas of the building's roof and floor systems. These buildings can also vary in height from two to four stories. Floor areas can contain anywhere from two to four apartments per floor, accessible from a common interior stair. Exterior walls of the building are made of brick and masonry, where concrete balconies will be accessible through sliding glass doors from each apartment. Floors and ceilings will be constructed of concrete, with a peak or flat mansard roof covering the building. The roofs are designed from factory-made lightweight wood trusses fastened together by metal gusset plates. In this area the attic space will often

Fig. 5-2 *Garden apartment type 2*

serve as a utility area, housing electric, gas, and heating ducts for the apartments below. Vent pipes for the plumbing (as well for the kitchen and bathroom exhaust ducts) can also penetrate this space.

Garden apartment type 3

We categorize this type of construction as a Class 5 wood frame structure with hybrid materials possibly found in the roof or floor supporting system. These buildings can vary in height from two to three stories with possible variations if built on a grade. The interior design is a platform wood frame with a peaked or mansard roof. Depending upon when the building was built, the roof supporting system may be of factory lightweight wood trusses or a ridge/joist design. The building's exterior can be a brick veneer, stucco veneer, or a combination of wood in the spandrel spaces with some type of masonry veneer surrounding it. The combinations will vary based on design and the aesthetics desired.

Fig. 5-3 *Garden apartment type 3*

Common construction concerns

What is common to all three garden apartment construction types is that all kitchens and baths will be stacked vertically in a line from floor to floor as well as placed back-to-back in each apartment. This was done for obvious economic reasons. Placing all utilities in a line for all the buildings' bathrooms and kitchens reduces costs associated with the installation of all the electrical and plumbing services, a significant savings when building a multi-family unit.

Fig. 5-3a *Light-weight wooden trusses can be found within the floor and roof assemblies.*

Firewalls may or may not exist in any of the three types of garden apartment construction. In most cities the installation of a firewall or fire separation will be required in a building that exceeds a designated length. A firewall is a freestanding wall constructed from the ground level through the roof to eliminate any fire spread through, around, or over it.

A fire separation wall, or as it is also called a fire partition, is a wall that is constructed from the basement floor to the underside of the building's roof deck. Both of these types of walls can be built using gypsum or masonry material and are supposed to be constructed to meet their intended design. However, past history has shown that defects in design, construction, or alteration of the firewall and fire separation wall have led to early and rapid spread fire into adjoining buildings. Never trust them!

Townhouses

Just as research showed different types of construction associated with the garden apartment, research for the townhouse showed the same.

Townhouse type 1

What is referred to as the older version of the townhouse is a building designed of Class 3 ordinary construction. The exterior walls of the building are brick over cinderblock, or a masonry veneer over cinderblock. Floor and ceiling joists throughout the building will be constructed of wood. The interior walls and ceilings are constructed of frame studding with gypsum wall and ceiling coverings. Sub-flooring within the building is normally plywood with a carpet or parquet floor covering. Roofs will either be of a peaked gable, mansard, or gambrel design with the possibility of more than one design represented within the row. Buildings of this type can usually range in height from one to two stories with the possibility of a garage under the unit.

Fig. 5-4 *Townhouse type 1*

Townhouse type 2

This type of townhouse is of Class 5 wood frame design and consists of frame construction on top of either a concrete slab or block foundation wall. Exterior sheathing consists of stained T 1-11 plywood, wood clapboard, aluminum/vinyl siding, or a stucco veneer over plywood or particleboard. Interior walls and ceilings are gypsum over frame studding with either a carpeted or tile floor over plywood subfloor. These units can range in height from one to three stories with the possibility of a garage under each unit. Roof designs will consist of asphalt singles over a peaked roof or mansard design.

Fig. 5-5 *Townhouse type 2*

Townhouse type 3

This is a true hybrid grouping of the classes of construction and a confusing one to visualize if you don't have one in your town or city. In this particular category of townhouses we are finding multiple types of construction all with different designs in one complex. As an example, in one row of buildings you may find what starts to resemble frame construction building methods. Here you'll find frame exterior walls, brick veneer exteriors, gypsum wall and ceiling coverings, and an asphalt-shingle peaked roof. However, as you undress the building further you will find steel girders for support in the center of the building, with TJI floor joists (Truss Joist I-Beams), metal C-joists, or wooden I-beams spaced on 24" centers. These larger townhouses can now accommodate any wall layout desired with this type of supporting system. In addition, due to the larger areas associated with the complex, the peaked roof supporting system will also use truss joist I-beams as roof rafters tied into what is referred to as a powerlam or microlam.

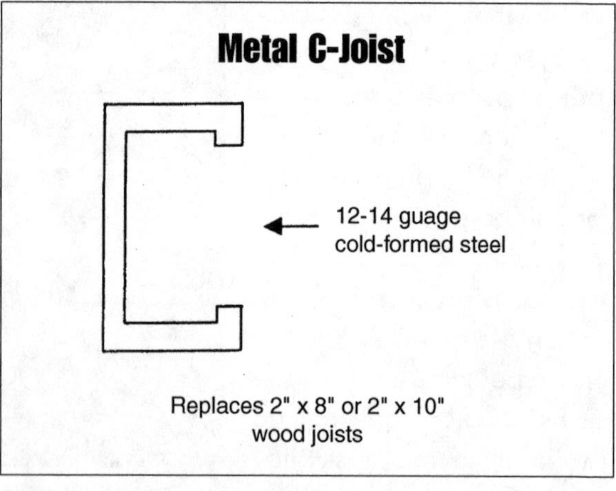

Fig. 5-5a *C-joist can replace a wooden floor joist.*

Directly across the street in the same complex you may find buildings built of Class 2 non-combustible construction. The buildings shell is a network of steel columns and girders with open web steel bar joist trusses used for floor supports all covered by corrugated steel decking and a nominal poured concrete floor of 3½". The walls in this building will be constructed of 2"x 4" and 2"x 6" aluminum studs covered with gypsum board. In this type of construction the gypsum wallboard will be applied to the inside and outside of all exterior walls. The exterior walls will resemble a masonry building from the many different masonry veneer methods used. Roof supporting systems used in this type of building use steel bar joists covered by steel roof sheets all at extreme pitches.

Also, in the larger type of townhouse complexes there is the possibility of a lower-level parking garage with individual areas for tenant/owner storage. All of this will be located at grade or slightly below. These are much larger than any other townhouse complexes described with heights ranging from five to seven stories.

Fig. 5-6 *Townhouse type 3*

Fig. 5-7 *Owner parking and storage located under the dwelling units*

No doubt that there is a lot to absorb here. What should be evident is the need to get out there and take a look. Listed are only a few categories of what is a growing and expanding concern. The bottom line as it relates to the construction concerns of garden apartments and townhouses is that there is much more than one format. Inspections and familiarizations within your own city will show this to be true.

Occupancy: Garden Apartments and Townhouses

COAL TWAS WEALTHS

Determining the number of occupants and their status is a primary concern at any incident we respond to. The number of occupants that firefighters may be confronted with at a garden apartment or a townhouse fire can be surprising and often overwhelming.

Garden Apartments

In one single building within a garden apartment complex the number of individuals can reach into the triple digits. The number of occupants within a given building will fluctuate depending upon the design and layout of the building. What is important to remember is that the same fire will affect all of these people within this one building. With a single stair serving as many as eight to sixteen apartments where, in many cases, there may be no secondary means of egress from the building, arriving firefighters can have an overwhelming number of occupants to be concerned with.

Contents

Building content concerns for the garden apartment are generally limited to the ordinary combustibles and furnishings of a residential occupancy. Barring any unique encounters, incidents normally do not present a difficult concern in this area until they involve tenant storage areas.

Townhouses

Townhouses can also be referred to as a condominium complex. For our efforts, the fact that they are referred to as a townhouse or condominium should be viewed by the fire service as nothing more than a method of ownership.

The number of people immediately affected by a fire in a townhouse complex will be much smaller when compared to the garden apartment complex. Most townhouses, namely those referred to as Type 1 & 2 within our construction classifications, are viewed as single dwelling units within a row of many. What allows us to consider the occupancy concerns to be somewhat less when compared to the garden apartment is the design and construction methods used within the building. Not having the numbers of people stacked on top of one another (as with the garden apartment design) eliminates the larger numbers of people firefighters will have to be immediately concerned with. What will dramatically change this thought is when we respond to a reported fire in a townhouse of Type 3 design. This type of structure will allow a much larger occupant concentration within a single building. This particular type of structure presents itself with many of the same difficulties as a large apartment building.

Contents

The content concerns for a fire involving a townhouse will come with a heightened concern due to the individual ownership of each property. With single unit townhouses, property owners may or may not be responsible for their own lawn maintenance, painting, plumbing, and any other building repairs. For the weekend carpenters, painters, and plumbers, the attached or under unit garage space becomes the storage area for all their household projects. As you would expect, these areas can be storage for numerous chemicals, solvents, paints, and thinners that will go unchecked until a problem arises. Whether in a private dwelling or a townhouse unit, members must always exercise caution when entering a garage storage area. The content concerns must be approached as if you're going to work in a commercial occupancy.

It is also important to note (as it relates to occupancy content concerns) that some of the more affluent, upscale townhouse complexes may contain their own grocery store, bank, hair salon, dry cleaners, doctor, dentist, as well as their own pool and spa. These different establishments, which will most likely be found on the lower-level of a series of buildings, will come with their own unique content concerns to the incident commander. Pool chemicals, cleaning solvents, and stored gases are just a few of the concerns that will add to the difficulties if they become involved in the incident.

Apparatus and Staffing: Garden Apartments and Townhouses

COAL TWAS WEALTHS

Engine Company Operations

Engine company operations in the garden apartment and townhouse complex will focus around the company's ability to establish a water supply, select the proper size hoseline, stretch the hoseline, and place it into operation quickly. Initially, these will seem simple to accomplish, and in many cases they may be. However, difficulties can arise when companies are not prepared to deal with the unusual, unsuspecting, and the unique areas the garden apartment or townhouse can present.

Apparatus placement

Some specific areas that can be considered challenging for the engine company at a garden apartment and townhouse complex may include long, narrow, dead-end, or cul-de-sac streets. This may be further compounded by buildings that are setback from the street or parking area, and buildings that can be surrounded by decorative landscaping, pools, ponds, fences, and sloped topography. These concerns are not new to the fire service; one or more have presented themselves in the chapters previously discussed. However, when they are all present in one type of building as they can be with the garden apartment and townhouse, company officers must be better prepared.

The key to the effectiveness of the engine company will come from the preplanning of streets, building layouts, as well as the accessibility and terrain familiarity associated with the complex. Garden apartment and townhouse complexes can come with streets that may make it difficult for engine companies to establish attack and supply positions. Fire departments accustomed to one way of establishing and obtaining water may find themselves hard pressed getting quick and sufficient water on the fire when confronted with an incident in one of these complexes.

Initially thoughts for arriving units must focus around investigation, and then direction when nothing is showing. Committing most or all of the alarm assignment to the reported street address may not prove to be the best course of action, especially knowing some of the mobility problems associated with these types of complexes. When nothing is showing, or when conditions do not lend themselves to quick identification of where that first hoseline is going to be placed, everyone other than the first due engine company should remain out of the street. Buildings that are fronted on two sides, whether by a street or parking lot, may allow a second due engine company to move to the opposite side of the complex to determine if that access will allow for a quicker and shorter hose stretch. With this approach, other arriving engine companies must stand by outside the congested area to await the location, and then direction on augmenting the specific engine company's operation. Allowing all your resources to go into the maze early, cannot only prove to be time consuming, but thoughtless.

Hose selection, stretch, and placement

When it appears that your engine company has the closest and best advantage at getting water to the fire, your concerns have only just begun. Proper hoseline selection, stretch, and placement will be critical for the effective operation of that first hoseline. Being the closest company may still require a 300′ hose stretch to the front door, let alone the additional lengths required to reach to the apartment. If your hose bed is not set up for long stretches, members must be educated and trained to put into motion the steps that will ensure that once the hoseline gets to where it is supposed to, it has enough volume to do the job. A 500′ stretch of 1¾″ hose from the apparatus will not allow that size hoseline to deliver its designed flow. Policies and procedures must be put into place to deal with this type of stretch before the incident occurs. Getting a hoseline in position for a quick knock down is going to be critical due to the early collapse potential some of these buildings can present.

In cities that normally respond to these types of buildings, engine company hose beds are pre-set to handle the long hose stretches. Pre-connected 3″ hose with a gated wye or manifold stretched to a front door or courtyard paves the way for a number of smaller and more mobile hoselines to be stretched into the building.

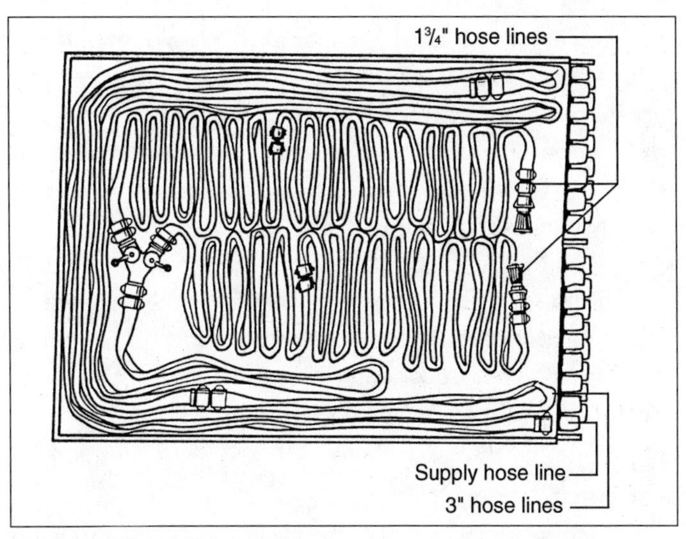

***Fig. 5-8** Top side view of a pre-set hose bed for long stretches*

Additional decisions on putting effective hoselines to work in these occupancies must be made early. When it is evident that the building involved will not allow the first due engine company to get close enough for a simple and short hose stretch, help from the second or third arriving company should become an automatic thought. Officers must realize the importance of getting that first hoseline to work. If we put the fire out, all of our other problems will go away.

***Fig. 5-9** Anticipate the difficulties*

Ladder Company Operations

Apparatus placement

Efficient ladder company operations for the garden apartment and townhouse will again depend upon the officer's ability to best use available resources and equipment. Garden apartment and townhouse complexes with narrow streets that are setback from accessible areas can trap poorly placed ladder companies. Ladder company officers must be concerned about street accessibility into these types of complexes well before the receipt of the alarm. Knowing where they can get in, and where they can't, is something that must be reviewed in the company pre-plan. Some fire departments responding to building units that are accessible from more than one side have the first due ladder company's apparatus remain out of the general area until the location of the fire and its best access route is confirmed. Buildings that are protected by fire departments with two ladder companies responding on the initial alarm may assign one ladder company to each side in an attempt to move people and equipment in that general direction in order to quickly handle the priority tasks.

When the building is set back from the street by decorative landscaping, sloped terrain, or by any obstructions that will prevent the apparatus from pulling close to the building, gaining access to the upper floors and roof will most likely come from ground ladders. Providing a secondary means of egress for the occupants as well as for the firefighters in this type of building is going to be a must.

Laddering

Placing ground ladders to specific areas at a townhouse or garden apartment complex may come with limitations. With sloped terrains and decorative landscaping it may require the placement of a 40'or 50' ground ladder to allow firefighters to access the roof of a three-story building. Placing a ground ladder up only to find that you can't reach the target height causes a serious delay in achieving the particular objective. The decision on prioritizing one type of ladder operation over another will be based on the officer's size-up that can be aided by viewing the obstacles well before the receipt of the alarm. Walking a ladder in from an extended distance because of grassy or sloped terrain is going to be frustrating when it can't reach what you hoped it would. Knowing what you can and can't do around these buildings before you need to do it, will again greatly affect the outcome of the incident.

Ventilation

Vertical ventilation considerations for the ladder company in both the garden apartment and the townhouse complexes will vary based on design and construction of the building. Knowing the design type and the construction methods used will greatly aid in any efforts to ventilate the building when fire is located on the top floor or extending into the roof space. What comes with particular concern to the firefighter in this area of the building is the construction method used in supporting the roof deck.

In the newer garden apartment and townhouse complexes you will find the ever-popular use of the truss design. This design will most notably be of the lightweight wood design with gang nail/gusset plate fasteners or the open web steel bar joist design. The effects of the truss under fire conditions have been well documented over the years. Committing members to the roof to remove ventilation obstacles or to cut the roof comes with great risk. Any attempt to work on these types of constructed roofs should only come from the afforded protection of a bucket or aerial ladder. With the truss design, firefighters must expect that there will be no ridge rafter running the length of the roof to give the firefighter the support needed if attempting to straddle the peak or place a roof ladder to distribute weight. A failure of a truss will drop the entire deck in the affected area from load-bearing wall to load-bearing wall. If any roof operations are going to be attempted at all, they must be done from the protection of an aerial device.

Depending upon the fire's location, and the size and span of the truss loft, providing a roof vent to draw the convection heat in this type roof design may be accomplished from the gable end(s) of the building. These areas will already have a manufactured gable vent in place. This louvered or screened vent will be part of the design of the roof space as it allows accumulated heat from the sun and any built up moisture to escape from the underside of the roof deck.

Another type of natural ventilation opening found on some garden apartment complexes that should be considered early when fire enters or threatens the roof space is the roof turbine. This wind/heat driven device is installed by building contractors to provide a means for removing the heat from an attic space. The curved blades of the turbine rotate freely from either wind movement or from the built up convection heat in the roof area. The movement of the turbine will create a negative pressure on the underside of the roof deck as it draws air in the roof space toward its opening and out through the blades as they spin.

Fig. 5-10 *Roof turbine*

What causes concern to the fire service with the roof turbine is a number of these devices that vent a large roof area. Fire or heat that enters the roof space at one end of the building can be pulled throughout the roof space by the movement of a turbine at the other end. When possible members should remove all roof turbines in the uninvolved areas to eliminate the negative pressure pull associated with their operation.

The roof turbine is an excellent device to remove convection heat when it's directly over the fire area. When they are outside the immediate fire area, remove them.

Life Hazard: Garden Apartments and Townhouses

COAL TWAS WEALTHS

Firefighters

The life hazard concerns for the firefighter are similar in many respects to the residential types of occupancies previously discussed. Whether we are discussing the private or multiple dwelling, many of the thoughts and concerns identified will also be present in the garden apartment and townhouse. There is at least one additional critical factor in your decision-making that must be identified in the garden apartment and townhouse complex. Anytime fire extends beyond the contents and begins to attack the actual structural members the risk to firefighters increases dramatically. Depending upon the type of dwelling unit and the materials used in their construction, many of the garden apartments and townhouses referenced will contain truss floor and roof assemblies. When fire involves one of these areas failure will occur within minutes. Firefighters and fire officers have to use a calculated approach when fire involves one of these areas. Using thermal imaging to identify the fire's location and extent becomes a critical first step. To put it simply, you need to know where it is and where it is going. Once identified, do not allow firefighters to operate on top of or below truss supported areas. Use the effective reach of a hose stream to dampen down the fire. Operate from a refuge area or supported areas around or within the building. When practical, breach walls in order to gain access and egress points around involved areas.

There is no doubt that many of the points previously discussed have to be calculated into your decision-making regarding the life hazard concerns to the firefighter. However, when truss supported areas are involved this factor moves to the top of the list.

Occupants

Occupant life hazard concerns in any type of residential occupancy will always be a great concern to responding firefighters. When viewing occupant concerns in both the garden apartment and the townhouse, firefighters must remember that there will be a difference in the initial difficulties of occupant rescues. The main difference between the rescue difficulties in garden apartment and townhouses is that the garden apartment complex is essentially a multiple dwelling capable of housing 45 to 60 people, whereas the townhouse, is considered a single-family unit averaging two to six people. The number of individuals within the building, in association with the fire's location and extent, are going to be the influencing factors.

The saving of life will always be the fire officer's first priority in the decision-making process. Members directed to remove any endangered occupants, especially those within the garden apartment complex, will be hard pressed to accomplish all these tasks in a safe manner.

Occupant removal and direction

When the building's occupants are unable to exit their involved or threatened apartments via the interior stair expect the apartment balconies and windows to become refuge areas as they await rescue. With numerous people showing at different levels and different locations, removal efforts must be prioritized based on the areas of greatest danger. Firefighters should make every attempt to complete their tasks based on this concept. However, problems will arise when those not initially helped believe that they should be, and then compound your problems by threatening to jump. Often those yelling or reacting the most may not be the ones in the most danger.

Occupants jumping to ladders not placed directly by them or onto ladders just placed below them have injured and killed a number of firefighters over the years. A person falling from a height onto a firefighter below must be a serious consideration when fire has forced the building's occupants to windows and balconies. When efforts are compounded by numerous occupants at different locations all calling for help at the same time, stern direction must be given to those being rescued as well as to those awaiting rescue in the hopes that your efforts can be carried out safely.

Another serious consideration in your occupant rescue attempts for those who have fled to a balcony is the path of least resistance they have just created. Sliding glass doors that lead onto a balcony will create an opening of significant size that will draw the fire. If we consider the layout and square footage associated with average garden apartments we will find that they are not only small, but also predictable. Often what you find is a small foyer or entranceway that leads into a small work-in kitchen. Opposite the kitchen will be a combination dining and living room with a bathroom and bedroom on either side. The balcony will generally access the dining/living room area. If the entranceway is blocked by a fire that has possibly begun in the kitchen area, the sliding glass door that leads to the balcony will most likely be the area sought for escape. Occupants awaiting rescue should be directed to close the sliding glass door behind them. These doors are constructed of very durable, tempered, double-insulated glass. Closing the doors will give the occupants some refuge from the heat and smoke that is being drawn to the opening. With a door left open, not only will the natural movement of the fire be drawn to its opening, but the conditions at the sliding glass door could be further accelerated as the engine company makes their movement into the apartment. Openings created opposite the advancement of the hoseline can quickly create blowtorch conditions.

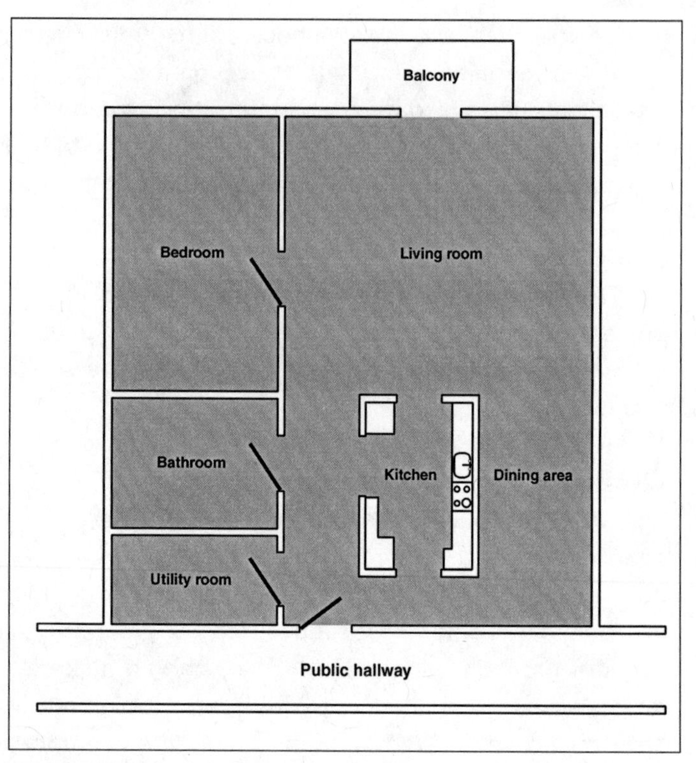

Fig. 5-11 *Typical garden apartment layout*

Terrain: Garden Apartments and Townhouses

COAL TWAS WEALTHS

Of all the different occupancies represented in this book, none will capture all of the terrain size-up concerns as well as the garden apartment and townhouse complex. From building setbacks and sloped topography, to accessibility concerns in multiple forms, the garden apartment and townhouse complex can present them all.

Setbacks

Building setbacks in the garden apartment and townhouse complexes can be very significant. With depths of 200′ to 300′ from streets or parking lots not uncommon, officers should entertain the possibility of identifying drivable versus non-drivable lawns and courtyards. Barring unusual weather conditions that make a lawn totally unusable, identify areas where you may be able to place an engine company or a ladder company. When hoseline stretches and their placement are going to be significantly delayed, or a number of ground ladders will have to be walked long distances from the street, consider bringing the resources closer to the troubled area by driving on the lawn. Building managers, even property owners, may be upset when you attempt this. They'll get over it.

Garden Apartments

Grade

Grade/sloped topography concerns are a common problem when dealing with the garden apartment complex. The term "garden apartment" was derived from the architect's design of the grassy decorative areas associated with the complex. These areas will often be sloped or raised around the building units to give a more pleasing environment to the tenant. With the sloped environment there often will be decorative shrubs, trees, and flowerbeds that can make fire department operations difficult. With these concerns, firefighters who are forced to walk ground ladders in for rescue or suppression efforts may have added difficulty finding any level working surface as they attempt to raise their ladder. On more the one occasion, members with shovels have dug out a section of sloped earth to allow one side of a ladder heel to rest level with the other side.

Another significant problem with the changes in grade that surround a garden apartment will be the fireground management of the building in what appears to be either a three- or four-story building. As previously stated, some departments will use varying methods to determine the designated areas. The important thing to remember is that a uniformed reference must be used.

In garden apartment complexes the grade concern can present further challenges when determining floor levels from one particular side. In some garden apartment complexes there may be occupied spaces that from the street side or address side of the building may give the appearance of being mostly below grade. These apartments (often referred to as terrace apartments) will occupy this below grade area. The word "terrace" is given to these particular apartments due to their accessibility and use of a rear yard from their sliding glass doors. This same below grade area may also contain utility or laundry rooms accessible from the same public hallway used by the terrace apartments. In the scope of our fireground management these areas must be identified. Referring to this area as either a terrace apartment on the building's first floor, or a basement apartment on the basement floor is what research from different departments has shown to be typical identifications. If your department has similar concerns, set the policy and the game plan well before you need to go to work here. It should be obvious that gaining the information about these areas, their accessibility, and what they will be referred to well before a fire is critical to an efficient, effective, and safe outcome.

Townhouses

Accessibility

When viewing general accessibility concerns for the townhouse complex, we find that these can also be numerous in scope depending upon their design and location. From narrow alleyways between building units to fenced-in yards commonly found in the rear, accessibility and a firefighter's ability to work here safely must be reviewed. One elaborate townhouse complex in Jersey City consisting of canals and boat docks situated to capture the views of the Statue of Liberty and Ellis Island, greatly hinders the firefighters' ability to operate on the water's side. Winding walkways along the water's edge in close proximity to the buildings make the stretching of a hoseline and the placement of a ground ladder a great concern to members attempting to go to work here. Even taking advantage of a marine company on this side comes with significant planning.

As unique as this situation may be, the townhouse complex will come with a number of different obstacles. If you don't take a look, you won't know what they are.

Fig. 5-12 *A townhouse nightmare*

Water Supply: Garden Apartments and Townhouses

COAL TWAS WEALTHS

Availability

Water supply availability for these structures will hopefully not be a concern. With most of the garden apartment and townhouse complexes being relatively new, adequate supplies should be available from a well-spaced hydrant system. What does become a concern for the fire service as it relates to water availability is hydrant location and hydrant accessibility. Location is a relatively easy task to master. Hydrant location maps should be easily available from local water departments or complex managers. Many departments will have the nearest hydrant location information for a specific address printed on a computer dispatch printout or, at the very least, a map in the apparatus to view as firefighters leave their quarters.

Accessibility

Garden apartment and townhouse complexes are notorious places for overcrowded parking lots and streets surrounding the complex. This concern will quickly impact fire hydrant accessibility. Conditions in many complexes have become so difficult that local fire departments have requested owners to identify hydrant locations by installing an indicating light on a lamppost or telephone pole near a hydrant. Hopefully the fire department will see it as soon as they arrive. An earlier practice used by some departments of painting a large H in the street indicating a fire hydrant location was quickly found to be useless after a short period of wear or if the road surface became snow covered.

Fig. 5-13 *A typical accessibility problem*

Requirements

The water flow requirements for fires in a garden apartment and townhouse complex will depend on a number of factors that will affect the decision-making of the company and chief officer. For the average room-and-contents fire most departments will lead off with their department's designated attack hoseline. Most often the choice hoseline will be either a 1¾" or 2" handline. Flows from 180gpm to 200gpm, accompanied with the mobility of these lines to get into and around an apartment or dwelling unit, are generally more than adequate to handle the average room-and-contents fire. However, it is important to note there may be circumstances in the garden apartment complex in which a well involved storage room(s) may call for a larger hoseline. Storage rooms, especially those that are below grade with limited or no ventilation options, can be very hot and difficult fires. When presented with these conditions, officers may consider using the extended reach and penetration from a larger hoseline.

Delivery

Water delivery concerns can be very different for the garden apartment and the townhouse. With most townhouse complexes being more accessible than the garden apartment complex, fire officers must be prepared for specific complexes that require long hose stretches. Orders given to firefighters stretching and operating a hoseline into a curbside occupancy are going to be very different from those given for one set back from the street. Bringing a larger hoseline to the front door of the building that can be branched into smaller hose lines is a concept that we have discussed previously and will discuss in later chapters with any building that is set back from a street or accessible area.

Auxiliary Appliances and Aides: Garden Apartment and Townhouses

*COAL TW**A**S WEALTHS*

Detection equipment

Auxiliary appliances may vary depending upon local jurisdiction and code. Detection equipment for the garden apartment can take on the presence of smoke detection of either the hard wire or battery design, with hardwire mostly being in the public areas. This system may be tied into a central station alarm company, or the alarm may only sound locally within the building.

Detection equipment for the townhouse will vary again based on local code. Individual units may only have battery operated detection equipment supplied by the property owner, wherein another complex in a different jurisdiction may have hardwire detection to a local monitor received at a security gatehouse.

Suppression equipment

Suppression equipment for both the garden apartment and townhouse will be even scarcer. In the garden apartment complex a sprinkler system may be found in storage or trash areas. This system will most likely be fed by the building's domestic water supply. In multiple floor townhouse complexes, standpipe systems are usually found in the public stairways, with sprinkler systems in the public hallways. These large townhouse/condominium complexes are of elaborate and often unique design. They may have a number of suppression and detection features that must be found.

Aides

People or aides who can assist the fire department with their efforts will be even more limited. In the larger and more affluent townhouse communities, people on site may include complex security, a building manager, building maintenance, and possibly a building engineer. When present, they could be of great assistance not only with a malfunctioning alarm system, but also with such information as the location of handicapped or non-ambulatory occupants, or control of any possible building systems.

Street Conditions: Garden Apartments and Townhouses

*COAL TWA**S** WEALTHS*

We have already mentioned in related size-up factors the numerous concerns with street conditions in and around garden apartment and townhouse complexes. In addition to those previously mentioned, research has also shown the inconsistency with the unit/building address numbers. Many firefighters, primarily in the urban areas, are accustomed to streets that are straight in direction, meeting and crossing other streets at perpendicular intersections. In addition, many of the types of streets mentioned are numbered with the odd numbered addresses on one side of the street, and even numbered addresses on the opposite side of the street. In many garden apartment and townhouse complexes this is not the case. Depending upon the complex size and its design, street layout and address referencing can be unique and confusing.

We know from some of our previous discussions there is no definite and easy geometric pattern to the street layouts in many of these complexes. Many will be winding and often narrow in an attempt to give some aesthetics or to capture views in the neighborhood. To add to this appeal, streets within the complex are often named after such things as trees, flowers, or U.S. presidents. Depending upon the building's arrangement with the street layout, one building may actually have two different street names. It might be Maple Street on one side and Oak Street on the other.

What can add even further confusion is the method used in the lettering and numbering of the buildings within the complex. In many of the complexes reviewed, there was no consistent pattern used between the different complexes. In one particular townhouse complex buildings are referenced to by letter. Individual units within this building encompass all four sides of a very large building. Each individual unit/apartment is assigned a number in a consecutive order with those numbered units increasing as they circle around the building. The building directly across the street will have a different letter with its unit numbers increasing in order from where they stopped at the neighboring building. This process will continue until the street ends and a new series of letters and numbers start on another street. Confusing? Even with extensive pre-planning, I still have to look at the map on the way in.

Weather: Garden Apartments and Townhouses

COAL TWAS __W__EALTHS

When we review the weather size-up concerns, the categories of wind, temperature, humidity, and precipitation will remain as universal concerns to responding and operating forces.

Precipitation

The weather concern that stands out most noticeably for the garden apartment and townhouse complex will be the precipitation category. In previous size-up factors, we listed the concerns with apparatus maneuverability and placement on dry streets around the different type complexes. If we were now to encounter a significant snowfall on the ground, or large accumulation of water after a storm, already known difficulties will become compounded.

Snowplowed streets will dramatically shrink in size greatly affecting what is an already difficult situation. Impassable streets from floods may require apparatus to seek alternate routes. With all of this in mind, hose, ladders, and tools may need to be moved in from longer distances by hand. Add the delay of members moving the equipment by foot in these types of weather conditions and the delays become even longer. First arriving officers should never hesitate in calling for additional help early even when fire conditions don't initially appear to be serious.

Fire, smoke, and heat conditions will not be delayed by the weather. Call for help early anytime a fire and extreme weather conditions present themselves at these types of occupancies.

Exposures: Garden Apartments and Townhouses

COAL TWAS W__E__ALTHS

The exposure concerns for the garden apartment and townhouse complex will vary based on their design and the methods used in the construction process.

Garden Apartments

When we view the garden apartment complex, it generally presents itself as one, large, multiple dwelling with fire extension primarily focused within the building itself.

What adds to the exposure concern of the garden apartment is the layout design used within the complex. In some complexes you may find a number of buildings separated from other buildings within the complex. The building separations and distances associated with the separation will depend upon the amount of land within the complex, as well the aesthetics sought by the developer. Other complexes with less land to build on, and minimal appearances sought, will be attached to other building units often giving the appearance of one long, single building. With this design, the exposure concerns will initially come from the construction methods used where the buildings are attached.

Firewalls

Firewalls, which separate the buildings in most cases, will be evident by penetrations extending above the roofline, as well as any penetration through the building's front and rear walls. The presence of this wall, barring any shoddy workmanship, will limit fire spread to the attached exposure.

Fire separations

The cousin of the firewall is the fire separation. In this type of design, it is rare during the construction process that any installed masonry blocks used within the wall will be cut to fit the pitch of the roof within a peaked roof design. Often there will remain a gap between the fire separation wall and the underside of the roof deck. Even with the methods used to cover the opening, fire officers must anticipate that once any fire penetrates into the roof space of the fire building, it has the opportunity to extend to the roof space of the exposure.

The concerns of a fire separation wall can be further realized when it is used as a party wall between two buildings. In this design, floor joists of both buildings will not only share the same wall, but they will also share the same joist pocket. Any fire extending to a void space may extend to a joist pocket and find its way through the fire separation wall into the adjoining building.

Townhouse Complexes

The townhouse complex has a number of the same concerns and similarities as the garden apartment complex. Initially, we view each townhouse unit as an individual dwelling unit with the consideration of the fire's location and extent within the unit similar to that of the private dwelling.

Roof Space

Exposure concerns for the townhouse complex will also come from the construction methods used, and their layout within the complex. Again we focus our thoughts on the building's roof space. Depending upon local code and jurisdiction, as well as the date they were built, a common roof area could cover a number of dwelling units. This will generally be limited by the number of units within the row, the design of the complex, and the type of construction methods used at that time. It's not uncommon to find a common roof space over two, three, or four dwelling units. Any roof area that is common to more than one dwelling unit immediately creates an exposure concern when fire extends to the roof space.

Additional attached buildings within the complex will contain fire separation walls or firewalls. Firewalls will not only be evident from the actual appearance of the wall as it penetrates the building's roof and wall line, but also from the offset design of the units within the row. In an attempt to give a row of building units some aesthetic design, building designers would offset, or stagger a number of units of one section from another attached section. Placing a grouping of two to three units slightly offset from the next grouping gave design to the complex as well as illusion to the appearance of an unsightly firewall.

Fig. 5-14 *Offset fire wall*

Area: Garden Apartments and Townhouses

*COAL TWAS WE**A**LTHS*

Area size-up concerns for the garden apartment and townhouse complex will focus around their square footage, their layout, and how they will affect fire department operations.

Garden Apartments

As we have mentioned, the garden apartment complex will generally follow a floor layout design of two to four apartments per floor, all accessible from a single interior stair. This stair will often be the only means of egress for those apartments. This design in most cases will be the same for all floors served off the interior stair throughout the height of the building. It's important to note that this design concept of two to four apartments per floor, all being served from a single interior stair, can be repeated two, three, and four times within one large building. One stair can serve as many as 16 apartments varying in size from a studio, one-bedroom, or two-bedroom unit.

Townhouses

Area square footage concerns for the townhouse can be very different when compared to the garden apartment complex. First, the townhouse must be viewed as a single dwelling unit consisting of multiple floors. Floor and room layouts will generally be the same as the private dwelling with bedrooms on the upper levels, the kitchen, dining, and living room on the intermediate level and a recreation, utility, or garage area on the lower-level of the dwelling unit. This may vary based on the height and design of the unit.

Larger townhouse/condominium type complexes can have duplex and triplex dwelling units grouped together within a large single building. The unit layouts with these larger buildings cannot only be unique, but dangerous. In one such building we found duplex units that were accessible from street level, and triplex units directly above them that were only accessible from a remote interior stair and hallway that traversed the building. In other units within a nearby building, upper-level duplex units only allowed access into the particular apartment from its second floor. Therefore, if there is a fire on the first floor of this two-story duplex, your only access into the unit would bring you directly onto the floor above the actual fire.

Not only will these large townhouse complexes be very confusing for firefighters attempting to find and extinguish the fire, but assigning firefighters to check for fire extension may also be delayed as they attempt to find access to adjoining or neighboring apartments.

Location and Extent of Fire: Garden Apartments and Townhouses

*COAL TWAS WEA**L**THS*

Garden Apartments

In reviewing the fire's location and possible avenues of extension in a garden apartment, we find numerous places for fire travel. The number of concealed spaces and voids in the garden apartment complex with the opportunities available for fire to travel through them will make this a challenging incident. Fire officers must be able to anticipate avenues of fire spread and prioritize resources into the suspected areas.

Kitchens and bathrooms

The most significant cause of early fire extension in a garden apartment will be from the practice of placing kitchens and bathrooms back to back, as well as from stacking them in a vertical line within the building. This is not the first time we have discussed the concerns with these two types of rooms. In previous chapters, as well as in some future chapters we will mention them again to reinforce the thought that, if there is more than one within the building, anticipate the fire spread.

With the installation of a kitchen and bath come the vertical arteries associated with their electrical and plumbing utilities. These voids can be numerous in size as well as number. Openings created to provide these services will run the height of the building often originating in the basement area and ending at the top floor or (in the case of ventilation ducts or piping) ending above the roofline. It is important to note in many of the installations the kitchen and baths will be located in the interior or middle of the building complex with no direct exterior access for their ventilation.

One of the earliest and easiest means for fire to travel once the kitchen is involved is from the cabinets and their soffit areas. With most fires starting in the kitchen, this small space with ample amounts of combustible and flammable

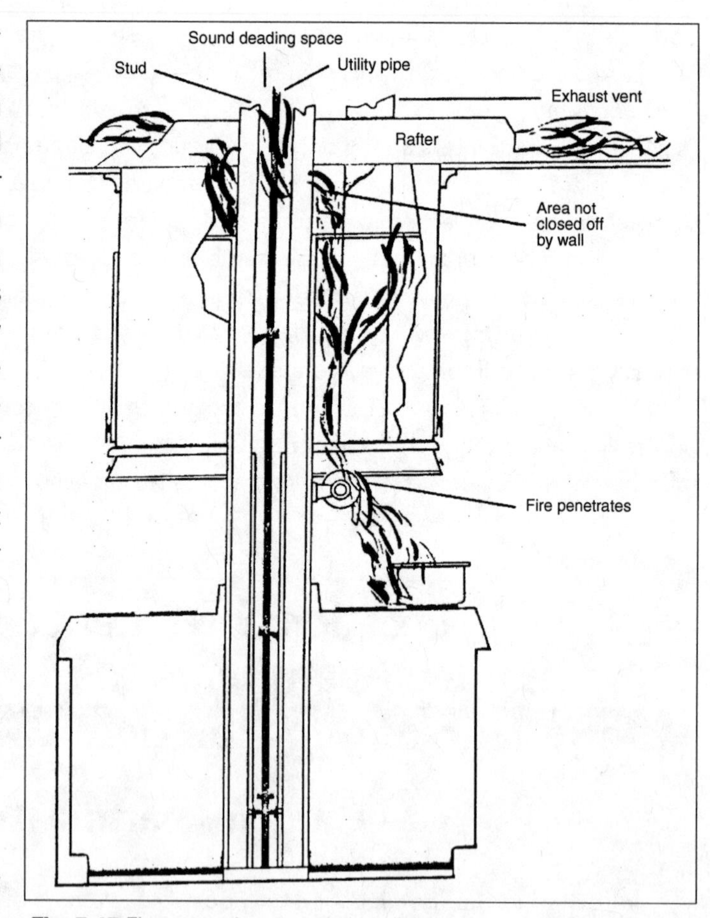

Fig. 5-15 *Fire extension probabilities from back to back kitchens.*

materials can create an intense fire that will quickly deteriorate cabinets and flimsy covered soffits.

The soffit area above kitchen cabinets is nothing more than boxed out void spaces that if not properly fire stopped, can allow uninterrupted access to wall and ceiling voids of the apartment involved, the adjoining apartment, as well as the apartment above. This boxed out space over the kitchen cabinets is required to have fire-rated gypsum board placed over its framing with tape and spackle covering the screws and seams. Often this is not the case. Many times during the construction process this is not done, or it may be poorly done because of its anticipated covering by a decorative material. The decorative material if applied, is often nothing more than a ⅛" piece of masonite. This material is decorative only, and will fail quickly during the early stages of the fire. The taping and spackling over nail or screwed sections of a sheet rock surface is an integral part of its fire resistance.

The kitchen cabinets themselves are a concern with the fire's ability to access the void spaces behind and above the cabinets. Depending upon the quality of the cabinet, thin particleboard may be used for the cabinet backing and ceiling. When fire enters the cabinet, any failure of the particleboard ceiling will allow fire to access the wall and soffit space.

Bathrooms in the garden apartment will also come with significant avenues for fire extension when given the chance. The fire extension possibilities from this particular room will be most notably from plumbing access panels and from recessed medicine cabinets.

Plumbing access panels are openings in a wall behind shower installations. These cutouts will often be found in a closet next to the shower wall. The opening will generally have nothing more than a ¼" piece of particleboard, or ⅛" piece of masonite covering the area for easy access in a plumbing repair. Any fire that penetrates this covering will again have the opportunity to extend into void spaces of the adjoining apartment as well as the apartments above.

Medicine cabinets recessed into the wall of a bathroom will be placed into the void space, often back to back with the medicine cabinet of the adjoining apartment, if you could believe that. This ridiculous past practice of placing medicine cabinets back to back allows for early fire extension into the adjoining bathroom, as well as into the vertical wall space.

With the continued use of plastic piping in the kitchen and baths as well the use of plastic in the tubs and shower stalls, fire attacking these materials not only accelerates the burning process, but allows plastic drippings to fall down into void spaces possibly starting fires below. Unsuspecting firefighters going to work in the original fire apartment may be surprised to find fire burning in a void space below them.

If there is any one building where the incident commander has to make sure he/she covers all six sides of the fire, it's going to be when a fire involves a bathroom or a kitchen area of a garden apartment.

Utility rooms

Another room or area within the garden apartment that will cause a fire extension concern is the utility room. This room is generally no bigger than the size of a closet and is usually located in or next to the bathroom or kitchen. This room will contain a hot water heater, a heating unit, washer and dryer, and possibly an air conditioning unit that will serve the individual apartment. Depending upon the workmanship within the room, inferior fire stopping can allow fire to extend into the void spaces.

Storage rooms

Storage rooms within the garden apartment complex can be located in a number of areas within the building. They can be very small, and serve each individual apartment. They may be consolidated and found on each floor, or they can be designated in an area on a lower-level, or an attic level within the building. Depending upon a storage room's location, each will present its own unique opportunity to allow fire to spread.

Basement storage rooms will house the building's utilities with gas, water, and electrical services throughout. These utilities will penetrate fire barriers. The methods often put in place to prevent fire spread from poke-through construction are often inferior or non-existent. Fire involving this area will also present the added concern of exposed gas piping that can fail early when subject to intense conditions. The storage room when located in either the basement or attic of the building may or may not be sprinklered, further allowing for an intense and difficult fire.

Auto exposure/fire lapping

Fire extending out to the exterior from a window or balcony is an additional threat for the garden apartment complex. Generally we consider fire venting out of an opening (such as a window or balcony) a plus from the standpoint of being able to get in behind it with a hoseline. However, fire venting out of a window or balcony of a garden apartment creates additional concerns that fire officers must be aware of. Depending upon the building design and the construction methods used, features such as wooden exterior sheathing and soffit/attic vents can play a significant role with the fire's ability to extend.

It was once believed that five feet of masonry on the exterior of the building between the top of window and the sill of the window above was enough distance to prevent fire from automatically exposing the floor above. This is no longer a true statement due to the increase in the fire load within the apartment and the amount of heat released when the contents are involved. Take this same spandrel space and install a wooden exterior instead of a masonry covering, and you must anticipate the possibility of fire running the exterior of the building with extension to the floor above.

Another serious auto-exposure threat will come from fire venting from a top floor window into a soffit overhang. The soffit vent is nothing more than an opening in the roof overhang covered with a wire mesh or soffit panel. An initial consideration with fire extension through this avenue is the location of the soffit vent in relation to where the fire is venting. In many buildings you may find that the vents have been unconsciously installed in direct line with a number of top floor windows of the building. With this type installation, fire venting out a window inline with a soffit vent will enter the roof space quickly.

Fig. 5-16 *Venting fire will be given easy access to the building's roof space.*

In similar situations where fire venting from a window that is not in direct line with a vented area, the concern must then be directed toward the integrity of the building material used in the overhang. Depending upon how the attic space is used and insulated, the building materials associated with the soffit can range from ¼" plywood to thin gauge aluminum or vinyl. In the latter case, there will be no resistance to fire as it vents from the top floor. With a plywood, particle, or pressboard soffit, you may get a few minutes before it, too, fails and allows fire into the roof space. Your recourse in this situation is to get a stream directed over the fire vented window to prevent fire from extending into the roof space. Selective and limited use of a hoseline to extinguish any surface fire of the soffit covering will prevent fire spread into the roof space. Tremendous control must be exercised with this stream so it does not interfere with interior operations.

Townhouses

The fire extension concerns for the townhouse will present a lesser concern of fire spread into adjoining dwelling units when compared to the garden apartment. This is primarily from the fact that the townhouse is viewed in many respects as a single dwelling unit. Each individual dwelling unit will have separate and more isolated utilities, with little concern of the inherent problems associated with the back-to-back or stacked kitchen and baths of the garden apartment complex.

Interior stairs

The most significant fire extension concern for the townhouse will come from the unenclosed interior stair of the dwelling unit. Any amount of fire originating on a lower floor of a townhouse will quickly expose upper levels of the home in a very short time. Depending upon the height of the unit, the interior stair will serve all floors of the living space with an additional stair from the basement area to the first floor. In many cases the basement stairwell will only have a hollow-core door at the first-floor level as the only means of limiting fire spread to the upper floors. This type of door has little resistance to fire and will fail early with any significant fire exposure. These large, open, vertical voids within a dwelling will continue to be the focal point for fire department operations.

Utility chases

Next in consideration as an avenue of fire extension in the townhouse is the plumbing and electrical chases originating from either a basement, kitchen, or bathroom fire. Any fire originating in an area that has vertical chases running from it will require the checking of all areas above, adjacent, as well as below.

Flue enclosures

Another source for possible fire travel in the townhouse will be from fireplace flue enclosures. As an added amenity in many townhouses, fireplaces are installed in a living room and/or bedroom. Design or installation flaws with the flue have been the source of numerous fires starting in the void space that surrounds the unit. Any doubt on the integrity of the flue will require a thorough check of all areas up to the roof deck.

Unit separation voids

Other interior areas of concern for the townhouse will come from the void space that separates the dwelling units. In townhouse units with fire-rated gypsum walls between units, a small void left between the partitions separating the units can serve as an artery for some of the building's utilities. A fire originating in this area or extending into this space will spread to both units.

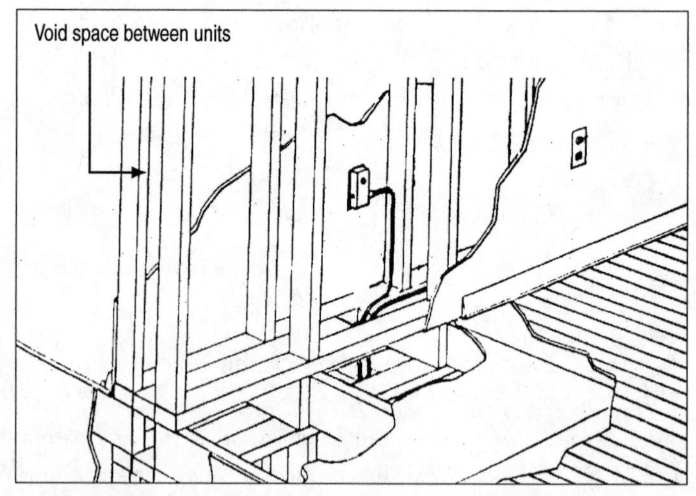

Fig. 5-17 *Space between units which was originally intended to provide sound proofing is also utilized for running building utilities*

Overhangs/extentions

In some townhouse units, aesthetic designs allowed for the second floor to protrude or overhang beyond the first floor a distance of two to three feet. The underside of this protrusion is normally covered with 3/8" plywood or masonite. Fire venting out of a first floor window can expose this covering and enter the void space between the first and second floors. The type of exterior sheathing on the building involved can compound this situation. With a wooden exterior, fire left unchecked will run the sheathing and enter the floor above as it works its way up the exterior to the roof soffit. If given the opportunity, fire will enter the roof space through this means where the tasks will become more intensive.

Fig. 5-18 *This window overhang can present a concern if fire accesses this area.*

It becomes evident early on that based on the class of construction, the hybrids associated with the classes, and the inherent design concerns of the different units, if you were ever going to put your thermal imaging cameras to work to assist with limiting the fire extension, it's going to be when fire involves a garden apartment or townhouse complex.

Time: Garden Apartments and Townhouses

*COAL TWAS WEAL**T**HS*

Time of the day/year

When we look at the factors associated with our time size-up concerns, the first category that should come to the mind of the fire officer with all residential structures is the time of the day.

During the day, the majority of the occupants who live in the complex will be out of their homes and at their places of business. Those that are home will be awake and hopefully more able to react when a fire erupts. For those incidents that take place in the late night or early morning hours when more occupants are home and asleep, the concerns increase significantly simply due to the larger number of individuals in the building and their status at the time of the incident.

The evening hours also bring the added street concerns we have mentioned previously. With most of the complex occupants home, parking lots and surrounding streets become a nightmare for moving and placing fire apparatus.

The concern with street congestion and the increased number of occupants will grow larger during weekends and holidays. It's during these times when individuals will be visiting relatives and friends not only adding to the congested street concerns, but also increasing the life hazard concerns within the apartment or dwelling.

Height: Garden Apartments and Townhouses

*COAL TWAS WEALT**HS***

Height ranges for the garden apartment complex will average around two to four stories with ranges for the townhouse from two to three stories. There are exceptions in this area with higher stories possible, but for the most part heights will fall between the two- to four-story range for both types of buildings. Our biggest concern on attaining height in the garden apartment and townhouse complex is going to be how to do so. With most of the height concerns of these buildings falling within the two- to four-story range, initially it may seem that gaining access to a roof or balcony will not be much of a problem. In a significant number of these incidents it will not be. However, when buildings are set back, surrounded by decorative landscaping all within a sloped terrain, gaining height will become much more challenging than you might expect.

Special Considerations: Garden Apartments and Townhouses

*COAL TWAS WEALTH**S***

Garden apartment and townhouse complexes create many significant concerns for the fire service that cannot be handled without information gathered from a pre-plan. The best way to handle the different concerns associated with these type occupancies is to get the companies out to see them first hand. If there was ever a need to get out and take a look at a specific grouping of buildings, it will be these. This is not to say that any of the other types of occupancies discussed in this book require any less a concern, but the unique concerns presented by each different garden apartment and townhouse complex will require a definite look.

Depending upon the design and the location, complexes can vary from the type of construction, location of fire separations, building setbacks, landscape and terrain obstructions, street and address layout, to the occupancy load. With each complex often taking on its own identity, members need to review what will cause concern for that specific complex. Information gathered on one specific complex on one side of town, may be totally different from another complex on the other side of town.

As chief officers go to work at these types of occupancies, they also need to review the pre-fire plan information in order to direct effective as well as efficient operations. This information coupled with the individual's education and experience should start to direct some thoughts on fire extension probabilities and possibilities once the fire's location is determined. Specific concerns for the incident commander upon arrival must initially include the following information.

Garden apartment and townhouses

1. *Accessibility.* Ensure that companies don't commit themselves to areas around the complex that they cannot get out of quickly. It is best in congested complexes to have the first arriving engine and possibly the first arriving ladder company investigate to determine the fire's location and best access, while the remainder of the assignment stages outside the immediate area. Holding resources and awaiting this critical information will prove to be valuable when you need to put additional companies to work.
2. *Additional resources.* Don't hesitate calling for additional help. When conditions indicate early on that your first alarm assignment is going to be committed quickly, start additional companies into the incident. Even if at that time you don't have specific assignments for them, get them coming in and establish a nearby staging area. This way when a need arises and a company needs to be assigned to handle it, the lead/reflex time will be short. Waiting to call them until you need them will get you caught in the playing catch up mode. Using this form of calling resources will guarantee the loss of the building.
3. *Information gathering.* Obtain reports early about:
   - The number of occupants in the building.
   - The occupants' location in reference to the fire.
   - The condition of the interior stairs of the building.
   - The progress of the first hoseline.
   - The conditions in the rear of the building.
   - Information about fire extension on all six sides of the fire.
   - The location of any fire separations or firewalls within the row.
   - The necessity to open the roof to relieve heat and smoke.

4. *Fire extension*. When fire goes beyond the contents and involves the structure of these buildings, the checking and opening up of concealed spaces becomes a critical priority of your ladder company members. Quick deployment of resources in vital areas around the main body of fire is a necessity on regaining control of the building. With the deployment of resources into these areas also comes the responsibility of opening up these spaces to check for fire travel. Fire officers who hesitate to open up because they don't want to cause unnecessary damage may quickly find that the fire is past them on its way up to the attic space. Keeping the fire from taking possession of the attic should be a strategic concern for the fireground commander. The only way to prevent this is to assign resources around and above the fire early.

When available, the early deployment of thermal imaging will greatly assist with efficiency, effectiveness, and safety of the operation. Thermal imaging should be put to work with the first deployed hoseline. Additional cameras should be deployed above, adjacent, as well as below as soon as possible. The images and their interpretation will reveal what needs to be done. Departments who lack multiple cameras or have none at all must approach the areas surrounding the main body of fire aggressively. Ladder company members must open vital areas with engine companies ready to direct water into the concealed areas. Hesitate at all, and you can give up ground you will never regain.

As you can gather by now, I'm a big fan of thermal imagining. But it is also important for me to state that a camera's use must not replace the education and experience of a firefighter or fire officer, it is only meant to enhance it. If an experienced and educated member has reason to open up and investigate an area or void space, they should do so. As useful and beneficial as thermal imaging is becoming for the fire service, education and experience of the firefighter and fire officer continues to be the fire service's greatest asset.

Questions For Discussion

1. Review the different types of garden apartments and townhouses mentioned, and identify how they may compare or differ from the types within your own city.

2. Identify the different occupancy concerns of the garden apartment as compared to the townhouse.

3. What is the key to an engine company's effectiveness at a garden apartment fire?

4. What is meant by term long stretch occupancies and how does that affect the decision-making of all engine company officers?

5. Describe the pros and cons of the roof turbine.

6. Identify and describe the different grade level concerns that may present themselves at a garden apartment fire.

7. What is the distinct difference between a firewall and fire separation wall?

8. What is referred to as the most significant cause of early fire extension in the garden apartment? List the related areas of concern.

9. What is referred to as the most significant cause of early fire extension in the townhouse?

10. What should be the main strategic concern for the incident commander at a garden apartment fire?

6
Row Frames and Brownstones

The row frame and brownstone type buildings are very different, and can be, unique and challenging structures the fire service must continue to deal with. Although you will find that we will be able to cross-reference some of the concerns with other types of occupancies, to include these two types of buildings in any of our previous chapters would not have given the recognition these buildings deserve, nor identify the specific concerns we need to fully understand.

Construction: Row Frames and Brownstones

COAL TWAS WEALTHS

Row Frames

As the name implies, the row frame is a row of wooden frame buildings attached together possibly for an entire city block. They can vary in height from two to four stories with each individual building ranging in width from 20′ to 30′, with depths ranging from 40′ to 60′. Within the row, the number of attached buildings could range from as few as three or four attached together, to as many as 20 to 25, all with a common cockloft, a common cornice, and in some cases a common basement.

Balloon frame

These buildings are built of Class 5 wood frame construction consisting of either the balloon frame or braced frame design. As previously discussed in balloon frame design, the exterior walls of the building will contain wall studs extending from the structure's foundation to the roofline with no fire-stopping in between. Any fire entering this void space not only has a raceway to travel, but also directly attacks the load-bearing members that support the floor, wall, and roof of the structure.

Fig. 6-1 *Row frames*

Brick nogging

The load-bearing walls of the row frame (also known as a division or party wall) is what separates the individual buildings within the row. What is often found within this wall is the use of *brick nogging*. Brick nogging was a concept of placing individual bricks in the division wall's vertical stud channels for their entire height. This idea of placing bricks and mortar together between the studs of the separation wall was to give strength to wall, as well as prevent fire from traveling through the wall into the adjoining building. Many of the old-time building framers felt that the brick, in addition to its intended design, also gave the wall some insulating value. What needs to be identified as an immediate concern is the increased dead load in the building's load-bearing walls. There should be no doubt that this type of wall design can collapse early, producing disastrous results.

Fig. 6-2 *Brick nogging*

Braced frame

In the braced frame design of the row frame our primary concerns must focus on the connection points of the system. Within this type of design there will be a framework of vertical timbers called "posts". Posts will be positioned at each of the four corners of the structure as well as intervals in between. These wooden structural members can vary in size from 4"x 4" to 6"x 6" posts. Further within the design there will be horizontal cross-members called "girts". Girts are wooden structural members fastened to the posts through a mortise and tenon connection at each floor level. Their sizes will average the same as the posts. The girts, which act as a vertical fire stop at each floor level, are cut down at their ends to form a tenon, which is designed to fit snugly into a mortise opening cut into the vertical post. The combination of post and girts is what makes up the wooden skeleton of the structure.

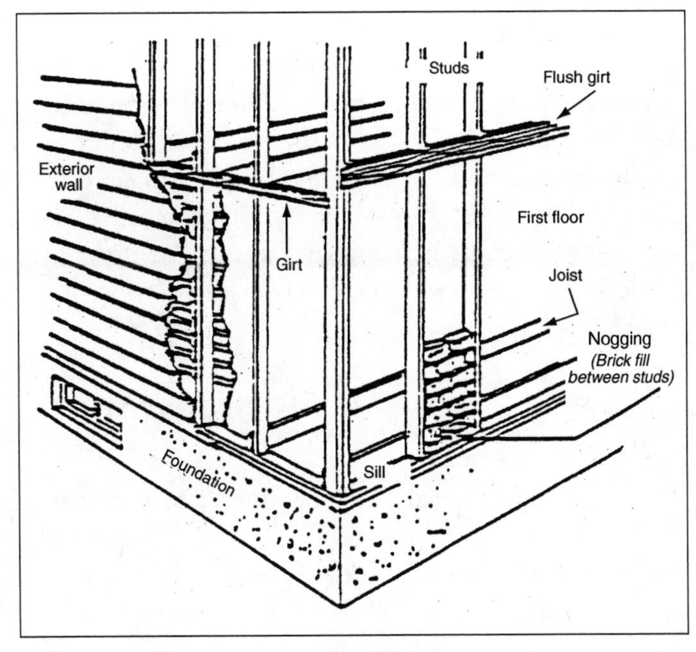

Fig. 6-3 *Braced framing*

A significant concern with this type of wood frame design is its age. Many of these buildings with this design and method of construction (most popular during the early 1900s) will now be more than 100 years old. Through the years the wooden structural members will have dried out and contracted from age and poor sheathing ventilation. The structural members also may have been subject to rot or building infestation. With the design of the building allowing a large structural member to be shaved down at the connection point into a tenon often no bigger in size than half of its original dimension, the concern must focus on the integrity of the connection point.

Many times the only support holding these structures vertically in line is gravity and the support of the adjoining structure. If one or

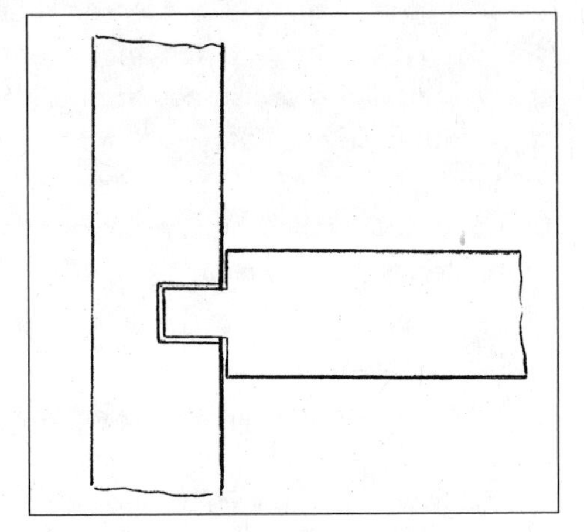

Fig. 6-4 *Mortise and Tenon joint connection*

more of these structures in the row has been removed from a previous fire or renovation the collapse concerns of the building increase dramatically to the unsupported side. If fire involvement affects the load-bearing wall, any shift in the load-bearing members can allow the building to fall to the unsupported side. Add the presence of brick nogging, and the slightest movement at a connection point can collapse the entire building. This same concern holds true for the last building at the end of the row. The last building at the street corner will lack any support of an adjoining building, making it more susceptible to collapse than when placed within the row.

Brownstones

Brownstone buildings are built of a Class 3 ordinary construction. The exterior components of the building are of brick or stone masonry with the interior structural members comprised of wood. Buildings will range from two to four stories in height with street widths averaging 20′ to 25′ with depths ranging from 40′ to 60′. With most brownstones you will find the classic stoop that accesses the second level from the sidewalk entrance. Below the stoop there will be an entrance to the lower-level, which is normally a few steps below street grade within a sunken court. Below that you may also find a cellar that is completely below grade and is only accessible from an interior stair. A tip-off to the presence of a cellar in a brownstone may be an old coal chute cover in the front sidewalk, or a metal grate within the sunken court.

It must be remembered that even though many consider brownstones strong and well-designed buildings, they still come with the inherent concerns of a Class 3 ordinary building, with a few additional concerns from the features associated with this building.

Fig. 6-5 *Brownstone*

Building features

Shafts that originally concealed an old dumbwaiter shaft or hot air heating duct will often still be present. These openings will be large in size and traverse from floor to floor. Buildings will also contain an undivided cockloft over its entire top floor that may or may not be separated from an attached building by a division wall. Hollowed-out spaces around electrical and plumbing voids within the building will still cause the same concerns with fire extension up into the cockloft. Overhanging cornices will be on the front of the building, with wooden Yankee gutters found in the back of the building.

Alterations

The interesting point to note here that further relates to the construction of these buildings is that in many of the country's urban areas the brownstone (more so than any other type of individual building) seems to be undergoing the greatest frequency of renovations. The reason for the renewed interest in these structures seems to be from the aesthetic value they present. From the brick exteriors, decorative entranceways, to the wrought iron gating, their popularity in many cities commands a high real-estate price tag. With these changes, the fire service must also anticipate building alterations, which in many cases could be extensive. The use of wooden I-beams and truss floor/roof supports to create large open spaces are just a few of the alterations that can be expected.

In many cases, today's incident commanders could be standing in front of a building that contains a 100-year-old brick facade, with all lightweight interior construction components behind its stone surface.

Occupancy: Row Frames and Brownstones

COAL TWAS WEALTHS

Classification

The occupancy classifications for both the row frame and the brownstone will usually be residential. Within the brownstone there may be the possibility of a mixed occupancy; you may find a small grocery, barbershop, or retail shop on the first floor.

Fig. 6-6 *Mixed occupancy brownstone*

Occupant load and status

Occupant load and status of both types of structures will generally depend on the amount of alteration done to the structure. Row frames will generally follow their original design of the railroad flat type of apartment with two apartments serving each floor. The brownstone, on the other hand, was originally designed to serve as a private dwelling. In many of the affluent urban areas you may still find the building being occupied by only one family within the structure's original design. However, for the majority you will find two or three apartments with large floor areas, vaulted ceilings, and the possibility of open stairways between levels.

Other occupancy concerns for the row frame and brownstone that may not be obvious is the possibility of SROs. Their presence will present a much larger occupant load than the building's original design intended.

Apparatus and Staffing: Row Frames and Brownstones

COAL TWAS WEALTHS

Early considerations

Fires reported in row frames and brownstones must demand significant consideration to apparatus response and placement. In some cities surveyed, due to the attached row concept of this type of structure departments have elected to send an engine and a ladder company within the response to the end of the block or possibly the parallel block to view and eventually obtain access to the rear of the building. A large number of buildings attached without direct access to their rear could quickly create difficult conditions if not planned for. If companies are given additional opportunity to access the back of the buildings early in the incident, large fire loss may be prevented. Committing all of your companies to the street side of a row of attached buildings may find you chasing a well-advanced fire in the rear.

Fig. 6-7 *Rear accessibility may be difficult*

Engine Company Operations

Apparatus placement

Placement of the first due engine company must consider leaving the front of the building open/accessible for the ladder company. For a fire in a row frame or brownstone many departments stress placing the first arriving engine company at least two buildings past the fire building in a attempt to increase the ladder company scrub area of the fire building and its exposures. Considering the average widths of the buildings, placing the first arriving engine company two or three buildings past the actual fire building only increases the hose stretch to the front door by approximately 50'.

Hoseline selection

Hoseline selection with both of these structures will focus around the hoseline's speed and mobility. Due to the life hazard, apartment layout, square footage, and high probability of fire extension, a well-placed hoseline with sufficient water is a must in both of these types of buildings. The hoseline of choice for most departments in these types of occupancies will be the 1¾" or 2" hoseline. Anything less will often prove to be insufficient, anything more than a 2" hoseline in a row frame or brownstone will lack the speed and mobility required to quickly place the hoseline into operation.

Hoseline placement

Placement of the first deployed hoselines at any fire will always depend on the life hazard and the location of the fire. Barring specifics, it is safe to say that the first hoseline in these types of occupancies, usually marches in through the front door to gain control of the building's interior stair. With this area not only being the lifeline of the building, but the area most vulnerable to fire spread.

Fig. 6-8 *Anticipation is the key to assigning resources.*

Additional considerations with hoseline placement will depend upon the fire's location, the building construction, and the exposure proximity. As an example, with a top-floor fire in the row frame, anticipation of fire spread into the cockloft is a must. Visual observations and information from the preliminary reports of first due companies can very well dictate deploying the second and even the third hoseline into the top floor of exposure buildings if a fire is taking possession of the cockloft.

To expand this idea, fires that involve basements and cellars of both types of buildings will generally require the first hoseline to go through the front entrance door to the top of the basement stair. This hoseline can advance down the stairs as an extinguishment hoseline, or if conditions restrict its advancement, it can act as a confinement hoseline with a second hoseline attacking the fire from a basement entrance. Even with those fires that involve the basement or lower-level of the brownstone, directing the first hoseline through the stoop area and into the basement (which at times can seem very inviting) without the protection of a confinement hoseline above, can allow fire to race up the basement stair out into the building's interior hallway.

Ladder Company Operations

Apparatus placement

Ladder company placement at attached structures must enhance their operation as well as anticipate the fire's spread. With the significant fire-spread problems associated with the row frame, the front of the building belongs to the ladder company. With only one truck company initially responding to the incident, this becomes a priority. This statement is derived not only from the need for quick and easy accessibility to the building's roof, but also for close and quick accessibility for additional ladders, tools, and equipment. In some cities researched, it is a common practice to assign two ladder companies to each reported building fire. With this type of response, officers should consider assigning the second arriving ladder company to the rear of the building. With attached row design of both the row frame and the brownstone, accessibility to the rear of the buildings with the requirements of tools, ladders, and hose must be given early consideration.

Fig. 6-8a *Ladder company members take advantage of a vacant lot for roof and rear yard accessibility*

Forcible entry

Forcible entry concerns for both the row frame and brownstone will generally follow the use of conventional options of forcing entry. Additional points of consideration for forcible entry in these structures is the presence of security gates over the windows at the first floor or basement level, as well as the possible presence of SROs in both type structures.

Fig. 6-8b *Fixed window bars*

Ventilation

Ventilation will focus around your knowledge of the building construction combined with information about the location and extent of the fire. As an example, with top-floor fires, especially in the row frame, knowledgeable and well-equipped firefighters are a must. Skylights and scuttles will be the two primary means of natural ventilation available. While both will provide vertical ventilation of the top floor, these two openings are not enough for a significant fire on the top floor. A large primary ventilation hole directly over the fire should be the primary goal. Due to the volume of fire that can develop and spread through the cockloft in the row frame, it is important to get the primary ventilation hole opened as well as enlarged to create a significant thermal updraft of the fire. Not only is it the intent of this large opening to draw the fire back from its horizontal movement, but it will also allow you time to open ceilings and stretch hoselines into attached exposures, as well as place any defensive cuts for an effective trench.

As we have described in previous chapters, the defensive measures associated with the trench cut are going to be staff and equipment intensive. Attempt to gather enough people to handle the tasks. If you can't, keep enlarging your initial primary hole in an attempt to halt the horizontal fire spread.

The scuttle openings previously mentioned in these buildings are nothing more than a 30"x 30" opening with a small vertical ladder leading from the top floor stairwell to the roof deck. These openings are usually tarred over and nailed shut. Not only do they allow the ability to vertically ventilate the fire building, but they have also proven to be a quick and easy access for the checking and halting of extension into the cockloft of the exposure buildings. By removing the scuttle hatch

Fig. 6-9 *Return walls*

of the exposure building(s) you can easily access the cockloft area by removing the return walls on the exposed side of the fire. The return walls of the scuttle are the boxed out space that spans the top floor ceiling and the roof deck. Its span will actually tell you the height of the cockloft. Usually a couple of pokes with a halligan tool and you have access to the cockloft.

Search

Search efforts for the row frame and the brownstone will follow normal residential search procedures. Concerns will increase significantly when your search shows the presence of an illegal multiple dwelling in either the form of a basement apartment or attic apartment, or the presence of SROs.

As we continue to focus on the need to access the rear of these buildings, it is important to note with both types of structures that there may be a rear door to a lower-level that is directly accessible from the rear yard. With a fire in the basement/first floor area, an early attempt to access this area should be made.

Another significant point of consideration for those firefighters assigned to the search of a brownstone is the fact that most will lack a secondary means of egress from the upper floors. With the possibility of two or more apartments in the brownstone, there may be no rear fire escape to serve the upper floor apartments. With ground ladder accessibility delayed or even impossible to the rear, removal of building occupants trapped at upper floors will be difficult.

Rear Accessibility Considerations – Row Frame and Brownstone

1. Through an adjoining structure(s) into the rear yard.
2. Through the rear yard of a structure on the parallel street.
3. Through the end of the block from the nearest cross street.
4. Up and over the roofs of nearby buildings.

Life Hazard: Row Frames and Brownstones

COAL TWAS WEALTHS

Firefighters

Fire spread

The row frame and brownstone's ability to allow fire to spread within the structure comes as the most obvious and most significant concern to the firefighter. From the concealed spaces and voids of both structures to the open interiors of a brownstone duplex apartment, the firefighter must be educated and prepared to operate within as well as around the building.

The seriousness of the open interior stair should be apparent by now from our discussions with the private dwelling. What will come as a less obvious, but serious, threat to the firefighter is fire involvement of the concealed spaces and voids. As fire burrows its way into the walls, ceilings, and cockloft, gases can build to the point where they can starve themselves of the necessary air to support combustion. With an unlimited amount of fuel and heat, we are reminded of the ingredients necessary for a backdraft. For years fire service educators have cautioned firefighters of the dangers of opening concealed spaces with a high heat presence within a room or area. Today with newer insulating and protective capabilities of a firefighter's gear, the heat conditions may be masked to the point where the unsuspecting firefighter may open a hole in the middle of a ceiling introducing a rush of air into the cockloft, exploding fire and the entire ceiling down on the firefighters.

With a fire in any concealed space, most notably the cockloft, firefighters must exercise extreme caution when opening up the space. Too many times you may have heard about, read about, or even seen firefighters being caught in an explosion of fire and debris after poking a hole into a concealed space. Because of the dangers of the large and numerous concealed spaces of the row frame and brownstone, listed below are concerns and precautions to take.

Concealed space precautions

1. Make sure others know of you and your partner's location. Fire department accountability at a fire scene should track member assignment and location. Many departments have standard operating procedures (SOPs) on member placement and objectives at certain occupancy fires. Anytime your assignment and location change, make sure more than one person knows.
2. Have a charged and staffed hoseline ready for use when you are opening up any suspected area.
3. Note heat conditions in the room or area where you are operating. A significant presence of heat in a room with no visible fire may very well indicate fire is occupying a void space. Be observant to blistering paint, smoke stained trim, and fixture openings.

***Fig. 6-10** Row frames and brownstones are known for their numerous concealed spaces and voids.*

4. *Always* operate and open up the space from an egress area. Examples are from a doorway off the hallway or from a window that leads from a fire escape or ladder. Only after conditions are acceptable can you continue to open up and move in from these refuge areas.

Exterior operations

Another unique and often-overlooked concern to the firefighters is the unsuspecting dangers on the outside of the row frame and brownstone. On more than one occasion firefighters have walked off, or fallen off the rear roof, or walked into a light and airshaft between the buildings. Barring how we should act when we can't see where we are walking, the fact that should be remembered is that there is no rear wall parapet or cornice if we walk toward or slide on an icy roof in the rear of these buildings. The parapet wall or raised cornice on the front of the building will give you a visual as well as physical barrier to remind you to stop as you work or walk in that area. At some incidents I have seen light and airshafts protected (and I use that term loosely) by a flimsy construction of wood and chicken wire in an attempt to form a fence around the opening. Although their integrity is questionable, their presence is better than nothing.

An often-present danger that usually cannot be seen by the unsuspecting firefighter assigned to the roof is the wooden gutter/*Yankee gutter*. In the chapter on private dwellings the concerns and dangers of their installation were discussed. Please note that they are also very common on the rear of both the row frame and brownstone buildings.

Collapse

Collapse concerns to the firefighter for the row frame and the brownstone will differ due to their classes of construction and the inherent design features within. As in any building frame, the most significant collapse concern will be failure of the bearing wall. As we previously described, the row frame is most notably constructed from either the balloon frame or braced frame design. With these structures originally designed within an attached row, the building age, rot, overall settling, and the effect from fire conditions will affect the structural stability of the building. With each building possibly relying upon its neighbor for support, any missing building or buildings from a previous fire or demolition could cause the new end buildings

Fig. 6-11 *Rowframe collapse*

to become structurally unstable early in the fire. When a fire extends beyond the contents to the structure, collapse concerns increase not only for members operating inside the structure, but for those members operating outside the structure.

Occupants

Number and location

Occupant concerns for the fire department in a residential structure will always focus on how many people are in the building and their location in relation to the fire. With both structures there will be a high occupant life hazard due to the number of people who may possibly occupy these buildings. Upon arrival, firefighters may be able to obtain some preliminary information about the number of apartments within the building by viewing the first floor vestibules for mailboxes and doorbells. Their presence may not lend information to the presence of a basement apartment or SRO occupancies within the building; however, observations in this area may help to guide your thinking.

Exposure buildings

The occupant life hazard concerns for the firefighter must also extend to the people of the exposure buildings especially when they are attached or very close. With the anticipated rapid fire spread of the row frame, the occupants of the attached exposures must receive early attention. With rapid extension possible due to an enclosed or open airshaft, exterior sheathing, and the common cockloft, searches must be assigned to these buildings early in the operation.

Terrain: Row Frames and Brownstones

COAL TWAS WEALTHS

Brownstones

Grade level references

The terrain concerns for the brownstone focus on the difficulties with grade level references of the basement and the first floor. The question often asked is "when is it considered a basement? A cellar? A first floor?" A reference by one individual on the fireground to the area in question as the basement, and the reference by another individual to the same area as a first floor may cause confusion and serious problems with the fireground operation and its management. This size-up concern needs to be addressed well before the receipt of the alarm.

When referencing the grade level areas within the brownstone, some departments reviewed always refer to this area as the first floor. Other departments reviewed will always reference this area as the basement; while still others use the words "fire floor" and "floor above" to reference their floors. What was interesting was that most departments don't have a policy on this below grade concern, whether it is in the brownstone or for that matter any building.

The concern that I have with always calling this area the "basement" or the "first floor" is that these below grade areas can differ significantly when viewing two different brownstones only a block away from one another. The concern with the reference of "fire floor" and "floor above" is that the distinct reference made by these terms may not be initially recognizable. Smoke showing from a particular area on arrival may not necessarily be the fire floor.

Department policy and procedure should dictate the reference to these below grade residential areas to eliminate a potential disaster. The last thing anyone wants to see is a firefighter call a May-Day from the area recognized as the basement, only to have rapid intervention resources misdirected to a different area because of confusion on what this level is called. This delay in the arrival of resources could cost a member's life.

For grade level concerns in residential structures, the following are suggested reference guides to eliminate confusion as well as enhance fireground operations.

Brownstone Grade Level References

Basement – Any area that has more than half of its area below street grade. With the average story height in a residential structure being 8′ to 10′, this would mean more than 4′ to 5′ below the street level.

Cellar – Any area that has its entire area below street grade.

Sub cellar – Any area below a cellar.

Brownstone/first floor reference – Any area that has more than half of its height area above street level grade, not court level. Court level is the area within the sunken court. With the average story height being 8′ to 10′, this would mean more than 4′ to 5′ feet above street level.

Quick reference – If it appears close in consideration to a basement or a first floor, call it the first floor.

Additional reference – Window size may help you identify the floor. With a full-size window in place within the below grade area and close to the half above or below reference used, consider the fact that it is probably a living space and call it the first floor.

The above guide is nothing more than that, a guide. Departments must use what is comfortable for their day-to-day operations. The important consideration is to use something.

Fig. 6-12 *More than half below street grade, refer to it as a basement.*

Fig. 6-13 *More than half above street grade, refer to it as the first floor.*

Water Supply: Row Frames and Brownstones

COAL TWAS WEALTHS

Requirements

The water supply requirements for fires in the row frame and brownstone are based on the fire load and your ability to confine it. With a potentially fast moving and well-fed fire in these structures, speed and mobility are once again the primary considerations. If we refer to the initial fire concerns of the Class 3 ordinary and the Class 5 wood-frame constructed buildings, two important inherent features of both types of construction remind us of the need for speed and mobility. One will be the concealed spaces and voids, and two, the building's open interior stairs. Each of these is very well represented in the row frame and the brownstone. From the stacked kitchens and baths to the three-story chimney with a banister, fire will have the ability to move from one level to another quickly.

When the company officer opens the cab door to give orders to his/her members on hose lengths, diameter, and placement at a fire involving a row frame or brownstone, the 2½" is generally not the choice hoseline to be pulled from the apparatus hose bed. Row frames and brownstones come with features that are not conducive to a big water stretch. Small square footages, cramped surroundings, railroad apartments, narrow passageways, and relatively low fireloads make the thought of the 2½" hoseline impractical at most offensive operations. This is not to say that there is no place for a larger hoseline at a fire involving one of these buildings. With the possibility of a commercial occupancy found on the first floor or basement area of the brownstone, a larger hoseline may be warranted from the factors associated with that type occupancy. A commercial occupancy with an open floor area, increased fire load, and a heavy fire condition threatening to extend to the residential floors above, warrant the capabilities associated with the reach, penetration, and knockdown power of a big hoseline. Barring an occupancy change, choose and stretch what will give the quickest and most effective results.

Auxiliary Appliances and Aides: Row Frames and Brownstones

*COAL TW**A**S WEALTHS*

Detection equipment

The thoughts and concerns of this size-up consideration for the row frame and brownstone are for the most part identical to those previously mentioned for the private and multiple dwelling. Prior to the code enforcement requirements of hardwire and battery operated smoke detectors in the hallways and apartments, the fire activity in the row frame and brownstone was at an alarming rate. Personal experience is starting to show in these buildings, as well as in many other residential buildings, that complacency is starting to overtake the maintenance of a previously installed detector. The most common removal of the battery and the direct removal of the detector is what many of us are seeing as we enter buildings and apartments on routine alarms. When observed, it is important to correct the danger.

Street Conditions: Row Frames and Brownstones

*COAL TWA**S** WEALTHS*

As we reference street width and its effect on apparatus movement and placement, we noticed little to no difference between the row frame/brownstones and many of the other multiple dwellings types we have discussed. Multiple dwellings are normally built on any type of street within any location of a town or city. You can find them on one-way streets, two-way streets, and streets with multiple lanes of traffic. The row frame and brownstone can also be placed in the same type of setting. They could come with the same difficulties presented with the multiple dwelling. Familiarization with the area you're assigned to protect is critical.

Gathering information about street conditions will greatly affect the efficiency and effectiveness of the overall operation when it comes to placing, positioning, and operation of the apparatus. This is a concern that must be addressed at the company level. The officer and the apparatus chauffeur (engineer/driver) must know the street width, traffic flow, their assignment on the alarm, and their responsibility before they commit themselves to an incident.

Fig. 6-14 *Here we go again...*

Weather: Row Frames and Brownstones

*COAL TWAS **WEALTHS***

Of the universal concerns, wind, temperature, humidity, and precipitation, the one that will cause the fire department the most concern when operating at a fire in a row frame or a brownstone is the wind. With exposure buildings attached to the brownstone (more importantly the row frame) even a mild wind condition can significantly enhance the fire's spread.

Row Frames

Wind

Wind speed and direction is an important consideration when fire has taken possession of the cockloft area of a row frame. Cocklofts in the row frame building can be as small as 18", or as large as 6' in height. This opening will extend over all buildings that are attached in the row.

As indicated in the tactical considerations of the apparatus and staffing section, fires that involve the top floor and cockloft area of a row frame will require a commitment of firefighters to the top floor of the fire building, a ventilation hole over the main fire area, with early assignments to both exposure buildings to pull ceilings and deploy additional hoselines. As firefighters open the ventilation hole and continue to enlarge it, significant winds that are moving across the roof deck can affect the thermal updraft of the opening, as well as drive fire throughout the cockloft. It's difficult to state how much wind will drive how much fire. Even with wind movement as little as 10 mph, firefighters and fire officers must consider how it can impact the fire.

With this in mind, the speed of the wind and its direction can aid you with the deployment of your initial resources. It's rare that many fire departments can deploy an army of firefighters to the fire building, the roof, and both exposures at the same time. The street-smart fire officer realizes the need to prioritize areas on the fireground and then delegate what is available to launch the attack. If you're lucky enough to have the end building involved, then the decision is simple, but when the fire is in the middle of the row, where do we send the complement of available resources? If it's a factor, let the wind be your guide.

If the wind direction is from the left or right of the fire building toward either one of the attached exposure buildings, assign a significant percentage of your initial forces downwind of the fire building. The objective behind this is not only put resources in and on top of the fire building to attack and suppress the main body of fire, but to also prioritize what will be the most seriously threatened exposure from the fire's anticipated movement.

If the wind direction is perpendicular to the front or back of the building, a wind driven top-floor fire will quickly create blowtorch conditions pushing fire equally in both directions. In conditions as described, it is best to split a portion of the available resources to both exposure buildings in an attempt to hold the fire spread as remaining resources attack the fire floor and open the roof over the main fire area.

Until the majority of the troops arrive, fire officers must prioritize areas and assign resources where the best overall results can be achieved. In some situations that we arrive and go to work at, the wind may influence the decision-making.

Fig. 6-15 *Top-floor fires + wind = problems*

Exposures: Row Frames and Brownstones

*COAL TWAS W**E**ALTHS*

The class of the fire building's and the exposure building's construction, as well as their proximity to each other, are the driving forces in your exposure size-up concern.

As we reference both the row frame and the brownstone, one type of structure stands out as more troublesome than the other—the row frame. The row frame will allow a fire many more opportunities to extend to its exposures when compared to the brownstone.

Row Frames

Light and air shafts

The division/party wall that separates the attached buildings will possibly contain a light or airshaft. In the multiple dwelling chapter, the fire officer's concerns with their presence and ability to allow fire to extend should be carried over when arriving at a fire in a row frame. When arriving and viewing the building from the street, it is difficult to determine their presence. You can sometimes get an indication that fire has entered the shaft from what appears to be a fixed column of smoke or flame coming from the roof area where the buildings meet. This can be used as a reliable indicator to initially assign your resources. Barring this visual cue, if you have information that the fire building's depth exceeds 40′, expect their presence and anticipate the fire spread. When you need to obtain specific information about the presence and possible involvement of any light and airshafts, firefighters assigned to the roof position or rear of the building must relay the information to the officer in charge of the incident.

Common cockloft

The common cockloft is a major contributing factor for fire spread into adjoining buildings. With spaces ranging in height from 18″ to 6′ possibly covering an entire city block, the complications can be staggering. Aggressive tactics, a comprehensive incident management system, and an army of firefighters are going to be a

Fig. 6-16 *Fire involvement of a shared light and air shaft, as well as a common cockloft*

prerequisite when fire takes possession of a common cockloft in a row of frames. Incident commanders must not hesitate in delegating officers to divisions or sectors in and around the building to coordinate what will be an intensive fight. Assigning a fire officer to a division or sector and providing a number of companies in an attempt to halt the fire's spread has to be a consideration early into the incident.

Fire spread from the rear

Another significant exposure concern with a fire that involves the row frame is the buildings that are exposed from the rear. As we have stated, accessibility to the rear in these types of buildings is a key concern, not only for the need to access the rear of the fire building, but also for the need to protect the buildings that are exposed.

In many areas within a city where these buildings are located, congestion seems to be the norm. In many instances rear yards of 20′ to 25′ is all that separates buildings that front on opposite streets. A well-involved building could quickly jump from the rear of a building on one street to the rear of a building on another street if left unattended. Accessibility to the rear of a building must not only include a compliment of ladders and tools, it must also include a mobile 2½″ hoseline that could knock down a sizeable amount of fire while protecting any exposures.

Brownstones

When we think of the brownstone type of building we are reminded of Class 3 ordinary construction. The brick and joist concepts of the brownstone eliminate some of the combustibility concerns associated with the row frame. Fire spread via the exterior is usually limited to auto-exposure to the floor above. Fire spread to attached exposures is usually limited due to the attached brick walls being carried up through the cockloft space. Cocklofts do exist in these structures, but are not common with the adjoining structure. This is not to say that inferior construction or decay will not allow fire to spread to an exposure, but the speed at which fire can spread will be considerably less than compared to a common cockloft.

Exposures require early attention in both types of structures simply due to the fact that they are attached. The class of construction and their inherent design features will cause you to act more quickly when operating at one building type as compared to the other.

Area: Row Frames and Brownstones

*COAL TWAS WE**A**LTHS*

Area size-up for the row frame and brownstone will focus around the unusual or unexpected layout and accessibility concerns for both of these types of structures. With both buildings originally averaging the same width and length, the concern will come from the interior layout and any unusual alteration. In the row frame, room sizes will vary based on the depth of the building. The most common arrangement of the rooms in this particular building will be the railroad flat, usually with two apartments per floor. On the other side of the street, the *brownstone* can be very unpredictable.

Brownstones

Originally, the basement or first floor area was comprised of the kitchen area in the rear, with a dining room toward the front of the building viewing the street from the sunken court. An interior stairway from this area will lead to the main entrance level, which is comprised of the front and rear parlor rooms with a bathroom and possible bedroom. The upper floor(s) generally consisted of an additional bathroom and bedrooms. Today, due to the high cost and demand for housing, it is difficult to find the layout described. Modern brownstones could contain a variation of basement and dorm apartments housing two and three families, to the possibility of a single family in a four-story building.

It seems to have become very popular to create duplex or multi-floor apartments within the brownstone. Building owners have found that by creating one apartment on the base-

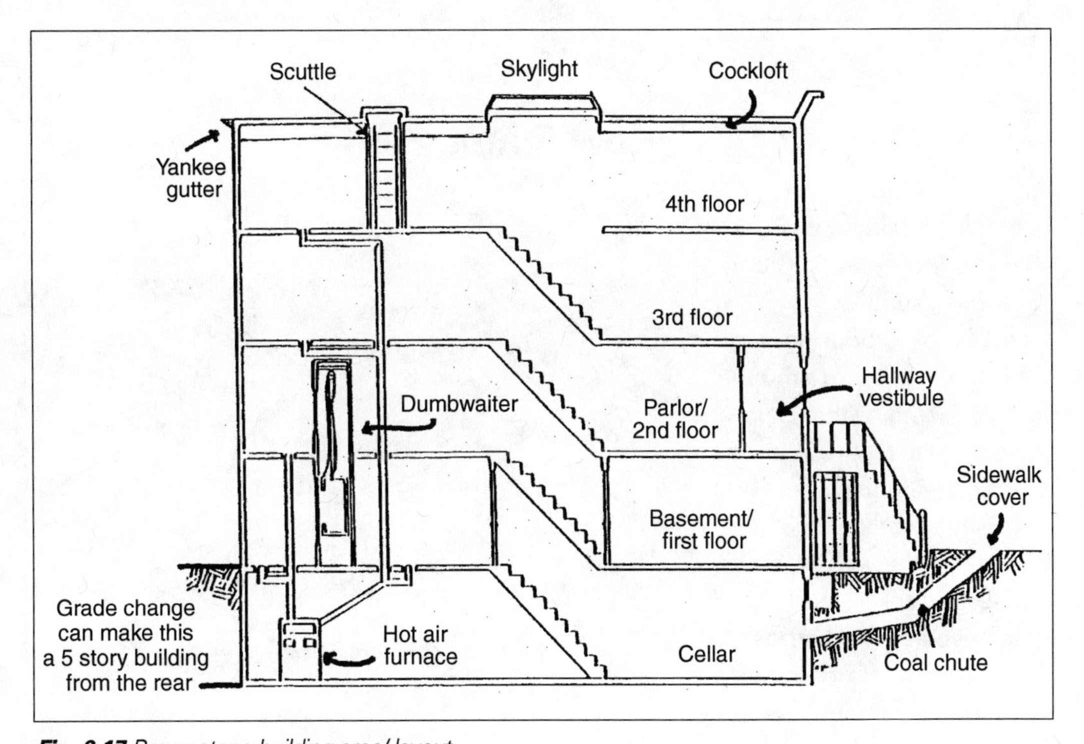

Fig. 6-17 *Brownstone building area/ layout*

ment and lower-level, usually with another more spacious apartment on the upper-levels, translated to good rent money without many of the code restrictions associated with the multiple dwelling. In this type of conversion the main entrance will generally access the upper apartment with the lower apartment accessible from the entrance under the front stoop.

Dormered brownstones can extend the height of the building to four or even five stories. Small, decorative windows built into cornice walls are usually followed by a sloped roof with an additional full story in the rear. Full dormered brownstones with full windows are an obvious indication to an additional floor. Look for additional signs that this area is being used as a living space, most notably curtains, blinds, and the presence of an air conditioner in the windows.

Location and Extent of Fire: Row Frames and Brownstones

*COAL TWAS WEA**L**THS*

The classes of construction and the inherent design voids of the row frame and the brownstone lend themselves to the anticipated means of fire travel from a fire below grade to a fire on an upper floor.

Row Frames

Below grade fires

The row frame is a combustible building that is capable of producing and concealing a large amount of fire. Within the below grade areas of this type of building we must first anticipate the probability of the balloon frame with its built-in ability to spread fire. Unprotected and exposed floor joists in the basement area will not only allow for rapid fire spread to the ver-

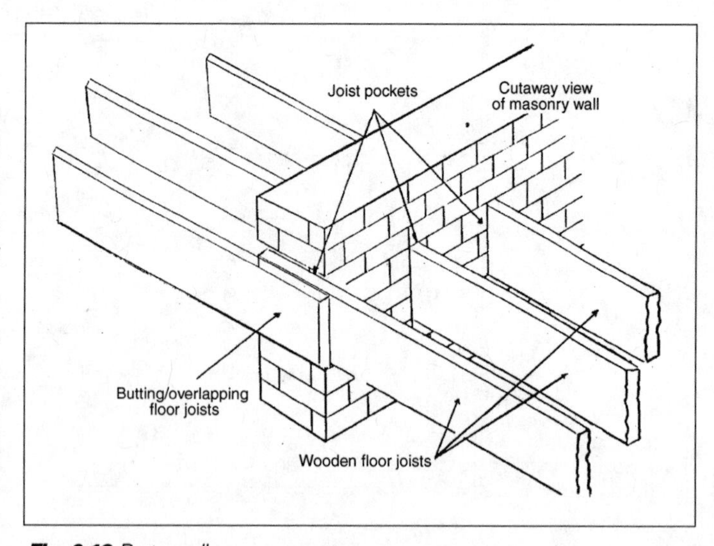

Fig. 6-18 *Party wall*

Fig. 6-18a *Shared joist pockets*

tical wall bays of the building, but they will also be subject to direct exposure of the fire. Within this area there is also the possibility to a common basement or cellar. When first built, this below grade area may have been common to two or more buildings with no separation between them. Although their presence is rare today, most cellars of this original design could contain separation walls of nothing more than studded lumber and clapboard.

In those buildings with separate basements and cellars, extension to adjoining buildings is not eliminated by the presence of a masonry foundation wall. There is the possibility of fire spread through wooden beams resting on a common/party wall. The partition wall, common wall, party wall, or division wall may allow floor joists from one building to butt up to joists from an adjoining building in the same joist pocket. Fire involvement of the wooden structural member on one side may allow fire to transfer to the wooden member on the opposite side.

Lower-level fires

Lower-level fires in the row frame will come with a variety of fire extension concerns that the firefighter and fire officer must be aware of. The unenclosed interior stairway of the building will always be considered the largest and the most devastating. Simply due to the size of the opening and its placement within the building, any extension into or originating in this area is to be considered a priority.

Extension via a light and airshaft is a real probability in the row frame. In most row frames you will find an enclosed air and light shaft with all four sides of the opening covered in either a wooden clapboard or a wooden clapboard covered by asphalt siding. The wood when present by itself will be very old and often untreated, creating a dried-out tinderbox of fuel. If the wood has asphalt shingles nailed over it the concerns become greater due to the combustibility of the covering. In either case, both will produce blowtorch-like conditions exposing the floors above in both the fire building and the attached exposure.

Still within the lower-level of the row frame is the concern from the stacked kitchen and baths of the building. By now their presence and inherent concerns in any multi-family building should be well understood.

Top-floor/cockloft fires

Top-floor/cockloft fires within the row frame seem to cause us the greatest concern. Extremely large cocklofts will provide a lumberyard for the fire to gain considerable strength. Add the concern of this area stretched over an entire city block of buildings, and you'll find your resources quickly depleted.

Fig. 6-19 *Fire spread within the row*

Brownstones

Below grade fires

Below grade areas for the brownstone may also share the same concerns of the common party wall and butting floor joists within a shared joist pocket. Though these areas do pose less of a concern from horizontal fire spread than areas mentioned for the row frame, this possibility will still warrant a check for extension. What fire officers must remember is that the below grade areas of the brownstone are often finished. Usually these shared joist pockets will go unnoticed as well as unexamined due to their covering. When fire involves this area, examination must be made to eliminate the possibility of fire traveling through this space. What becomes useful in any concealed space examination is the use of a thermal imaging camera. When the image shows the possibility, open it up.

As we stated earlier, the below grade area of the brownstone was originally the kitchen area of the home. Today, chances are this will be the first level of a duplex apartment. Stairways from the original basement to the first floor now become the interior stair for the two-story apartment. Being open, they will allow any fire quick and easy access to the floor above, further exposing additional living areas.

Brownstones are also known to have old, hot air furnaces in the basement /cellar area with the unit's floor ducts running throughout the building. If the furnace is still operational, then concern of fire spread from below is obvious. For those that are no longer in operation, the ductwork has most likely been left in place and covered over. Any penetration of the covering and the duct network will allow fire to move throughout the structure.

Lower-level fires

Lower-level location and extent concerns for the brownstone are initially viewed from the area layout of the apartment(s). Those brownstones that have duplex apartments with living space on more than one floor will have the same initial concerns of the open interior stair. In some of the older brownstones it is important to note that depending upon any renovation from the original design, you may also find a second set of stairs that is accessible from this lower-level area. This stair when originally constructed served as a servant stair. It was constructed in the rear of the building and only accessed the kitchen area to a bedroom on the floor above. In many buildings this small, remote stair may still exist. Any fire on a lower floor could quickly move to the floor above, trapping occupants and unsuspecting firefighters. Their presence, as in the Queen Anne, must be anticipated and found.

Just as yesteryear brought the use of a servant stair in the rear of the basement/first floor area, we must also be aware of the possibility of a dumbwaiter that serviced the same floor areas. The dumbwaiter was originally designed to serve the kitchen and parlor areas of the residence. The opening to the shaft has been in many cases, sealed over. However this void, and its ability to spread fire, will still exist. If the covering for the dumbwaiter shaft is compromised, it will provide a very large void for fire to spread.

Other structural features that in many cases will still be in use, most notably from their aesthetic appearance are the *pocket door* and the door and window *transom*. In earlier chapters we discussed the concerns of the pocket door and the fire's ability to spread in this area. The pocket in which the door slides allows a void in the wall that will expose structural members both horizontally as well as vertically. Fire, if given the chance, will find its way into this area without too much effort.

Transoms are those small windows just above a door or window opening. Their original intent was to allow light and possibly air into a room. When present many homeowners restore their appearance to add to the home's decor. In most cases, all that is preventing fire from penetrating the opening is a single pane of glass. Those that were designed to allow airflow will be hinged, permitting smoke to penetrate the opening sooner.

Fig. 6-20 *Transom*

Top-floor/cockloft fires

The threat of fire spread from the top floor and cockloft of the brownstone to its exposures is of much less of a concern than that of the row frame. Brownstones can be, and often are, attached in a row for an entire city block. As we have stated, what allows us the decreased concerns of fire spread is the division wall that may penetrate up through the roof deck separating the cockloft areas. This is not to say that these walls may not be compromised. The fact is we must always check attached exposures; the brownstone just allows us to breathe a little easier.

Time: Row Frames and Brownstones

*COAL TWAS WEAL**T**HS*

As we reference the time of day and its effect on the row frame and brownstone, we are reminded of the same concerns of the multiple dwelling. With any residential structure your concerns with the life hazard will be 24 hours a day with more emphasis during the late night or early morning hours. With a fast spreading fire associated with the row frame, usually occupants in the exposure buildings are unaware of fire next door until the fire is upon them.

Height: Row Frames and Brownstones

*COAL TWAS WEALT**HS***

The height concerns of the row frame and brownstone are going to be limited to buildings of two to possibly five stories. These heights are well within the reach of a fire department's complement of ladders. The use of fire department ladders for gaining access to the roof of the building is a viable option when accessibility is not a problem and they are not needed for the

Fig. 6-21 *Ladder use may be limited*

removal of building occupants. When fire department ladders are not immediately available, firefighters must use other available options in their attempt to gain access to the roof.

In the row frame and the brownstone, the scuttle ladder of an attached exposure becomes an early consideration route to the roof of the fire building. The scuttle ladder and its approach path, however, do not come without some significant concerns that need to be mentioned to the responding firefighters.

Row Frames

Scuttle ladders

First is the location of the scuttle ladder. In the row frame they will be found at the top of the interior stairway. It is important to note that this is the area within the building that is often a collection point for the building occupant's junk. You can find anything from bicycles, mattresses, to old furniture cluttering this area. Members must exercise care when going to work in these areas to avoid entanglement with these obstructions.

The next concern for the scuttle ladder is its use. Specifically, which one do we use? For the row frame, the fire's location and extent will be the deciding factor. If the fire is located on the top floor, gaining access to the roof via the scuttle ladder of the attached exposure is not an option. With the inherent building characteristics of the row frame allowing for rapid horizontal spread in this area, an unsuspecting firefighter could be caught in the scuttle/top floor area from an advancing fire when attempting to force open the hatch. The hatches associated with any scuttle opening are often nailed shut or tarred over often causing a delay on opening the cover. It is a recommended procedure not to use the scuttle approach of the attached building when the fire is on the top floor. Put some distance between you and the fire. Move down two or three buildings to gain access to the roof areas of the row frame if you choose the scuttle approach.

Brownstones

Scuttle ladders

In the brownstone, the scuttle openings can be found in a closet or hallway. Depending upon the amount of renovation, they may have been removed all together.

The approach via the scuttle area of the brownstone is less of a concern due to the division wall that penetrates the cockloft of the buildings; however, the concern still exists. As a reminder, the division wall may have been compromised by decay or faulty workmanship. In either case, members should always exercise caution when using this area due to its uncertainty.

Special Considerations: Row Frames and Brownstones

*COAL TWAS WEALTH**S***

As in all chapters we continue to save this section for those special thoughts, strategies, or concerns that can aid the incident commander's decision-making.

I enjoy this section because it puts the entire chapter into perspective, and it becomes a checklist of strategic and tactical thoughts to consider.

Row frame considerations

1. Has the fire entered the cockloft? If so, to what extent? Any time fire has possession of the top floor and is extending into the building's cockloft, assign resources to the top floors of both exposures as soon as possible.
2. Are there any shafts in between buildings? If yes, and fire is threatening this area, you again need to assign resources into exposures.
3. What are the fire conditions in the rear? Combustible siding will not only enhance the exterior fire spread to the building's upper floors, but also to attached exposures and exposures across rear yards.
4. What are the life hazard concerns in the rear? Building occupants may be forced to rear windows and rear fire escapes. Note: with brownstones, fire escapes may be non-existent.
5. Rear yard accessibility. How can we get resources back there? This may involve moving people and equipment through exposure buildings, rear yards, or from the end of the block. Depending upon resource availability, give early consideration to assigning an engine and ladder company to the rear of the building.
6. Incident management of a row frame - Break the incident down into geographical areas as soon as practical. This will involve the interior of the fire building, the roof, both exposures and the rear of the affected building(s). Assign a firefighter or fire officer in charge of each area with a complement of resources to handle the tasks.

Brownstone considerations

1. How many apartments are there in the building? Renovations can turn an originally designed one family home into a two- or three-unit occupancy.
2. Are there any additional floors to the property in the rear? Dormered areas may be a clue to the presence of an additional story in the rear of the building. Members assigned to the roof and rear should be able to help you here.

3. Has fire entered the dumbwaiter shaft, heating ducts, or light and airshafts? Thermal imagining can help here.
4. If fire has entered the top floor and cockloft, check the integrity of the cockloft of both exposures, and check for possible fire extension through common cornices on the front of the building as well as butting or shared Yankee gutters in the rear.
5. How are basements/first floors referenced? This is one is up to you.
6. Rear yard accessibility? Again, how do we get resources back there?

Questions For Discussion

1. Identify and review the concerns associated with brick nogging.

2. What are the concerns about the rear of attached buildings?

3. What is a return wall? How and where can it be used by a fire department?

4. What are some possible visual indications that fire is within a void space?

5. Review the references of the first floor, basement, cellar, and subcellar of the brownstone. How do the listed references differ, if at all, from references used within your own department?

6. List, identify, and then describe the fire extension concerns of the row frame.

7. List and describe the fire extension concerns of the brownstone.

8. Upon arrival of the first due companies or chief officer, what observations may indicate that fire has extended or is threatening a light and air shaft between the buildings?

9. What concerns do you have with party wall construction within the row frame and brownstone?

10. What is a transom, and how can it affect fire spread within a building?

7 Churches

Firefighters will respond to churches that will come in all shapes, sizes, and ages. They can be a 200-year-old, gothic-style, heavy timber building or a modern contemporary building, composed of all lightweight building materials. For the purpose of this chapter, we will be focusing our attention on the older masonry gothic-style churches, and the size-up concerns associated with this particular style of church.

Fires involving large, old-style gothic churches will be some of the most difficult, resource intensive, and challenging fires anyone will ever respond to. When a fire involves one of these large structures, there is nothing easy about them. From the age of the building, their size and construction, to the large unreachable void spaces, fire can gain possession of the entire building in a short period of time. The important point to remember when going to work at these fires is knowing what you can and cannot do in the little time you have to work. Limitations come quickly and it is important for a fire officer to recognize them.

Construction: Churches

COAL TWAS WEALTHS

Gothic-style churches derive their name from the style of architecture developed in Western Europe, anywhere between the twelfth and sixteenth centuries. This style of church appeared in the United States during the nineteenth and twentieth centuries, and is still a well-represented symbol in many towns and cities throughout the country today.

These structures, which will often cover an entire city block, consist of masonry exterior walls with large wooden interior timberwork, all covered by a steeply pitched slate roof. Buildings of this style, although considered one-story in height, can have roof peaks reaching 60′ or more, with steeple heights 150′ or more.

The construction class associated with the gothic-style church is Class 4 heavy timber. Large wooden interior timberwork will make up a significant portion of the building's skeleton. Large columns and girders of wood will support the floor and roof spans, while the timbers used to support the roof system can be erected in a number of truss designs. The triangular design and the scissor-type truss are some of the most common designs used to support the peak area over the church nave.

Fig. 7-1 *Gothic style church*

Hanging ceiling space

The space that makes up the roof support members is often referred to as the hanging ceiling space, the attic, or the cockloft. This area between the nave/auditorium ceiling and underside of the roof deck can be quite sizable. Heights can range from 12′ to 18′, and will often cover the entire church auditorium. If accessing this space is required, it will be extremely limited. Often the only entrance to access this area will be through a small trap door that is only reachable by a narrow staircase, or access ladder. Furthering the concern of this space, flooring in the attic space will be for the most part nonexistent. Anyone venturing into this space will only find a catwalk of wooden planks to traverse the area. Stepping to either side of the wood planking could cause a person to fall through the plaster ceiling of the auditorium. If that isn't enough, the attic space will lack any suitable lighting. Lighting, when present, is gen-

Fig. 7-2 *Hanging ceiling space*

erally limited to a 100-watt bulb or two, if you're lucky. Any smoke within the space greatly hinders what is already a difficult place to see in.

Any attempt of a firefighter to enter the hanging ceiling space will come with great caution, and should be only considered for small incipient fires at best. Fires of any magnitude are too dangerous to be fought from this area.

Vaulted ceilings

Vaulted plaster ceilings will make up the underside of the roof space. These heavy ornamental ceilings could extend 50' or more above the church auditorium floor. Supporting columns for the ceiling are constructed from large wooden timbers or cast iron columns. It is important to note with cast iron columns within this design, that their appearance may not be immediately visible from within the church auditorium. Often they will be framed out, and then covered in ornamental plaster. If within the building, their presence will be most likely seen in the undecorated areas of the basement or cellar.

The main concern for the firefighter with the heavy ornamental plaster ceiling of the church auditorium is the probability of large sections failing and dropping to the floor as fire attacks the attic space. Sections weighing a hundred pounds or more can drop into the church auditorium, bringing large lighting fixtures down with them. Anyone struck by sections of the ceiling will be seriously injured or killed.

Roofs

Roofs associated with the gothic style church will most often be slate tile over wood planking. These roofs will often date back as the building's original roof. Slate is a material that is virtually unaffected by its exposure to the weather. The design of the roof, with pitches of 45° and greater, will allow these roofs to last a lifetime. Often the biggest structural concern from a slate tile roof will come from any loose and dislodged tiles during fire department operations.

Wood, wood, and more wood

Wood, the chief combustible concern with these types of structures, is not limited to the columns, roof deck, and supporting system. Interior walls will often be constructed of a heavy plaster over a wood lath. Walls within the church can be hollowed out anywhere from 16" to 20" to accommodate heating ducts for what will most likely be an antiquated heating system. These large void spaces are highways of fuel that will allow fire to move up to the attic space, as well as possibly drop down into the basement with little effort. In addition to the wood lath behind the walls, interiors of the church will contain wood trim, wood wainscoting on the walls, wooden pews, wood benches and balconies, as well as wooden choir and organ lofts.

Floors

Floors in a gothic-style church can be constructed of marble, terrazzo, or wood, supported by wood or cast iron columns over a large, open basement. Depending upon the floor material, fire conditions below the floor surface can be masked.

Windows

Gothic-style churches are also known for their large oversized windows. These windows, in most cases, will contain ornamental stained glass. The glass, in addition to being very expensive, is often irreplaceable. This fact has often caused reluctance to break the windows in order to ventilate the building. These windows, because of their value, may also have wire mesh or a metal grate to deflect an occasional rock or bottle as well as eliminate any perching birds. *Any* obstruction covering the window will affect your ability to ventilate the opening.

Thermal imaging

As we begin to venture deeper into the difficulties associated with these structures, you will quickly agree that thermal imaging will be the chief officer's best friend when attempting to commit forces to the interior. Large void spaces, high ceilings, and marble floor surfaces will not give firefighters the heat indications they need to make early assessments on the fire's location and extent. Thermal imaging cameras deployed from multiple points by educated operators can give vital information to the decision makers about a complicated structure of this size.

Occupancy: Churches

*C**O**AL TWAS WEALTHS*

Load and status

At first glance, occupancy concerns for churches will focus around the concept of a place of worship, where the building is only occupied at certain times of the day, week, and year. Further investigation of this association will often prove otherwise. Although we can expect the gathering of parishioners on Saturdays, Sundays, and holidays, we must also be aware of other uses, as well as those people who may be occupying the church and any other attached buildings.

Churches today take on many additional roles other than Sunday service. From day care centers, homeless shelters, banquet halls, to food kitchens, these additional occupancy concerns can take place any day of the week, at any time of the day or evening. Their location within the building and the number of people associated with the function will be your primary concern. This is obviously where your preplanning of the building proves to be invaluable. Children, the elderly, and parishioners can be anywhere within the building. From back rooms behind the altar, below grade kitchen and basements, to adjoining annex buildings or attached schools, the occupancy load can overwhelm an unsuspecting first due company. Obtaining information about the events scheduled for the upcoming year, the presence of a day care facility, or the number of classrooms and student enrollment for Sunday school are just some of the pieces of vital information you need to know and plan for well before the receipt of an alarm.

Residents

The occupancy concerns of your local church will not be limited to those who visit or who occupy the building on a periodic basis. Most congregations require that the minister of the church have living quarters on the premises for himself and his family. This may be a separate residential property on church grounds, a building attached to the church itself, or a couple rooms located somewhere within the church proper. This can again surprise the unexpected firefighters when arriving at an incident at 2am.

The added occupant concerns could extend even further. Depending upon the size of the congregation, there may also be a nun's convent on the property or within the church itself. Their location can be an attached structure, a separate structure, or as we have also noted, somewhere within the church itself.

You cannot wait to gather the information associated with this concern from your on-scene size-up. What was initially viewed by many as an occasionally occupied building can become a building that has a residential occupant load 24 hours a day, 365 days a year.

Fig. 7-3 *Attached residence and office*

Content

Content concerns for the church will vary slightly depending upon some of the additional usages associated with the property. Those that house a kitchen will have added content concerns from the gases and liquids associated with the operation. Depending upon the size of the grounds and the number of buildings associated with the property, there may be rooms, buildings, sheds, or garages that may house anything from paints and thinners to gasoline and propane.

High value contents

Additional content concerns that will be found in the church proper, will be those items of irreplaceable value. Such items could include scrolls, robes, chalices, and statues. These items, when possible, do warrant considerable attention. When conditions allow, and resources are available, consideration should be given to removing irreplaceable items. This action however, must come with significant thought. We all know that items of this type are extremely valuable, and of great concern to parishioners. But committing a complement of firefighters into a well-involved building to retrieve an object is a no-win situation. It comes down to the risk versus gain consideration. With the inherent dangers associated from an advanced fire in a church, the risk far outweighs the gain.

Apparatus and Staffing: Churches

COAL TWAS WEALTHS

As one might imagine, a fire in a church, especially one that is large and old, will tax the resources of even the largest departments. Fires of any significant size will require multiple alarms, bringing large numbers of firefighters and equipment. If there is any opportunity to control a fire in a gothic-style church, it will obviously be in the early stages of the fire. Responding and attempting to control a fire with a skeleton assignment of people and equipment will help contribute to the loss of the building. Fire officers must anticipate heavy fire loads, large area buildings, as well as numerous concealed spaces and voids with extreme difficulty in accessibility. All of the mentioned concerns will support a fast moving fire. Underestimating the speed and size of fire these buildings can produce will prove to be disastrous. I think all would agree, that the call for help has to go out early if you want to afford yourself any opportunity to make an effective and safe effort on controlling a fire in this type of structure.

Offensive considerations

Committing resources to an offensive attack on a church will come with great thought and concern from the fireground commander. Only with a significant number of firefighters to mount the attack while the fire is in its early stages can an attempt be made in controlling the fire. The key to this strategic decision on whether to launch an offensive or defensive attack is your ability in obtaining immediate information on the fire's location and extent. In large area buildings with high ceilings and numerous concealed spaces and voids, this can be extremely difficult. As difficult as this may seem, today's technology in the fire service can provide us with the information. Thermal imaging can give information on the fire's location and extent that will greatly aid the incident commander in decision-making. Being able to scan vital areas of the building, interpret the image, and then inform the incident commander of the observations, will help determine whether the chief officer reinforces the initial attack, or withdraws forces from the building.

Fig. 7-4 *Determining fire involvement of the building is not only critical for effective extinguishment, it is also critical for the safety of the firefighters who may be operating inside.*

The key to the use of thermal imaging is that it is deployed early, and used by trained operators. If you're going to rely on the individual(s) assigned to the camera(s) to aid you in your decision-making, then the information has to be accurate. As a good friend and fire officer from the City of Atlanta once told me, "The information obtained by thermal imaging will only be as good as the firefighter's level of training with the camera."

Thermal imaging deployed early in this type of building may not only give you information on the fire's location and extent, but it can also help prevent serious injury or death to firefighters attempting to go to work inside the building.

Engine Company Operations

Hoseline selection, stretch, and placement

When the decision is made to commit arriving engine companies to an offensive attack, actions must focus on establishing a sufficient water supply, with the stretching of large hoselines to the seat of the fire. The requirements for a successful outcome for a fire in a gothic-style church will be the speed, accessibility, and amount of water that can be delivered to the fire. Hoselines of adequate size, preferably 2½" in diameter, must be given early consideration when arriving at a church fire. The amount of fuel this type of building can produce, the sizes of the void spaces, the reach, penetration, and volume required not only to slow the fire spread, but to extinguish it, will require the capabilities of a large hoseline.

There is no doubt that the movement of a large hoseline will require an increased number of people. If we took the resources that we were going to assign to the two smaller hoselines and teamed them up to stretch one large hoseline capable of delivering a more effective stream, the ability to not only slow the fire spread, but extinguish it, would greatly improve.

Based on the above concept, upon recognition of a difficult hose stretch, it must become the responsibility of the engine company officer to request additional help to stretch the first hoseline. With this request in mind, the chief officer must assume that the stretching of the backup hoseline (as well as any additional hoselines) will require the same, if not more; so plan ahead.

Having enough people and equipment there or already on the way in will cut down on the lead/reflex time associated with this or any other task. Until the rest of the troops arrive pool your resources to deliver the biggest punch on what is known to be a very difficult building.

Ladder Company Operations

Apparatus placement

Ladder company operations in churches will also focus around the fireground commander's decision of an offensive or defensive attack. In either position, the tasks will be difficult. A consideration that must come early in the operation is the placement and position of the apparatus. Company officers must direct apparatus drivers/chauffeurs to place apparatus with both the offensive and defensive modes in mind. Keeping the front of the building open for the placement of the truck company will become an initial consideration.

The front of a gothic-style church will offer openings for ventilation and stream placement. Gothic style churches will contain an oval shaped window in the front gable, which is referred to as the *rose window*. This decorative window allows the colored light of the stained glass to shine

Fig. 7-5 *Vent and Rose windows*

into the church auditorium. At a fire incident, this window can be penetrated to vent the church proper, it can allow access for a strong stream from a tower ladder into the church auditorium, or it can be used by a tower ladder to penetrate the ornamental ceiling in an attempt to get water into the attic space.

Another window that can be found within this same space that is much smaller, and often higher into the gable is the *vent window*. This window, or louvered opening, allows direct access into the attic space over the church auditorium. The vent window will give the ladder company another opportunity at venting or directing a stream into the attic space. This opening, when available, is a much smaller window than the rose window. Its small size will definitely limit the maneuverability of an elevated stream. Nevertheless, it presence will warrant early access.

Ventilation

Ventilation options for a fire in a church will come with limitations, as well as with some hesitations. Stained glass windows as mentioned earlier, are expensive, and often difficult to replace. The hesitation on breaking them and providing a large opening for ventilation has resulted in the loss of entire buildings. Consideration must not only be given to removing them, but also coordinating their removal with respect to the fire's location and potential spread.

Windows of this type can be removed one at a time to see if conditions improve, before *all the glass* is taken out with no added effect. Even when orders are given to remove a stained glass window, efforts may be delayed by wire mesh, metal grates, or lexan coverings that were installed in an attempt to protect the window.

Fig. 7-6 *Window with protective covering*

Ventilation efforts by a ladder company to open the roof of a gothic-style church are going to be impossible to extremely difficult, as well as dangerous. Architecture during this period was not only decorative and ornamental, but it also commanded a great presence by constructing buildings of great height. With this design, towering steeples and roof heights of six, seven, and eight stories designed with steep pitches and valleys is what the fire department will be forced to deal with. A roof pitch of 45° or more is a common site for a gothic-style church. Not only from the height, but also more importantly from the extreme pitch of the roof deck, ventilation is only to be attempted from the protection of an elevating platform.

Natural roof ventilation options for the ladder company may come from roof turbines, louvers, or scuttle hatch openings that run along the roof deck. Depending upon their design within the roof space, they may only provide ventilation into the church auditorium and not the attic space. Any natural openings that are placed high on the roof deck, close to the ridge, or directly on top of the peak, will most likely vent the attic space. The only way to be sure of what your options are, and what they will directly affect, is to get out and take a look well before an incident occurs. Information associated with this concern should be part of the incident preplan. Knowing what you can and cannot do prior to the actual incident is critical for an effective outcome.

Fig. 7-6a *Ventilation options for attic space*

Fig. 7-6b *Ventilation options for attic space*

It is without a doubt that the two main reasons fires cannot be controlled in large, old churches are our inability to ventilate the roof, and our inability to quickly open concealed spaces and voids. When conditions allow, great efforts must be made to create openings ahead of the advancing fire. Whether it is from the removal of stained glass windows, or the aggressive opening of ornamental plaster walls and wainscoting, our objective is to save the building. To ensure that these options are pursued during the early stages of the fire, it may be necessary to put someone in charge of these tasks to not only ensure multiple and coordinated efforts in this area, but to prevent any injuries to firefighters as they go to work. In some departments researched, a vent coordinator or a vent group leader is used to direct the tasks in this area. This individual's direction, in coordination with an officer directing the interior operations of the building, may make the difference.

Salvage

As mentioned earlier, there may be expensive, as well as irreplaceable, items within the church that must be given early attention. When conditions and resources allow, information from a pre-plan or on-scene parishioner can be used to locate and remove items from the building. It is important to know where these items are before you commit anyone to the building.

Life Hazard: Churches

COAL TWAS WEALTHS

Firefighters

Interior operations

Life hazard concerns at churches for the firefighter will be numerous. The greatest concerns to the firefighter will occur when members attempt to conduct interior operations. Advancing into these buildings with a hoseline or to conduct a search will prove to be the most dangerous activities for firefighters who attempt to go to work in the interior. When an offensive mode is in place, firefighters must obtain immediate information on the fire involvement of the void spaces. Fires involving these spaces will burn intensely as they travel throughout and make their way up to the roof space. Those that don't will produce conditions favorable for a backdraft. Whether in the free-burning or smoldering phase, fire involvement of the concealed spaces and voids can create fireball conditions when exposed to the open air of the church auditorium. Not only have these conditions forced firefighters to scramble in an attempt to find an exit, they have also brought down large sections of the ornamental ceiling, lighting fixtures, duct work, and structural roof members that can all trap and kill a firefighter. Any attempt at putting people to work inside a church must be accompanied by a thermal imaging camera(s). Information gathered from a number of images about the fire's location and extent can save firefighters' lives.

It should be a rule that if the upward extension of fire cannot be controlled, or if indications and observations show that fire is gaining control of the hanging ceiling space, a defensive attack is necessary.

Exterior operations

Exterior operations associated with church fires will also come with significant life hazard concerns for the firefighter. As we have already mentioned, gothic-style churches are known for their architectural style and design. Slate tile roofs that are commonly found on the gothic-style church were not only installed to add to the aesthetics of the structure's design, but they were really the only durable roofing material during that period. Slate tiles are large and heavy. Individual tiles can vary in size and in weight. Their installation is similar to that of a conventional roof with each tile laid side-by-side starting at the roofs edge, with overlapping rows up to the roof's ridge. When originally installed, tiles were nailed to the roof through pre-drilled holes with copper nails. Installation today follows the same design, but with galvanized nails.

Fig. 7-7 *Large caliber streams will dislodge slate tiles.*

What occurs with an older slate tile roof, is that the tile will most likely have loosened at its nailed points from the continued expansion and contraction of the roof deck. This primarily occurs from the large roof surfaces associated with this type of building, and the inferior construction methods used at that time to ventilate the roof or attic space. As the wood deck attempts to breathe, nails can become loosened. As the deck contracts, the loosened tiles might not retighten or reset themselves. With a tile no longer held rigidly in place, any additional movement could allow a tile to unseat and drop, seriously injuring a firefighter. The concerns with slate tiles are further realized from fire department operations. Tiles can be swept by a high caliber stream and dislodged. The force of the water can throw tiles extended distances and can severely injury a firefighter.

Fig. 7-7a *Slate tiles that have fallen from the building.*

Another concern that must be considered during the early stages of the incident is the large sections of glass windowpanes and their method of removal for ventilation. These large panes of stained glass are often lead trimmed. Individual sections can weigh up to a hundred pounds or more depending upon their size. Large sections of this glass can severely injure a firefighter if care is not taken when they are removed. Large windows must be removed by breaking them at the top and then working down. Breaking out a portion of the window near the bottom will allow heavier jagged pieces of glass to rain down onto members. Even from extended distances this will cause a severe threat. At a recent multiple alarm fire in a church, the tip of an aerial ladder was used to remove a large rose window on the front of the building. The operator placed the tip of the ladder through the window at the bottom of the window. As the window broke a large section of glass landed on the aerial rails and rode one of them down the beam forcing the operator on the turntable to literally jump out of its way.

Collapse

As the incident grows in size and intensity, fireground commanders will be forced to operate in defensive modes. The life hazard concerns for the firefighter must not lessen in this mode; they must change in perspective. Within this mode will come extended operations with established collapse zones. Zones must be established to include the full height of the building, including any steeples or bell towers on the exposed side. This consideration must also include any attached or closely spaced buildings. Aggressive stands to hold fire from traveling into a building annex or attached residence will come with great concern as fire and the earth's gravity attempt to drop the church building or any portion of it to the ground.

Firefighters operating inside one these buildings can be severely injured or killed if a section of the church crashes through the roof of the exposed building. This concern has to be added to buildings nearby or across the street. In congested suburban and urban areas, steeple heights may force the fireground commander to evacuate civilians and firefighters from buildings that may have initially seemed remote from the main body of fire. With steeple heights reaching 140' to 150' or more and bells that can weigh 500 pounds or more each, collapse concerns must include the building's height in its entirety on the exposed side.

Fig. 7-8 *Defensive operations require well-established collapse zones.*

Occupants

Congregation

Life hazard occupant concerns for a church will notably be a concern when the congregation is present. Firefighters must be aware of the possibility of large numbers of people during these times, all attempting to flee the building through the same means they have entered—the front door. The large numbers of parishioners attempting to leave at the same time may cause a panic. Firefighters cannot add to this panic by blocking or obstructing exits with hoselines and equipment. Assisting with the orderly removal of occupants and the seeking of alternate access points for hose, tools, and equipment may be some of the initial concerns for arriving forces.

Social events

Church social events are an additional occupant life hazard concern to responding firefighters. Events usually in church basements, school auditoriums, or building annexes may house large numbers of people. They, like the parishioners, will attempt to flee through the same means they entered. The reason for this is obviously based on their limited familiarity with the building. Those attempting to flee by an alternate means may find locked or blocked exits, further compounding the fire department's efforts at entry and search.

Residents

As we mentioned earlier in our occupancy concerns additional groups of people, priests, ministers, rabbis and their families, the possibility of nuns' quarters, a building caretaker, and even an attached school will require the responding fire department assignment to ensure that all areas are searched, and that all occupants are accounted for. These people may be numerous as in a full time church/school, or may be smaller as in a full-time residence. In either case, incident commanders will have to assign resources to search all affected areas.

Your occupant life hazard concerns must extend further to those churches that house the homeless. It seems that many churches will house and feed the homeless with no notification of their activities to the local fire department. People who may be occupying a church at any given hour of the day can number well into the double digits. This is crucial information I think we would all like to know.

Terrain: Churches

COAL TWAS WEALTHS

Setbacks

Churches will come with limited concerns in your terrain size-up. Most notably we find some problems with building setbacks. Depending upon the property size and their location within the city or town, a church may have a considerable setback from the sidewalk or accessible area limiting elevated use from an aerial or platform. Depending upon the setback as well as any added obstructions such as trees, overhead wires, and parking lots, this could further affect placement of your aerial equipment.

As you can imagine, building setbacks will also affect the engine company's hoseline selection and stretch. Officers must anticipate longer stretches and develop a plan to ensure adequate flows.

Setback information and how to deal with it, should be another piece of information placed within pre-incident action plan. Information concerning the difficulties will be easily identified from your first visit to the complex. Identifying where you can, and cannot place and set up your apparatus is going to be extremely useful information when you go to work at that specific address.

Water Supply: Churches

COAL TWAS WEALTHS

Availability

Water availability is yet another piece of information that must be sought out during your pre-incident gathering. Seeking a hydrant to provide necessary flows at the time of the incident is too late. In any pre-incident plan, water supply availability is a must.

Delivery

With accurate information on water availability, next will come the concept of delivering it to the fireground with adequate volume. Large diameter hose with the positive pumping concept is just one way of delivering large volumes of water to the fire building. In a number of previous chapters that address large buildings, we have introduced the concept of operating and supply pumpers to deliver sufficient water volume. As you can gather, I'm a real big fan of their use.

This concept uses an engine company, usually the first due company, to act as an operating or attack pumper seeking a position nearest the building for the stretching of their hoselines. Another engine company, usually the second due company, acts as a supply pumper, dropping a large diameter supply hoseline to the operating pumper as they seek a water supply. The concept of placing a pumper on the hydrant to push water to the fire scene is referred to as positive pumping. This concept allows the delivery of large volumes of water to the scene where it is needed. If necessary, additional engine companies on the assignment, as well as those coming in with multiple alarms, can pair up using this concept in an attempt to deliver large volumes of water to the fireground.

The key to a fire department's water supply efforts will be its ability to deliver water on the fire. With an offensive attack, this will come by way of cautious, yet aggressive, placement of handlines to the seat of the fire. Depending upon the fire's location, options may be limited in directly stretching to the seat of the fire. Fires in basements or concealed spaces may require the use of a distributor or cellar pipe to hold the fire from advancing. Piercing nozzles can be used to get quick water into a void space, as the walls and ceilings are further opened.

Decisions to go defensive with large water applications will generally follow after fire has a significant hold on the concealed spaces of the church. With the known difficulty in getting effective water into these spaces, as well as the general inability to create an effective roof opening, fireground commanders must be prepared to go defensive as soon as information or observations dictate its need. Knowing when to withdraw your people from the building is critical.

Defensive delivery of water will come in the form of big water in any manner available. From engine company mounted stang guns, to elevated streams from tower ladders and ladder pipes, the application of water into concealed spaces and onto exposures is crucial. Master streams should use smooth bore nozzles to provide the best reach, penetration, and extinguishing capabilities of the streams put into place. Anything less will most likely turn to steam.

When decisions require defensive operations, sector/division officers must plan and provide for extended operations with big flows.

Fig. 7-9 *Big fires…*

Fig. 7-9a *…require big water.*

Auxiliary Appliances and Aides: Churches

COAL TWAS WEALTHS

Detection and suppression equipment

The absence of automatic alarms and fire suppression equipment is a major contributor to the large fire loss in churches. Many churches, primarily those of the gothic style, were built before codes were written, and (depending upon jurisdiction) are not required to conform to existing codes. Often the only time fire officers may find alarm or suppression equipment in any area of a gothic style church is when the church has undergone some type of renovation. Even with renovations, the alarm and suppression requirements will be based on the amount and type of work performed.

Other instances where an alarm or suppression system is present on a limited basis may be in a basement or storage area. Knowing where these systems are, and the location of a possible fire department connection is crucial to the first few minutes of the operation. Statistically, the basement and heating room, as well as those areas used for storage, are where most fires will start in a church. If they are protected, the local fire department will have a fighting chance.

Newer buildings associated with the church complex, primarily those of residences, schools, and building annexes will most likely contain an alarm and suppression system. When present, fire department personnel, namely those in the first due response district should know the type of alarm or suppression system and how to augment it. This will obviously prove to be valuable for any incident involving these properties, as well as any attempt to protect these buildings from the main fire. When a fire from an attached or nearby church is threatening the above-mentioned buildings, the threatened exposure should have their auxiliary appliances augmented. If the building has a sprinkler system, supplying the fire department connection to augment the systems supply may help with limiting any significant fire extension into the building.

Street Conditions: Churches

COAL TWAS WEALTHS

Street size-up concerns will be specific to each building's address. As we mentioned in the apparatus and staffing section of this chapter, it is important for companies to get out and seek key areas for spotting and placing their apparatus. Churches and the surrounding properties will come with their own unique obstacles and access concerns. Depending upon the street width and traffic flow surrounding the church, fire officers should review the placement possibilities during the street's most difficult times, as well those times when the street and surrounding areas are free of parked cars. Seeking apparatus placement areas on a Saturday morning might be quite different from the next morning when the streets and the parking lot are full with vehicles for the Sunday service.

Weather: Churches

*COAL TWAS **W**EALTHS*

Weather and its related categories will plague any incident commander with additional fireground considerations, regardless of the type of building. However, we need to again focus our thoughts for a fire in a church, notably a very large one.

Wind

The wind and its effect on a fire in a church will greatly increase what is a already a difficult situation. We know that winds as little as 10 mph will significantly affect the fire's intensity within a building, as well as cause a significant concern to any exposures downwind. The fire's intensity within a building will be directly affected by the wind's velocity, the size of the openings, the building's compartmentalization, and the buildings fire load.

As you can imagine with this association, gothic-style churches are known for their large, wide-open spaces, which could be present in the building in many forms. From the church nave or auditorium to the attic loft space, once they become influenced by the wind, fire conditions can be quickly accelerated to blowtorch conditions in a short period of time.

The large open spaces we often associate with the church auditorium or nave may be as small as 50'x 100', to as large as 100'x 200' or more. Cathedral ceiling heights can be expected to reach at least 50' . These large open spaces and high ceiling heights of the church auditorium can create tremendous drafts that will intensify a fire once it begins or enters the church auditorium.

In many of these areas, the failure or removal of large stained glass windows and the building's rose window or vent window, has to be factored with the wind's speed and direction. The premature ventilation of an area could promote fire growth and spread, not only involving additional portions of the building, but quickly overwhelming forces operating opposite the openings.

With the heavy fire load in the building from the large wooden interior timberwork, additional concerns from firebrands must be considered. High winds combined with a fire involving a gothic style church will create a tremendous firebrand problem. Add the possibility of the church's steeple becoming involved, and the hazard of flying brands will spread and land over a larger area. When feasible, and when available, incident commanders must give consideration to requesting a brand patrol downwind of the incident. Establishing a minimum of an engine and ladder company to patrol areas downwind, to observe the affected area from building rooftops, or from the height of an aerial ladder, may help to eliminate any additional building fires.

Exposures: Churches

*COAL TWAS W**E**ALTHS*

Radiant heat

Exterior exposure concerns that surround a gothic-style church will again vary, based on the specific address of each incident. What will become quickly obvious to anyone who operates a well-involved church fire is the tremendous amount of radiant heat that can be produced from these types of buildings.

With gothic-style churches being of Class 4 heavy timber construction, the large wooden interior timber work in addition to being old and dried out, will produce tremendous fire conditions once ignited. If additional buildings within the complex are attached or nearby, fireground commanders must anticipate early fire spread and plan to deal with it. Structures within as little as 50′, will become early exposure threats requiring resources to protect them from the radiant heat. What must be noted with this thought is that the exposure threat seems to be most dangerous to any higher, nearby structures. With the roof area often being the first area within the building to fail, the opening created will allow a more concentrated and intense source of radiant heat to expose any higher buildings nearby.

Attached buildings

Buildings that are attached will pose the earliest threats for fire involvement. Whether from radiant heat from the main fire area, or fire spread through openings between both structures, these buildings will require resources to protect them from fire extension. All surrounding areas will require an early commitment of resources all applying

***Fig. 7-10** Exposure D (right side of photo) only sustained minor damage thanks to aggressive protection efforts.*

water to vital areas. Water application requirements can be as obvious as the protection of exposure roofs and walls of nearby buildings, to the stretching and operating of a hoseline in a tunnel or passageway from the church proper to an attached building. On more than one occasion fireground commanders have been surprised by a underground tunnel or passageway that extended from the church proper to a rectory or church residence under the street. Here again, having pre-incident information on all possible avenues of spread may help eliminate fire spread to an adjoining, or nearby buildings.

Area: Churches

*COAL TWAS WE**A**LTHS*

Square footage in a church and its surrounding properties will come with a number of concerns to responding and operating firefighters. The concerns will not be limited to the large open spaces and their potential involvement, but firefighters also must be concerned about any maze-like areas in and around the building where companies may have to operate.

For the most part, we view churches as large open areas. We gather this thought from our past experiences in these buildings, which generally come from attending church services. Unless you have been actively involved in the church, or have been there from a previous incident, these buildings and their surrounding properties could present areas that would be difficult to find and then get out of.

As examples, there may be small and narrow stairs behind a closed door in the front vestibule that may lead to a choir and organ loft. Behind the altar, there also may be rooms used by church members for meetings. Locked storage areas for sacred documents and materials, as well as a kitchen, bath, and study for the church clergy may be found in the rear of the church. Basement areas, although often large open areas, may contain a stage area for church performances. It's in the basement again where you may find areas that contain numerous rooms for storage, classrooms for Sunday school, as well as a kitchen for church functions. The point that should be gathered here is, although we note the difficulties with the large open spaces of a building of this size, depending upon the fire's location and extent and the difficulties that surround it, the small and difficult to reach areas will often present us with more.

Location and Extent of Fire: Churches

*COAL TWAS WEA**L**THS*

Determining a fire's location and extent is information that all fireground commanders seek in order to deploy and assign resources. This holds true for all buildings that we arrive and go to work in, but what makes this information gathering process difficult in churches, especially older churches, is the numerous places where fire can start and travel through. In an attempt to determine the fire's location, we must look back statistically at where a majority of fires start in a church.

Ignition areas

For a time, during the late 1990s, arson became one of the major causes of fires in churches. The number of incidents that occurred during this time were motivated by individuals for one reason or another. Thankfully, it seems that in recent years, the number of these types of fires has since declined.

Barring the incendiary causes, most church fires can be contributed to defective furnaces, inferior electrical service, poor housekeeping, carelessness, and the overall age of the utilities and their components.

When we look at the furnace or boiler room of a church of this size and age, a number of problems will arise. The first being, the poor housekeeping. Many times due to carelessness and complacency, the furnace room becomes a collection point for the storage of combustibles, paints, varnishes, etc. The problem obviously starts with the use of a furnace room as a storage room. It is only a matter of time before the room's contents come in close contact with a heat source.

A second concern is the integrity of the furnace or heating room to prevent any amount of fire from spreading. Fires that start in a furnace room can spread quickly to other parts of the structure. Initially, the fire can spread to other areas of the basement area through flimsy or non-fire stopped partition walls. More often than not, all you will find is a framed out partition wall covered by wooden clapboards or gypsum wallboard. Partition walls associated with these rooms will only extend up to the underside of the unfinished floor joists. With an unfinished ceiling, a partition wall installed at a right angle to the floor joists allows large horizontal openings for fire to spread. With the high probability of an unfinished ceiling in this area of the basement, it also has to be expected that fire will have the opportunity to spread from the furnace room up through openings in and around the heating ducts. As we have mentioned, walls and ceilings throughout the church can be hollowed out as much as 16" to 20" in order to accommodate the old heating ducts in these buildings. These areas offer expanded raceways for fire to extend throughout the building. The openings will extend from the basement up into the sidewalls of the church auditorium, on up into the hanging ceiling space over the entire church.

With the concern of the heating system ductwork and its concealed space, its age and installation has also been a contributing cause of many fires over the years. The forced hot air systems used in the old churches can become very hot trying to keep the building warm and comfortable during the cold weather months. When a hot duct has been in contact with a wood structural member over the years, or if there is a defect in the duct itself, the hot air leaking or coming in contact with the wood will break the fibers of the wood down. This charcoal effect or *pyrolysis* of the wood lowers the ignition temperature of the wood until the next time the heat is turned on, the wooden member ignites. This same problem can occur in the heating systems chimney. A defective or inferior flue with loose or missing mortar has also been the cause of a number of fires in churches. Their installations in most cases will be over 100 years old.

Inferior electrical service is another major fire cause in churches. Many of the electrical services originally installed in churches are still in place and functioning today. The wiring network in a church, and the arteries within a building it may travel are extensive. Any alterations and additions made to the original installation or any attempt to plug in equipment with a high electrical demand may overload an already old system. Just the fact that the system is old, is a factor that could result in a failure of a component of the electrical system that in turn could cause a fire.

Below grade areas

Within a church of this size, there will be no easy area to deal with an uncontrolled fire. Below grade areas with their many possible points of ignition previously discussed, large accumulation of combustibles, and numerous avenues for fire spread, are going to create difficult (often impossible) conditions to work in. Often all you will find is a narrow staircase accessible from the interior of the building. Ventilation of this area will be limited, if available at all. If the building is built on a grade there may be a rear entrance, if not, possibly a few small windows around the perimeter of the building. The hoseline stretch will be difficult due to the concerns previously mentioned. Direct water application to the main body of fire could be time consuming, and resource intensive. Anything less than a well-coordinated movement of water and ventilation will not only waste valuable resources, but could also affect the safety of the members operating.

Fig. 7-11 *Large wooden interiors*

Church auditorium

Fire involvement of the church proper is definitely easier to access when compared to other areas within a church. However, this area will also come with its own challenges and concerns as it relates to fire extension. Most notably is the combustibility of this area. Most of the interior of the church will be made of wood. Columns, walls, benches, balconies, and lofts will be as old as the church itself. With these mentioned areas, comes the exterior finishes. Over the years, countless applications of varnish and polyurethane will have been applied to every possible square inch of exposed wood. The aesthetic look the finish may give will be a major factor in the rapid growth of a fire. Heated surfaces containing these materials will contribute to a much hotter and faster spreading fire. Consider the draft potentials of these large open areas with this factor, and the rate of spread is faster than one can even attempt to predict.

The church auditorium does not come without its own confined and difficult to reach spaces. Choir and organ lofts are notorious spaces for fire to start and travel from. Also utilities that service these areas and the numerous and inaccessible void spaces that leave these spaces are critical areas. If you don't control the upward spread of fire in the wall spaces with the first deployed hoseline, it's on its way up to the hanging ceiling area.

Hanging ceilings

The hanging ceiling space over a church is the most difficult and most inaccessible space in the building. Access to this area is generally limited to a narrow staircase at best, to a wooden ladder leading to a trap door. Fire can originate in this area or enter this space from numerous areas, most notably the large and numerous wall voids around the structure. With the large wooden interior timberwork of the ceiling and roof support system, accompanied with heights in these spaces averaging 12′ to 18′ over the entire length of the building, chief officers must be prepared for defensive operations once fire enters this space.

Time: Churches

*COAL TWAS WEAL**T**HS*

Time of the day/week and year

The fire officer's concern with time will focus around the universal thoughts of the time of day, time of the year and the day of the week. As a general consideration, a church's occupancy load will generally be at its highest when church services are in session. This is normally on Saturday evenings and Sunday mornings. With services, there is also the probability of Sunday school. As we have mentioned before, this will bring an untold number of children to an area in the church basement, or church annex.

In church basements, access and egress will be extremely limited. Often all you will find is one-way in and one-way out. If a church basement has a secondary means of egress from it, it will generally be a remote area behind a stage or through a kitchen, and the teacher and children are probably unaware of it. When is the last time a fire drill was conducted for Sunday schools in your district or city?

The day of the week also allows us to focus on other church events that generally occur on specific days. Weddings most often occur on Friday evenings, Saturdays, or Sunday afternoons. Holiday or special religious events and their dates will vary based on the day of the week and the time of the year. These events will bring large numbers of people to the church and its surrounding properties.

Holiday times also bring additional concerns with the higher than normal fire loads. The Christmas holiday, being the most obvious, brings an abundance of Christmas decorations throughout the church auditorium. Garland-laced entranceways, large Christmas trees, and a larger concentration of lit candles is what you could expect.

Incident time

A positive factor from the viewpoint of time is that most church fires occur during the unoccupied times, most notably during the late night or early morning hours. The negative factor from the late night or early morning time frame is that there is often a significant delay from the time the fire first starts until the fire department is actually notified. This is associated with the building being unoccupied during these times, and minimal (if any) alarm detection in the building.

Height: Churches

*COAL TWAS WEALT**HS***

In the beginning of this chapter we mentioned that a church roof peak can reach 60′ or more, with steeple heights known to reach up to 150′ or more. These high, often inaccessible, areas cause great concern for any members operating on or near the building. Attempting to gain access to a roof peak for any ventilation options is a limited and risky task. As we have mentioned, those roofs that contain roof or ridge vents should only be accessed by reach of an aerial device. Attempting to gain height to the steeply pitched and remote areas of a gothic-style roof by any other means is extremely risky, and in most cases not an option.

Collapse considerations

Steeple heights present another concern for those members operating nearby. Not only does their design and construction (which is similar to a giant chimney flue) enhance their fire growth, but the total height of the steeple must be considered in the collapse zone when operations go defensive. In many neighborhoods across the country, the church is the largest and tallest building in town. These buildings will often be the focal point for the town village or Main Street area. With this in mind, there will usually be many smaller, and sometimes closely spaced buildings. When church fire operations go defensive, fireground commanders must consider and establish collapse zones around the entire building. With the church auditorium reaching the height of a six or eight-story building, and steeple heights equaling the height of a 15-story building, all surrounding areas must be evacuated to encompass the collapse zone for that affected height.

Anticipating the direction and path of a steeple collapse is often an unpredictable event. Often the best course of action when collapse of a church steeple is anticipated is to establish collapse zones for its full height in all directions. To aid in determining the path of a collapse, the setting up of a surveyor's transit on a fixed point on the steeple may help to indicate a potential collapse area. In Jersey City at all defensive operations, the safety officer and his crew will setup a surveyor's transit on a corner or fixed point of a building to detect any early movement not easily noticed by the naked eye. A movement of as little as an inch can be detected and relayed to the incident commander and division or sector officers. Even with companies operating in a flanking position on a well-involved church, information that the church steeple is starting to lean their way will warrant repositioning.

Fig. 7-12 *Steeples can reach to great heights.*

Fig. 7-13 *Transits can detect early movement.*

Special Considerations: Churches

*COAL TWAS WEALTH**S***

In this section, we will again review specific thoughts, ideas, or concerns that will help any fire officer guide his or her thinking. Barring specific information as it relates to an address, the following is a list of considerations the fire officer should review when going to work at a church fire.

Churches

1. *Fire location and extent.* Gather information about the location and extent of the fire from the use of thermal imaging cameras. Because of the numerous concealed spaces for fire to travel undetected, obtaining early information about the fire's location and extent will not only help with fire extinguishment, it may prevent a serious injury or death.

   In addition, attempt a perimeter view of the involved building and any buildings threatened. Information gathered from information in the rear, may be totally different from the information gathered from the front of the building. This reconnaissance of the exterior will not only lend information about fire and smoke conditions, but also information about building layout, any exposure concerns, and accessibility to the building involved as well as those threatened.

   If observation by a thermal imaging camera indicates fire has possession of the hanging ceiling space, or if exterior observations show visible fire and/or smoke from roof ventilators, pull companies out of the building, conduct an accountability roll call, and go defensive!
2. *Ventilation.* When conditions and information allow us to operate in an offensive position, efforts should focus around establishing a vent ahead of, and when possible, above the fire. This as we know is a difficult task. When personnel allows, fireground commanders should assign the coordination of this task to one individual as the vent coordinator or a vent group leader. Allowing the fire to vent and to keep it moving in one direction are key concerns for limiting fire spread.
3. *Salvage consideration.* An area you might want to add, or make a specific note of in your pre-plan information, is the location of church valuables. These items can be given consideration if you know where they are.
4. *Defensive operations.* When conditions do not allow us to go offensive, or the situation forces us to change to a defensive position, fireground commanders must be prepared for large water supplies and elevated streams, all from safe distances. Exterior operations must take into consideration the protection of exposures, the probability of structural collapse, as well as the possibility of a secondary collapse

onto smaller structures. A strong incident management system with established division/sector officers ensuring that priority areas are covered, members are accounted for, and all precautions are taken to ensure the safest possible operation, must be strictly enforced.

5. *Collapse.* When available, safety officers should set up a surveyor's transit(s) to detect the slightest movement of any portion of the building. This tool can be a great aid in anticipating collapse direction.

The occurrence of fires in churches is not an everyday incident. The infrequency of working in these types of buildings may invite us to be complacent, and ignore some of the historic difficulties when advancing in to find the fire. These structures present the firefighter and fire officer with unique challenges that must be calculated with every step of our operations. Ignoring any one of them may lead to injury and death of a firefighter.

Fireground commanders must set realistic goals. It is your experience, but more so your education with these buildings that will make the difference.

Questions For Discussion

1. Identify the possible occupancy load and status concerns within a church.

2. Identify the location and purpose of a rose and vent window within a church and how their presence may affect your decision-making.

3. List and describe all your ventilation options at a church.

4. What are the two main reasons fires cannot be controlled in large, old churches?

5. What is considered a critical salvage concern at a church? What factors must be considered before firefighters can be assigned to this task?

6. List the occupant life hazard concerns that may be presented in a church.

7. List the firefighter life hazard concerns that may be presented in a church.

8. Where do most fires start in churches?

9. List and describe the exposure concerns presented by a well-involved church.

10. What will be the most difficult and inaccessible space to reach in a church?

8 Factories, Lofts, and Warehouses

What some consider as one type of building in this chapter may actually be a number of different types of structures with different types of construction, all associated with an untold number of occupancies.

Factories

Factory is a word that represents a building used for the manufacturing of a product. Buildings associated with this process can vary in size and type. A factory building can be either a large, 10-story, heavy timber building that is more than 100 years old, or a modern one-story noncombustible or fire resistive building within an industrial complex.

Fig. 8-1 *Factory*

Lofts

Loft buildings or tenant factories are non-fireproof structures built more than 150 years ago. They average four to six stories, and may consist of varying shapes and sizes. These buildings usually housed a number of different tenants or manufacturers, all within one building. Loft buildings are still found in large numbers within the mix of many older urban centers throughout the country.

Fig. 8-2 *Loft/tenant factory*

Warehouses

Warehouse is another word that represents a building that is used for the housing and storage of a product or products. These buildings will also vary in size, shape, and construction. You may find construction methods using lightweight materials spanning the size of a number of football fields, or a four- to five-story, heavy timber building that encompasses an entire city block.

There aren't too many cities in the country that don't have these buildings represented in one form or another. Buildings associated with this chapter have produced large losses of life, as well as large content losses throughout the years. Many of these historical fires have changed not only the way buildings are designed and protected, but also the way we fight fire in them.

Fig. 8-3 *Warehouse*

Construction: Factories, Lofts, and Warehouses

COAL TWAS WEALTHS

Newer factories

Factory buildings can come in many shapes, sizes, and classes of construction for us to consider. One of the ways factory buildings can be represented is by the large industrial complex built of Class 2 limited combustible construction. These types of buildings are generally one story in height, and can span areas equal in size to a number of city blocks. Buildings of this type are built with masonry walls, concrete floors, enclosed stairwells, and metal deck roofs supported by open-web steel bar joists.

Older factories

Older factory buildings can still be found throughout many towns and cities today. The type of construction most closely associated with this type building is Class 4 heavy timber. These older structures, which range in height from four to six stories will consist of brick exterior walls, large wooden interior structural members, wooden plank floors, and built-up roof systems. Depending upon their size and square footage, these larger heavy timber buildings may also contain firewalls. These self-supporting masonry walls are designed to act as a fire barrier. When present, their openings must be protected by self-closing fire doors in order to prevent fire from passing through to other areas of the factory.

Regardless of their contents, these buildings are highly combustible. With most being well over 100 years old, fires that are allowed to advance beyond the incipient stage, will consume large sections, if not the entire building in a short period of time.

Fig. 8-4 *Fire doors*

Lofts

Very old factory buildings also made of brick exteriors and wooden joist members may also be referred to as loft buildings or tenant factories. These structures built in the early to mid-1800s will range in height from five to ten stories, with building dimensions averaging 50′ to 100′ in width to 100′ to 200′ feet in length.

Generally, any building of this design will have incorporated into the framework of the structure a center steel girder that spans the length of the building from the front to the rear wall. This center girder acts as a support for the building's floor joist system. Supporting this center girder is a series of columns made of either large dimension lumber or cast iron. Loft buildings are known for their use of cast iron inside and outside the building. Inside, cast iron columns were used to support girders that held up the floor. Cast iron, which has a poor tensile strength, is often reported as having a poor record in fires as well. With this flaw, their failure will be devastating, especially when their use is superimposed within the building. Remember that in this framework design, lower floor columns of cast iron will not only support the floor girder system above, but all the columns on all the floors above as well. Should a column on a lower floor collapse, it will cause an additional loss of support for the columns on all the upper floors. A loss of a cast iron column on a lower floor of one of these buildings could produce a catastrophic collapse, bringing down the entire building.

Fig. 8-5 *Cast iron column supporting a series of wooden girders*

Outside the building, some lofts may also contain a cast iron front facade. These fronts were a popular style used by architects in the nineteenth century to give their buildings added character. Under fire conditions, the cast iron once heated and then cooled by exterior hose streams presents the added possibility of fracturing and pulling away from the front of the building, adding to the collapse hazard.

The cast iron facades are an appealing feature. During the renovation phase of one of these buildings, the original facade can be recreated by fiberglass or plastic, giving the same appearance as cast iron.

Fig. 8-6 *Girder and column integrity is already questionable.*

Fig. 8-6a *Cast iron supported facade*

Older warehouses

Warehouses are another building type that can vary in size and construction. We find that most buildings are built of either Class 4 heavy timber, or Class 2 limited or non-combustible construction. Older warehouses of the Class 4 design will contain exterior brick walls and large, wooden, interior timberwork within an average height of four to six stories. In older structures, steel shutters may be present over existing windows. Their initial intent was to prevent burglaries of building stock. Their presence can be looked on as a plus as they could delay fire from entering through the window, but they may also be looked upon as hindrance when entry and ventilation need to be achieved.

In addition to the structural features mentioned you may also find spreaders or anchors with the Class 4 or Class 3 designed warehouse. A series of steel rods and anchors, often with a decorative washer on the exterior of the building, were installed to spread the load among a number of structural members. Their installation may have been part of the original design, or possibly added later due to question with the structures stability. When viewed internally, their placement will tie an exterior wall to a wooden interior structural member or possibly from one load bearing member to another. When viewed from the exterior, a uniformed presence may very well indicate that their installation was part of the original design of the building. When viewed from the exterior to be sporadic in placement, chances are that additional spreaders were added due to a concern with the building's structural integrity. Regardless, their presence must send up a red flag when fire involves the structural members they are tied to. Failure on either side of the anchoring point or directly to the spreader itself could result in a collapse of a section of the building.

On a personal note, their presence is not limited to the Class 4 or Class 3 warehouse. They can be found within in any Class 3 or Class 4 designed building, notably the Brownstone, Multiple Dwelling, or Factory. So take an extra look!

Fig. 8-6b *Steel rod/anchor decorative washers*

Newer warehouses

Class 2 designed warehouses, while newer and more modern, could be constructed of masonry walls or tilt-up concrete panels covered by open steel bar joists supporting a metal roof deck, all in a building referred to as one story in height. What is critical to note with a warehouse constructed of tilt-up concrete panels is that the panels will be dependent upon the roof for their support. They remind me of the houses we would build out of a deck of cards when I was a kid. They don't stay up very long.

What will be universal to all warehouse buildings regardless of their construction is that they will all have large floor spaces. This is an obvious concern due to the nature of their business—the storage of large amounts of combustible stock.

Occupancy: Factories, Lofts, and Warehouses

COAL TWAS WEALTHS

When viewing the occupancy load for the different buildings mentioned, concerns will vary based on whether it is a factory, loft, or a warehouse.

Factories

Factory buildings can have a significant occupancy load 24 hours a day. Depending upon the type of product or products being manufactured, workers may be present all hours of the day on second and third shifts. It is not uncommon to open a door to a factory building at four in the morning and find dozens of employees all at work. Obviously this is another area where your pre-incident information is valuable. Without any pre-incident specifics, a building with all of its interior lights on, as well as a nearby parking lot that serves the factory full of cars during the early morning and evening hours may be a couple of clues.

Lofts

Loft or tenant factories will have one or more tenants per floor. In the past, there seemed to be a tendency for manufacturers of similar products to congregate in one building, especially those related to the garment industry. Today, fire officers can expect multiple tenants producing different products, all occupying the building 24 hours a day. Unless the fire department keeps close tabs on these types of occupancies, the number of people that occupy these floor areas can be alarming. Overcrowding, poor working conditions, shoddy housekeeping, all combined with a very old building can make for a disaster.

Warehouses

Warehouse occupancy loads will differ greatly from the factory and loft type buildings. Since the building is used primarily for storage, there is generally no need for a high occupancy load. People most associated with the operation of the warehouse will be responsible for the loading and/or removing of stock. Their presence usually follows normal business hours. In the evening hours, often the only occupant is a night watchman.

In the newer and more modern versions of the warehouse, there is the possibility that no one will be present at all. Automated handling and removing systems will stock, remove, load, and control all inventory within the warehouse. These self-sufficient operations obviously eliminate the need for employed humans, since a robot is now doing their job. We were amazed at a recent tour of one of these facilities in the battalion where I'm currently assigned. A building the size of four football fields had no employees in it; it was all automated.

Occupancy conversions

The occupancy load and status concerns for each of these three types of buildings are based on their original or intended design. However, as many of us know, things can change. Any one of the three buildings, but most notably the factory and loft building, can have a portion or its entire interior converted to residential occupancies. When presented with this conversion, your size-up considerations change most notably in the areas of the life hazard, the location and extent of fire, and the strategy and tactics associated with those factors. Many of these buildings will give little indication to their presence from the exterior. Upon entering a door off the interior stair, firefighters could find a floor space divided into a number of apartments. Everything from residential occupancies, an art studio, to a roof top restaurant could reside behind the front door of one of these buildings.

Fig. 8-7 *Warehouse converted to residential units*

Contents

Occupancy content in both the factory and warehouse will come with major concerns to the responding and operating firefighters. Depending upon the occupancy(s) within the building, these concerns can be enormous. From heavy fire loads, water absorbent stock, to hazardous materials, the concerns will seem endless. This is just another example that reinforces the need for some type of pre-incident information. Knowing that you're going into a no-win situation will not only save wasted time and poor deployment of resources—it will also save lives.

Information about the building's content, storage methods, and automated handling systems can come from a number of sources. Periodic inspection done on the property will not only ensure that the building and its operation are done to code, but familiarity with the occupancy and its layout before a fire is extremely valuable information that every firefighter should have. Specific information (once gathered) can be either entered into a department Computer Aid Dispatch (CAD) system or pre-plan book so it can be reviewed as the alarm is transmitted. In this inspection it is also critical to determine the location of any Material Safety Data Sheets (MSDS) in the building. In most jurisdictions, it is required that they be located at or near the main entrance to the building. Depending upon the occupancy and its operation, their review could be a lengthy task. MSDS can number into the hundreds, if not thousands, for just one occupancy. Difficulties and the time associated with their review must be anticipated.

Apparatus and Staffing: Factories, Lofts, and Warehouses

COAL TWAS WEALTHS

A fire in a factory, loft, or warehouse building is going to be resource intensive. With department staffing and equipment limited, many departments must rely on assistance from other surrounding communities to mount any type of effective attack. Larger departments with more staffing and equipment must rely on additional resources from within to accomplish the same objectives. In either case, it's important to recognize the fire potential, and request help early.

Fires in these types of occupancies generally come in two sizes—small and large. Fires that are small on arrival will only stay that way for a short period of time. Efforts by the first arriving companies must focus around a well-executed and coordinated attack if success is going to be achieved.

Engine Company Operations

Hoseline selection, stretch, and placement.

Initial decisions by the engine company officer will focus around what will be the quickest and easiest way to deliver the largest amount of water to the seat of the fire? The response to this question is often not an easy one. It generally starts with the number of people to do the job. Factories and warehouses are large square footage occupancies, often on multiple levels. Their fire load, stock configuration, square footage, and possible life hazard, all accompanied with the fire's location and extent are the concerns that will affect the fire officer and firefighters selection, stretch, and placement of the first hoseline.

Hoseline selection in the factory, loft, or warehouse buildings must be governed by the following:

1. *The building's fire load.* From the class of construction to the amount of combustible stock, you must expect it to be heavy.
2. *BTU generation.* From the association to the above, you will need the quenching capabilities of a large hoseline(s).
3. *The building's square footage.* Large open floor spaces with high ceilings will give a fire the opportunity to take possession of the entire floor area quickly.
4. *Volume, reach, and penetration.* From our association to all of the above, you will need to provide a stream that is capable of reaching, penetrating, and quenching the fire. Medium size streams will not do it.

The above considerations will best be handled by deployment of a 2½" inch hoseline with a smooth bore nozzle. Streams capable of delivering 225gpm to 250gpm from a distance of 50' or more are pre-requisites for a fire that involves these types of buildings. The use of a smaller hoseline, though deployed much faster than the 2½" hose line, will not have the reach, knock down capability, and cooling effect of the 2½" hoseline.

To further aid with this decision, it is also important to remember when considering the use of a larger diameter hoseline that the square footages commonly associated with this size building will greatly affect the friction loss of a smaller deployed hoseline. Even when stretching from a standpipe outlet, especially in a very old building, the age of the system will have piping most likely full of sediment, rust, scales, and possible debris further restricting the hoseline's flow. Even if this weren't enough, it must be further anticipated, especially in the warehouse, that getting sufficient water to the base of the fire can be compounded from narrow maze-like aisles, wall partitions, and varied stock arrangements. When these conditions present themselves, deflecting a heavy caliber stream off the ceiling not only attempts to control the flashover potential, but also allows a large curtain of water to rain down on top of and behind obstructions like a giant sprinkler head slowing the fire's development.

All of the above will require you to use your largest hoseline. When conditions lend themselves for mopping up, or you encounter a much smaller fire than originally planned for, the smooth bore nozzle can quickly be adapted to fit a smaller, and more maneuverable hoseline. So stretch the big line.

Sprinkler/standpipe options

Another consideration of the first due engine company officer when arriving at a fire in a factory/loft or warehouse has to be with the possible presence of a standpipe or sprinkler system, and the need to augment those systems. Sprinkler systems in these types of occupancies may quickly become taxed. The fire loading, the stock arrangement, and the ability for a sufficient amount of water to control the fire will be factors affecting this decision. With sprinkler systems only designed to deliver adequate flows to a percentage of the heads operating, the need to augment the system becomes great.

Standpipe systems in multi-floor and large square footage occupancies will also need to be supplied. Multi-floor buildings or those buildings that are extremely large in square footage will create long and difficult hose stretches. When a dependable system is present, augmenting and using the system will hopefully expedite the delivery of water.

Ladder Company Operations

Salvage

Ladder company operations for a fire in a factory, loft, or warehouse will take on a number of tasks, with additional emphasis in others. Salvage, a critical element of the ladder company functions in all buildings, gets specific attention here. With the large water flow requirements often necessary to extinguish fires in these types of buildings, early consideration has to be given to protecting the stock on the floor(s) below the fire. Depending upon the stock content, floor spaces could hold thousands to millions of dollars of merchandise and equipment. There may even be instances where some of the stock is priceless and irreplaceable. It is not uncommon to find in many of these buildings that the value of the stock far outweighs the value of the building where they are stored.

There is no doubt to the educated fire officer that salvage is a key concern at all building fires we respond to, but with factories, lofts, and warehouses, it becomes a major size-up consideration.

Ventilation

Ventilation responsibilities will focus on the fire's location within the building and the location of any endangered occupants. Roof and interior stair ventilation must still be achieved in an attempt to control the fire's spread, as well as remove smoke from the building's stair shaft. Removal of smoke from the stair shaft through bulkhead doors, skylights, and freight elevator shafts not only attempts to rid the smoke for fleeing occupants, but also aids firefighters in making their advance.

Freight elevators in some buildings can present a significant concern. The concern with freight elevators is from the possibility that they may have open/screened doors within the shaft at each floor level. Failure to achieve any type of vertical ventilation above this opening will allow smoke and heat to mushroom out onto floors above the fire.

Buildings built of Class 2 limited combustible will warrant serious consideration when firefighters are assigned to the roof. Roofs associated with this class of construction will be made up of combustible materials over a metal deck. As described in previous chapters, the involvement of the roofing materials may create a combustible gas fire below the deck, a surface fire on top of the deck, to say nothing of how the unprotected steel supporting the deck will be affected. If there is a serious enough fire below the deck of a Class 2 building that requires you to "create a ventilation hole over the fire," you shouldn't be up there.

Horizontal ventilation in factories and warehouses must be attempted from exterior fire escapes and fire department ladders. Due to the large, open area floor spaces of these buildings, it becomes necessary to create a favorable path for the fire and smoke to move toward. Creating these openings must take into consideration the building's lack of compartmentalization and the speed and direction of the wind. Ventilating on the windward side of the building will compound conditions on the fire floor, and all openings created opposite this side.

Of the three types of buildings listed, warehouses will offer the smallest number of opportunities to establish any type of horizontal ventilation. Many warehouses, due to the nature of their business, will have limited windows (if any) to assist the fire department with smoke removal. As one building manager said to me, "Windows are not necessary in a building that is used for storage." Options for smoke removal will be limited to loading docks, truck bays, and any vertical ventilation that can be established. With the above options, positive pressure ventilation may be another consideration for the incident commander to use.

Positive Pressure Ventilation. PPV is the planned and systematic removal of smoke, heat, and fire gases from a structure. By using a complement of fans or blowers, the natural ventilation process can be dramatically assisted or even replaced by forcing the movement of air through pre-selected and/or controlled openings.

The concept of positive pressure is similar to blowing up a balloon. As a balloon is inflated, the pressure inside the balloon becomes equal at the top, bottom, and sides. When an opening is created, the air is exhausted to the outside. The concept used in smoke removal in a building is the same. By pressurizing a floor, space, or entire building, and then creating an opening opposite the pressurization opening, contaminants will be exhausted to the outside of the structure. As easy and as practical as it may sound, its use must come with great consideration and considerable planning.

Forcible entry

Forcible entry in factories and warehouses has historically been difficult and time consuming. Members must expect heavy swinging doors, roll-down steel doors, screened gates, and steel shutters just to mention a few of the obstacles. Depending upon the fortified design, conventional forcible entry and hydraulic forcible entry tools may not be enough. It's not uncommon in buildings of this type to order a couple of oxyacetylene torches to work.

Fig. 8-8 *Forcible entry difficulties can be numerous.*

On a personal note, firefighters should not hesitate forcing more than the one or two doors needed for initial entry. Due to the square footages, stock arrangements and possible maze configurations presented, there may be a need for firefighters to exit, or a rapid intervention team to enter from a different exit/entrance point than the one or two initially created. Although it may be time consuming and task demanding, chief officers should not hesitate to call and assign additional resources for this task. Once additional exits have been created, advise interior forces of their locations.

Fig. 8-9 *Search rope operations in large area buildings must be well organized.*

Search

Searches in these building will also be very demanding and time consuming. Use of search ropes with timed operations is a must to ensure members return with sufficient air in their cylinders. Entering large square footage buildings without a procedure for establishing direction, pattern and distance, let alone operating time, is extremely dangerous for firefighters.

Search rope operations should include a main search rope that is knotted at specific intervals to aid the firefighter with distance traveled as they proceed in or exit out. Additional personal ropes should also be included with this operation so firefighters can search off the main rope to enhance the distance and area to be covered. With a search of this type underway, a firefighter or fire officer should remain outside the immediate search area to maintain radio contact with the search team advising them of their time in the area, and the time they need to return to ensure they can get out before their air supply runs out.

Life Hazard: Factories, Lofts, and Warehouses

COAL TWAS WEALTHS

Firefighters

Life hazard concerns to the firefighter are significant. Statistically more multiple firefighter deaths will occur in commercial occupancies, than when compared to any other type building we respond to. The concerns have been well documented over the years, but unfortunately they still seem to reoccur.

Disorientation

The most common concern encountered is disorientation. Large buildings, some with open spaces, others with narrow aisles, maze-like configurations, and highly piled stock that might fall on a firefighter or behind him blocking his exit, are just some of the many problems we face when going to work inside a factory or warehouse. In the apparatus and staffing section of this chapter, I mentioned the need to ensure that members searching in a building of questionable size or configuration do so under the guidance of a search rope. To assist, they should also use a thermal imaging camera. These two pieces of equipment are a must in any building or area of questionable size and configuration. Too many times before, we have all heard about firefighters entering a building with no guidance equipment at all. In others, firefighters have entered a building with no search equipment under the assumption of what appeared to be a small fire, only to become lost as conditions deteriorated later.

These are big buildings. More often than not, they will not show their true fire conditions until well after we arrive and commit ourselves. They can trap even the most seasoned firefighters into believing "it doesn't look like much". Use of a search rope and thermal imagining camera is a must, regardless of what is showing. They need to be deployed together. The last thing you want is for firefighters to become overconfident with a camera's use, and then find themselves out in the middle of the abyss with no reference home once a camera fails. Trained procedures with rope usage, accompanied with thermal imagining, are two excellent tools to use when advancing into the unknown.

Fig. 8-10 *Disorientation is a major concern in large area buildings.*

In these types of buildings members should also be reminded of how to determine direction when lost and then encountering a hoseline. Referred to as deployed hoseline escape, the procedure has been taught for years on how to exit a building if you come across a hoseline. By finding a coupling and determining the difference between the male and female ends of the hose can give direction to a firefighter attempting to exit a building. With the male end always pointing toward the fire, firefighters can crawl along the hose away from the male end to exit the building.

Fire spread

Heavy fire loads and the rapid fire growth associated with factory/loft and warehouse building is another life hazard concern for the firefighter. There should be no doubt in anyone's mind about the fire load of a factory and warehouse and the conditions they can produce. From large amounts of combustible stock, to flammable liquids, greases, and gases, the fire will burn intensely when given the opportunity. In the past, firefighters have been overwhelmed by unsuspecting fire growth in these types of buildings.

Fig. 8-11 *Conditions upon arrival may dictate your actions.*

High ceiling occupancies

One factor that contributes to the above concerns is the unusual height of the roof deck above the floor space. Factory and warehouse buildings are characteristic of this concern and must be referred to as high ceiling occupancies. In buildings that have this type of association, conditions at the ceiling level can be producing roll over conditions, yet not be observed or felt at the floor level. In many instances firefighters have been able to walk into a building with a light smoke condition at eye level, only to be caught in a violent flashover over the entire floor space minutes later. With ceilings heights in some of these buildings at levels 30′ and higher, conditions above may be drastically different than conditions below. Any building that fits this description must be entered with caution, regardless of what is showing at the floor level.

There are a few clues that can aid you in identification of the *difficulties* in high ceiling occupancies.

High ceiling occupancy clues

1. *Visually scan the ceiling area.* Any glimpses of yellow and orange dancing through the blackness is a reliable indicator of rollover.
2. *Listen to the radio.* If you hear conflicting radio reports from the interior of the building when compared to the roof of the building, something is wrong. An example would be, "Roof Division to Command, all natural vent openings removed, with heavy smoke and fire starting to show through the openings." Compared to "Interior Division to Command, we have a light smoke condition, still unable to find the fire, we are continuing in."
3. *Thermal imagining.* Observations from a thermal imaging camera pointed up toward the ceiling/roof level will indicate conditions well above you.

Building size, stock, and arrangement

The building's size as well as the stock arrangement and amount, are additional considerations that will also add to the life hazard concerns of the firefighter. The building's *size,* specifically the fire floor, will affect the engine company's time and ability to get water to the seat of the fire. The longer it takes to get water directly on the seat of the fire, the more opportunity fire has to travel.

The stock arrangement, whether it's placed horizontal or vertical, and its method for packing can either enhance fire growth, or retard it. Items that are stacked vertically will enhance the rate of burning by aligning the fuel with the convection column. Stock packing, and the ability for air to penetrate its arrangement can also enhance or retard fire development.

Stocking considerations can be furthered by observing how high the packing is stacked and if the arrangement can be knocked down by a hose stream or swell after it has been absorbed by water. Anyone of these mentioned concerns could possibly fall on firefighters or block their egress.

Collapse

The building features and building loads are another consideration. Firefighters must be aware of any heavy machinery, large tanks, safes, water tanks, hoppers, bins, and air conditioning units that may fall through burned out or weakened areas onto unsuspecting firefighters. Firefighters must also exercise caution when climbing on or working under loading dock canopies. Their design is intended to shield workers and stock from the weather as deliveries and pickups are made at a loading dock. With any fire exiting a loading dock opening, their integrity must be questioned.

All observations must continue, not only during the extinguishment phase, but also more importantly during the overhaul phase when our sense of danger becomes somewhat lessened.

Fig. 8-12 *Exterior loads must be noted.*

Fig. 8-12a *Be observant to any roof mounted equipment.*

Occupants

Occupant life hazard concerns will be more significant in the factory and loft building when compared to the warehouse building. We say this simply because of the numbers of people associated with the different buildings. As we initially mentioned in the occupancy section of the chapter, warehouse buildings will have a much smaller workforce associated with them as compared to the factory and loft building. Although there can be a limited number of people within a warehouse at any given time, the factory buildings with its workers will number much higher, often at all hours of the day and night.

Even though the occupant death rate from fire remains low for all three types of buildings, in years past there have been specific incidents that have resulted in major losses of life. The incidents of the past have led to radical changes in the protection of the factory and warehouse worker, but many of the concerns of the past still seem to plague fire officers today.

People and the building

The initial concern in these buildings that should immediately come to the mind of the responding fire officer is the number of people in the building and their ability to get out. Often factories are overcrowded with employees. A lack of exits within the building will create difficult and possibly panic conditions when everyone attempts to get out at the same time. In older factory/loft buildings there may only be one interior exit via an interior stair, and only one exterior exit via a fire escape in many cases. Depending upon the occupancy load and the square footage of the floor area, this number of exits is not enough.

During an inspection of a factory or warehouse, firefighters may find blocked exits, locked exits, inadequate access to an exit, unmarked exits, and poor housekeeping, just to name a few of the difficulties. An unfamiliar worker may not know if an unmarked door is a door to an office, bathroom, storeroom, or staircase. Routine visits to these buildings and our responsibility to correct these concerns will hopefully eliminate the lethal problems.

Terrain: Factories, Lofts, and Warehouses

COAL TWAS WEALTHS

Setbacks and accessibility

Concerns in this category, as in most, will be address specific. Buildings that are set back from the street or accessible side will be an obvious concern for forces attempting to go to work there. We find that in many factory and warehouse buildings, a significant number will be accessible from more than one side. Those accessible sides being the front, or street side, a possible parking lot or employee's entrance side, and sides that serve as a truck/cargo loading area.

Factories or warehouses that are served by a railroad will become an added concern. Fire department units and their members that are assigned and able to access this side of the building must be protected from any rail activity. Requests to the railroad to shut down train traffic must be made early to train dispatchers. It is also important to note with this consideration that more than one railroad could share the same or multiple tracks that are in the area of your operations. Notifying only one railroad to shut down train traffic may not be enough. Preplan information has to be specific here.

Depending upon the area or jurisdiction where the building is located, many older factories may have been built on waterways. Waterways once (and maybe still do) serve as a method of providing shipments to and from a building. In a number of older factory buildings, the waterway may have also served as method of powering or cooling a part of their production and processing. In either case, the ability for the fire department to gain access to this side, as well as to operate from this area, could be extremely limited. Those waterways that are accessible by marine traffic, will warrant the requesting and operation of a marine fire company for the possible use of fireboat streams, or for the supplying of water to land based companies. Those areas where a fireboat or marine based operation cannot take place, the fire will take possession of in a short period of time.

***Fig. 8-13** Buildings with minimal accessibility will require you to create additional access points.*

Water Supply: Factories, Lofts, and Warehouses

COAL T<u>W</u>AS WEALTHS

Availability

When first designed, water supply systems and the number of hydrants in a factory, mill, or warehouse district were planned to provide large water supplies with a sufficient number of hydrants that could be accessible by the fire department. This generally seems to be the situation in most cases. However, firefighters and fire officers must remember that these are the areas of your city that can be the oldest. Inactivity in the area, as well as the age of the system may have created debris, rust, scales, and sediment in the water system that could restrict what was originally considered sufficient flows.

Well before the incident occurs, water availability from the hydrant type and location, their accessibility, and flows they can deliver, must be determined. Complexes served by yard hydrants or private hydrants must be identified and distinguished from additional source locations. Yard hydrants will often be served by the same system that serves the building's auxiliary systems. Placing engine companies on yard hydrants may rob essential water from a building's sprinkler system.

Delivery

During your review of water availability, larger water mains must be found. By referencing water maps or simply having the size of the main painted on the hydrant's barrel will aid with this information request. Once identified, officers must know the gpm capabilities of their engine companies and place those within the water system that will provide the best results. Larger gpm pumpers must be placed at water sources in order to deliver the maximum volume allowed. The use of multiple or large diameter hose from source or supply pumpers to operating pumpers, is necessary in order to deliver large volumes of water to the seat of the fire. Everything we must consider and do in this category must focus around the concept of establishing and delivering large volumes of water from multiple sources.

Fig. 8-14 *Establish large water supplies.*

Delivery of water to the seat of the fire will initially be determined by a number of factors. Conditions allowing fire officers to initiate an offensive attack must be fought with 2½" hose with smooth bore nozzles. Nozzle tips of a 1¼" or 1½" must be accompanied with this size hose in order for it to provide any volume, reach, and penetration to the seat of the fire. Hoselines capable of reaching 50′ to 75′ and delivering 250gpm is a requirement for an offensive attack at a fire in a factory/loft or warehouse.

Exterior operations

When the first arriving engine company officer is confronted with a significant volume of fire on arrival, initiating exterior operations can take on varying forms that may or may not prove positive results. One such method that is considered and used by some is referred to as a *blitz* attack. This is a procedure that uses the availability of the on board tank water of an arriving engine company through an apparatus based deluge set or stang gun. This procedure of dumping a large volume of water through the windows of an involved area of a factory or warehouse has successfully darkened down a number of the fires in the past, ultimately allowing the stretching of interior hoselines to extinguish any remaining fire. This procedure also in the past, has produced adverse effects from uncalculated or haphazard use of its operation.

With the deployment of any hose stream, you must realize that air will be entrained into the stream. From our education and use of the different types of streams, we know that the wider the pattern, the more movement of the air. Solid streams, even with their tight continuity will entrain some air in and around the stream. This factor must be considered in the possible use, and the period of time a blitz attack operation is to be used.

Blitz attack use considerations

1. It cannot be used if occupants are in the building or immediate area. Use of this operation may push fire, smoke, and heat onto escaping occupants.
2. It cannot be used if firefighters are in the building or immediate area. However, interior and exterior officers can coordinate its use. If members operating on the inside can retreat to an enclosed stairway or behind a firewall, an exterior operation can be used to darken down a specific area.
3. It should only be used for a limited period of time. Continued, or sustained use of this operation may push the fire to uninvolved areas. Once the fire has darkened down, exterior operations should be shut down. Extended use of an exterior stream will push fire that was not reachable.

Street-based operations

With any street-based operation, use of exterior streams will be effectively limited to the second or third floor. Stream reach and penetration will not only be limited by their distance, but also from their accessible angle. With this operation, officers should give consideration to placing their apparatus on the opposite side of the street to provide a better angle into the floor area. Any effective use of elevated streams should come from ladder pipes, or streams from a tower ladder (or some type of elevating platform). Streams from a tower ladder are very versatile, providing hundreds of gallons of water at pinpoint precision. When conditions dictate the use of exterior streams above the second floor, chief officers should make every attempt to leave room for equipment that provides the best means of delivering the water to the seat of the fire.

Nearby buildings

When exterior operations from fire apparatus cannot obtain a vantage point to place their streams into operation, officers should consider use of nearby buildings. Operating streams onto the fire building from nearby buildings not only provides multiple vantage points to operate into building, but also provides exposure protection to the building from which they are operating. Depending upon the nearby building and its systems, this procedure may require supplying the building's standpipe system, with firefighters bringing standpipe hose and equipment to the designated floor areas. Buildings without an auxiliary system will require extensive efforts in stretching and placing hoselines at designated levels. However, the efforts would be beneficial.

Auxiliary Appliances and Aides: Factories, Lofts, and Warehouses

COAL TWAS WEALTHS

Detection Equipment

Detection devices, when present, will vary based on occupancy, the areas they are designed to protect, and their location within the building. Use and location of smoke detectors, heat detectors, and infrared detectors will vary. With a detection system we would hope to find a fire control panel or an alarm enunciator panel in these large-area buildings indicating not only the type of detection, but also its location. When an alarm company monitors the system, specific information is often available indicating the type of alarm and the source.

Suppression Equipment

Sprinkler systems

Suppression equipment in a factory or warehouse building is going to be the firefighter's best friend. Sprinkler systems in these types of buildings must be supplied by the fire department. Even if the building has a wet system, the building's sprinkler system can become quickly overtaxed from the fire load and its arrangement. Tall stockpiles and poor housekeeping practices will reduce the effectiveness of a discharging sprinkler head. Note that sprinkler systems are designed only to handle a specific percentage of involvement. In many instances they can become over taxed when their designed spray cannot control the fire. In situations where the systems fail to control a fire, it may be associated with a change in a building's fire load, and possibly a change in the arrangement and storage of the stock within. It is for these reasons that the fire department must augment the system to provide an adequate flow in an attempt to hold the fire, while hoselines are stretched and put into operation to finish the job.

Fig. 8-15 *The locations of the FDC and control valves are a important part of your pre-incident information.*

Standpipe systems

The building's standpipe system will require the same considerations mentioned relating to augmenting a sprinkler system. Large square footage, and multi-floor factory/warehouse buildings will require the augmenting and use of the building's standpipe system. Unless the fire is on the first floor or just inside the building's front door, use of a standpipe system becomes critical. Eliminating the needless stretching of hose and the cutting of the water flow lead-time is an essential step toward an effective outcome. With the use of this building system the same concerns about providing adequate volume to extinguish a fire remain. Wet standpipe systems in these types of occupancies cannot be relied upon to provide larger than normal flows. Heavy fire loads, stock arrangement, and the necessary reach and penetration requirements of a deployed hoseline in these occupancies, are reasons to supply the fire department connection and stretch your biggest hoseline.

Sprinkler vs. standpipe

When both a sprinkler and standpipe system are present, officers must augment both systems. Care must be given to ensure the supplying of each system. With this statement, follows the often-asked question of, "which one do you supply first?" What may help with this decision is the location of the fire. If the fire is on the upper floor of the building, supply the sprinkler system first. By the time the engine company reaches the location and connects into the standpipe outlet on the floor below, valuable time may have been wasted supplying the standpipe that could have been better used augmenting the building's sprinkler system.

Assistants/aides

When responding and going to work at a fire in a factory or warehouse, an incident commander may have the opportunity to use some on scene assistants/aides to assist in his/her efforts. People or aides who can assist the fireground commander could include a plant manger or plant engineer. People associated with this title will have a better understanding of the different

Fig. 8-15a *Seek assistance from on-site personnel.*

products and hazards within the building. There may also be a fire brigade, hazardous material response team, or a first aid team on the premises. These teams are made up of employees within the factory that have been trained to handle incidents as first responders.

In many buildings, management will seek out employees with outside interests as a volunteer firefighter, first aid squad member, or a person with knowledge of hazardous materials within the complex and give them a monetary incentive to join an on-site response team. This responsibility comes as an added duty to their normal work assignment where it allows management a chance to control a situation as soon as it happens. Their ability to aid in a situation should not be overlooked. Often these people will be of great assistance. However, problems can arise when the on-site response team attempts to control a situation without notification to the fire department. This delay in notifying the fire department could significantly compound the department's efforts once they do arrive. The importance of an early notification must be stressed to managers and response team leaders to ensure an efficient and effective outcome.

Street Conditions: Factories, Lofts, and Warehouses

*COAL TWA**S** WEALTHS*

Accessibility

Concerns within this size-up category will vary from building to building. A parking lot, an accessible loading dock, and wide spacious streets will most likely surround newer and more modern versions of the factory and warehouse. Older versions of the factory/loft and warehouse building that have been around since the turn of the century will not be so fortunate. Those that are found within congested urban environments will mostly be found clumped together in certain sections of the city. Multiple dwellings and other commercial type occupancies will surround other sections. In the latter, many of the factories and warehouses were built before many of the other buildings that now surround them. In an attempt to provide housing, stores to shop, and places to socialize, adjacent landowners would build apartment buildings, erect stores, and provide all the necessities to keep the factory workers in the same neighborhood that they were employed. During this process, little regard was given to parking, the street's width, traffic flows, the loading and unloading of shipments, and how this would eventually impact the area.

In many cases, street width in this area is not going to be any wider than in any other area of the city. Parking for employees, building tenants, and nearby store patrons is going to be anywhere they can find it. Factory loading docks for buildings placed in this type of setting will be right at curb level, forcing truck drivers to either double park, or block the street entirely while a delivery or pickup is made. Other truck drivers will park at the first available spot, then transport their goods down the street by dolly, buggy, or rack. What this all means for the fire department is that any attempt to get down the street and set up is going to be a nightmare. Double-parked cars and blocked streets can only be handled by law enforcement. Regardless of how many signs you request to be posted forbidding these practices, it always seems like it happens more after they are posted. Besides greasing the sides of the apparatus and pinning the accelerator to the floor to get down the street, double-parked cars and blocked streets must be expected and dealt with in these areas.

Weather: Factories, Lofts, and Warehouses

*COAL TWAS **W**EALTHS*

Fires in large buildings, even under the best weather conditions, are going to be taxing to your members. The difficulties these buildings can present will require frequent relief and rehabilitation for your members. Add high temperatures in the middle of July, or the very cold temperatures in the middle of February, and firefighters will need to be rotated and relieved more frequently.

Wind

One significant weather concern that will haunt incident commanders with large building fires is the wind. This weather consideration has been mentioned many times throughout this book and needs to be mentioned and stressed again, due to the size of these buildings. This re-emphasis concerning the wind simply stems from the fact that there is now more to burn. From the building's height, area, fire load, and likelihood of surrounding buildings of the same concern, a fire of any magnitude accompanied with a significant wind will produce large fire conditions quickly. There is no wonder as to why conflagrations are born

Fig. 8-16a *Wind will make difficult situations worse.*

from these types of buildings. When winds approach 25 to 30 miles per hour, conflagrations will become a real threat. Whether it is from fire spread to adjacent exposure buildings, or from flying brands downwind to remote buildings, buildings left unprotected can become involved in fire.

Fires approaching these magnitudes have also been known to produce their own winds. Firestorm conditions will be produced by the tremendous thermal updraft of the large amounts of combustibles being consumed. This updraft of super heated air will create a negative pressure, which pulls additional air into its core from surrounding areas. Under these conditions, the fire will produce tremendous amounts of radiant heat, literally lighting anything nearby.

Fig. 8-16b *Large fires can create their own wind.*

Exposures: Factories, Lofts, and Warehouses

COAL TWAS WEALTHS

Your education and experience with these types of buildings will quickly remind you that they can produce heavy fire conditions with tremendous amounts of radiant heat in all directions in a relatively short period of time. Depending upon the height and area of the building and the flame frontage presented, the radiation area source can be large. Due to the amount of British Thermal Units (BTUs) that can be produced, buildings as far-removed as 100 feet are not considered safe. As floor after floor of the fire building becomes involved, the radiant levels will multiply. In these types of situations where floors of the fire building can be literally igniting within minutes of each other, chief officers cannot hesitate in ordering companies into positions to protect exposed build-

Fig. 8-17 *Numerous large caliber streams are required to protect nearby exposure buildings. This is not an example of one of those streams.*

ings. Obviously the closer the exposure, the more serious the threat; however, buildings capable of producing this type of radiant heat can easily ignite buildings 100 feet away.

The protection of nearby exposures will not come as an easy task. In congested settings, the chief officer will be required to make some quick decisions regarding what exposure(s) receive priority protection. Barring a life hazard of any adjoining properties, exposure priority decisions regarding factories and warehouses will focus around the following considerations.

Exposure considerations

1. *Flame frontage.* With only one side of the building involved, the obvious concerns would initially be with any exposures to that side.
2. *Exposure distance.* The closer the exposure, the more serious the threat. Buildings that are attached or separated by an alley on the flame frontage side will require immediate attention. However, it is important to remind members that what may appear to be an adequate distance at that time, may become threatened moments later as the fire advances. Remember to anticipate this!
3. *Wind direction.* Although it is said that radiant heat will travel equally in all directions, the heat on the leeward side of the fire will always be more severe than the windward side. This occurs for a number of reasons. First, convection heat will be influenced by the wind causing an increase in temperature to that particular side. Second, radiant heat is related to temperature and the area of fire involvement. The flame frontage side will contain higher concentrations of heat. Third, the smoke given off by the products burning will absorb radiant heat. Smoke (being opaque) will further absorb the radiant heat given off by the fire making conditions and any positions downwind difficult.
4. *Exposure construction/features.* Nearby buildings of combustible construction will absorb radiant heat faster than those built of noncombustible construction. This will obviously create an earlier ignition concern. Window construction of the exposure and any covering it may have must also be considered. Although exposed buildings may be of non-combustible construction, the building's windows may contain single pane glass in wooden frames. In these situations, officers must anticipate the ignition of the window's wooden frames, as well as the early failure of the glass panes. This type of window construction will fail early allowing radiant and convection heat to enter the exposed building. Those windows protected by steel shutters or windows with wire glass reduce this concern, but not for very long.

After consideration has been given to exposure priorities, then tasks must be assigned to the protection of those exposed buildings.

Exposure protection

Fig. 8-18 *Exterior protection streams have to be directed onto exposure buildings.*

1. *Bathing/wetting the exposed surfaces with water.* Wetting the exposure building's surface helps reduce the transmission of radiant heat to its exterior. Water when directed against an exposed surface will absorb significant amounts of radiant energy. With this procedure, exterior hoselines could also extinguish any window frames and building sheathing that have ignited. Care must be given with the stream not to break any glass windowpanes of the exposure building. Doing so will allow radiant heat to pass through the opening. If we remember from probationary school, water has little opacity. It is only twenty five percent effective when used as a water curtain as compared to coating the exposed surface with water.
2. *Stretching hoselines into exposure building(s).* The use of exterior streams is an excellent tactic to use in order to protect exposures; however, it cannot be used alone. When safe, hoselines must be stretched and operated within the exposure building to provide additional protection from stopping the fire from extending into its interior. Fire officers assigned to the exposure building must make every attempt to get hoselines to all floors threatened. Once in place, these hoselines may also make great vantage points to extinguish fire in the main building.
3. *Supplying the exposure building systems.* Those exposure buildings that have a standpipe and/or sprinkler systems must be supplied. Supplying the building's sprinkler system augments the water distribution in case the heads are activated by extending fire. Supplying the building's standpipe system eliminates engine companies from making long, difficult, and timely hose stretches. Usually all an engine company needs is their standpipe hose and associated tools to put a hoseline in place.
4. *Remove combustibles away from exposed windows.* This is a sound practice in any exposure building. Remove boxes, blinds, curtains, or any combustibles that might cause a threat.
5. *Ventilation of the exposure.* Opening windows and roof accesses opposite the exposed side helps to eliminate a build up of smoke and gases within the building. Super heated smoke and gases allowed to accumulate, may ignite combustibles on a floor within an exposure building, as well as make overall conditions difficult to operate in.

Area: Factories, Lofts, and Warehouses

*COAL TWAS WE**A**LTHS*

Square footage

Fig. 8-19 *Big buildings = big fires*

There is no doubt that this size-up concern is a key player in the fire officer's decision making. As one would surmise, the bigger the building, generally the bigger the problems that go with it.

The size of these buildings will vary from as small as 25′ wide by 60′ in length, to buildings the length and width of a number of football fields placed together. The first consideration of the initial arriving officer in this category is the square footage involved, as well the square footage threatened. Determining the amount of involvement is not always an easy task in large area buildings. Often what is showing may only be the tip of the iceberg. Companies reporting information pertaining to the area of fire involvement is critical to assigning and calling additional resources.

Irregular shaped buildings

As we further reference this size-up factor, we must also consider building configuration. Knowing that a building is L-shaped and wraps around the back of the exposure on the *bravo* or *number 2* side, is critical information that will affect your decision-making. Irregular-shaped buildings, those other than a normal square or rectangle shape, have caught many unsuspecting incident commanders by surprise. In many congested urban areas, attached or closely spaced exposures will not allow the incident commander to gain a perspective on square footage, or the building's configuration from his or her street view. Information about irregular-shaped factories or warehouses hopefully would be available to responding forces from pre-incident information. Barring the opportunity to gather the information from this source, incident commanders must seek this information from reconnaissance reports and roof reports.

Large area buildings must be mapped out in order to increase the efficiency, effectiveness, and safety of your operations.

When building construction and fire conditions allow firefighters to go to work on the roof, members assigned to this position should aid the incident commander in the size-up of these large area buildings by relaying the following information.

Roof radio reports to command

1. Smoke and fire conditions other than the command side; be specific.
2. Occupant life hazard observations at the sides and rear.
3. Building layout and square footage.
4. Accessibility to surrounding sides.
5. Roof mounted equipment concerns:
   - Water tanks
   - Air conditioning units
   - Dust bins/hoppers, etc.
6. Surrounding exposure concerns.
7. Fire extension into building shafts.

Floor areas

Fire officers can safely assume that in both the factory and the warehouse building, the floor spaces will be large and open. There may be on occasion flimsy partitions separating work areas or officers. You may also find a number of rooms actually constructed within the open floor area, however, these areas will usually be large and undivided allowing smoke and fire to occupy the floor space with ease. It is from this association that we can further justify the use of large caliber streams. Large undivided floor spaces will allow firefighters to deliver water from safe distances. Having an effective stream reach of 50′ or more into an involved area affords members that extra dimension of safety.

Fire walls

Large area structures may also contain firewalls. These self-supporting masonry walls are designed to act as a barrier to fire. Openings within these walls must be protected to prevent fire from crossing through. Self-closing doors must be checked to ensure that they are actually doing what they were intended to do. Often they may be blocked open to aid with the passage of stock and building personnel. Knowing their location within the building or complex, and the checking of their integrity during an incident will greatly aid in your confinement efforts.

Location and Extent of Fire: Factories, Lofts, and Warehouses

*COAL TWAS WEA**L**THS*

Regardless of where a fire begins in these types of buildings, the tasks will be difficult. Below grade areas will be a maze and not offer any ventilation. Lower-level fires may be deep in the structure requiring a long stretches. Upper level or top floor fires will have the same concerns associated with the lower-level, with the added need in many cases to stretch up.

Below grade fires

The most difficult by far will be the fire that originates below grade. Below grade areas associated with both the factory and warehouse will most likely be used for some type of storage. This area will be large, dark, and full of combustibles and possibly hazardous materials. In addition, members can expect these areas to have maze-like configurations, no ventilation options, and limited access points for the fire department to conduct their operations. This will not only be the most difficult area within the building, but also the most dangerous. Any success with extinguishing a fire in this area will be when the fire is small and in its early stages (and easily reachable).

Advanced fire conditions, or those that are approaching these conditions, are going to be extremely difficult to advance on. Officers must make some hard pressed decisions on whether to put all efforts behind an offensive push, or to use a holding action while additional companies attempt the use of cellar pipes or cellar distributors through openings made in the first floor. In either of the cases, this is going to be a dangerous operation. When conditions negate any attempt at interior operations, large volumes of water through deluge sets, monitors, or tower ladders can be used to flood the first floor in an attempt to allow water to flow through cracks, pipe recesses, or other utility openings, in an attempt to act as a giant sprinkler head over the fire.

Lower-level fires

Fires that originate in below grade or lower levels will have a number of extension concerns to deal with. If the building is of Class 3 or 4 design, floors could be built of wood or concrete. With this probability, fire can extend through cracks, poke throughs, or directly through wooden members igniting combustible stock on the upper floors. In some instances this may not be that obvious, especially when you have a large amount of combustibles stored throughout the floor space. A cardboard box may smolder for hours before it shows signs of fire. This is especially a concern during the overhaul phase when the area of involvement was spread over a large area. Leaving the scene prematurely may have you returning later to a much larger fire. Barring the physical removal of each box, a number of thermal imaging cameras scanning up and down the aisles may help.

Service and utility openings. Depending upon the type of operations within the building, there may also be a conveyor belt system that serves not only one, but a number of floors within the building. As one could imagine, the presence of this type of opening will allow rapid fire spread to all areas quickly. Fire spread via floor openings can be further complicated from utility openings that serve workstations. Depending upon the type of occupancy and the manufacturing done, work tables or work stations may have electrical services and plumbing services for individual units or a series of work tables on the floor that may allow fire to pass up through them.

Skylights. Older factory/loft buildings may have large skylights on each floor, one right over another on up to the roof. As unbelievable as this may seem, their installation was an attempt at allowing light onto floor spaces from above. Deep buildings within a city block did not allow natural light into the middle of the building. In order to a create better working environment, sunlight was allowed in through these openings to not only add visual light to the floor area, but to also give the worker a better sense of environment as they worked.

Light and air shafts. Any factory/loft building that is attached on both sides, and greater than 60′ in length, will most likely have light and air shafts serving all floors on both sides of the structure. Depending upon the age of the building, and any renovations that have taken place, windows serving these shafts may be of single pane glass in a wooden sash, which is known to quickly fail under fire conditions. Fire entering into this shaft will extend to any adjoining buildings, as well as re-enter on floors above. In most factories, lofts, and warehouses I have personally experienced, windows that expose the shaft are constructed of wire glass. However, never rule out the extension possibility when fire enters this area.

Fig. 8-20 *Wire Glass*

Steel shutters. Steel shutters may protect the exterior windows on a factory or warehouse over glass. Although the shutters were originally installed as a deterrent for burglars, their installation also delays firefighters from ventilating and accessing the building. A positive factor of their presence is that they will delay fire from auto exposing into the floors above, or fire advancing from an adjoining building. However, with any flame impingement or radiant heat on the shutters for an extended period of time, the rate of heat being conducted through the shutters will be very high. This will warrant companies to check for heat extending to wooden window sashes, as well as heat penetrating the glass to interior combustibles.

Interior staircases. Open interior staircases will be a major concern in older factory type buildings. Depending upon the size of the building, there may only be one interior stair that serves the entire building. Smoke entering this artery will cause difficulty for anyone attempting to exit down, as well as firefighters attempting to enter and climb up. Building occupants not being able to flee down may attempt to climb the staircase and exit onto the roof. Many times in an attempt to prevent burglaries, these areas will also be locked and secured, indirectly preventing factory workers from escape. Just as in the multiple dwelling, firefighters assigned to the roof must open the bulkhead and check for the presence of trapped civilians.

Top floors

Fires entering shafts or top floors will require firefighters to check penthouses, bulkheads, and hoppers for fire extension. With the volume of fire associated with these types of buildings, fires on a lower floor can accumulate enough heat to ignite upper areas of a shaft with little effort. Even remote areas must be checked for fire extension. The volume of heat that can be produced from the building's fire load, the combustibility of the structure, along with the age of the building, will warrant a thorough check of all vertical arteries.

Time: Factories, Lofts, and Warehouses

*COAL TWAS WEAL**T**HS*

Fire officers will have different time concerns when comparing factories and lofts to a warehouse fire. We say this primarily from the nature of each building's business.

Factories and Lofts

Due to the nature of their business, there may be dozens to hundreds of workers on shifts all around the clock. These buildings will present a more significant concern with the time size-up factor simply from the fact that time of the day may not have a direct relationship to the number of people in the building.

Time of the year

The time of the year can be equally significant for the factories, lofts, and warehouse building, but it seems that the factory and loft building will again take center stage. During the holiday season, the supply and demand of the shopping season will cause the greatest concern. Factory owners may put on extra help or allow workers to work double shifts to meet production demands. During this time frame, normally crowded working conditions can become compounded as additional workers occupy the same space.

Also during this time frame, firefighters must expect higher than normal stock loads. In anticipation of the high production demands, factory managers will over-order material and equipment needed to manufacture their product. Storage of this stock will be found in any space available within the building. The storage will most likely be in total disregard to the life safety of the people within the building. From the building's loading docks, hallways, stairways, corridors, to the actual floor space, exits will become blocked and obstructed, to say nothing of the increased fire load in the public areas. It is during these critical time frames that fire companies must get out into their districts to spot check any concerns that might compound their efforts, or the efforts of the people who may have to evacuate the building.

Poor housekeeping, overcrowded conditions, and an increased fireload all in a building that has a poor fire record is an obvious recipe for disaster.

Warehouses

Warehouse buildings are built for the purpose of storing products. With this type of operation, only a small work force of people will be required to load or remove stock. Although their hours of operation may vary, most will occupy the building during the normal business day, with the possibility of a skeleton crew present in the off hours for security or miscellaneous work.

Time of the year

Warehouse buildings will come with the same concerns of overcrowded stock that a factory building will have during the holiday season. It seems, however, that the overcrowding of stock in the warehouse starts much earlier than the holiday season itself. Depending upon the types of products being stored, warehouses that service local stores and malls can have larger than normal accumulations of stock as early as August and September, all in anticipation of the needs of the holiday shopping season. With this in mind, stock loads can be extremely light following a holiday season. One local warehouse visited, had a drastically different fire load after the Christmas season.

With warehouse buildings being nothing more than a storage depot for local establishments, responding firefighters can expect large fire load all year long, but with the holiday season approaching, expect the concerns to increase in this size-up consideration.

Height: Factories, Lofts, and Warehouses

*COAL TWAS WEALT**HS***

Heights can vary between the factory and warehouse buildings. *Factory and loft* building heights will normally average four to six stories, with heights of eight or more possible. Smaller heights associated with the factory type building are not uncommon. Factories constructed of one-story in height designed with saw-tooth roofs to take advantage of the natural light, are also well represented in many areas.

When "one" becomes two or three

Older warehouse buildings still found throughout many areas of the country will commonly be four to five stories, with the newer, more modern versions of this type building only one story. What is important to note with the newer more modern warehouse is the confusion of referring to it as a one-story building when in actuality for laddering and access purposes it is not. Many of these one-story warehouse buildings may actually be 30' to 40' above street level. Although these structures are referred to as one-story buildings, their actual height to the roof or parapet line may equal the height of a two, or three-story building. To better our height size-up references when approaching these types of buildings, officers in Jersey City began referencing these buildings in their initial radio reports as a one-story equal to the height of two. This added reference not only gave members who may need to access the roof a better understanding of what was going to be confronting them, but it also painted a picture about the square footage and the high ceiling concerns that are commonly associated with these types of buildings.

Fig. 8-21 *A one-story building that is actually equal to a two-story height.*

Accessibility

Gaining access to the roof of a factory or warehouse building will be, for the most part, within the reach of a fire department aerial ladder. Depending upon the type and age of the building, as well as its square footage, roofs may also be accessed by one or more fire escapes, or by accessing an adjoining building. As in any roof operation, members must be cautioned when access to upper levels is limited. Establishing more than one way off is a vital element for the safety of members operating above.

Special Considerations: Factories, Lofts, and Warehouses

*COAL TWAS WEALTH**S***

The concerns and difficulties of a factory, loft, or warehouse fire, will test the resolve of any responding fire officer. What will obviously aid in the initial decision-making of all officers is any pre-plan information about the building that you are responding to. Information relating to the class of construction, the height and area of the building, the building's fire load, the occupancy load and hours of operation, the presence of any auxiliary appliances, and the local water supply are all vital pieces of information any responding fire officer would like to able to review while en route to a reported fire.

As extensive as the above may sound, it is not enough information for sound decision-making. When arriving on the scene more specific pieces of information must be sought out to aid in the decision-making, and more areas need to be addressed for a safe, efficient, and effective outcome. Some of those will include the following areas of concern.

Factories, lofts, and warehouses

1. *Determining the location and extent of the fire.* This is easier said than done. Large area buildings will make this difficult. First arriving companies should seek and give this information, as well as information relating to the type and location of the interior stairs, as soon as possible. Information relating to whether the stair(s) are open or enclosed, as well as which one will be closest to the fire, will be beneficial. With more than one stair available, occupants can be directed down one stair, as the fire department goes to work in the other.
2. *Determine occupant life hazard.* Obtaining information about the possible number of people on the fire floor and floors above is critical. Depending upon the type of occupancy, these numbers can be overwhelming. All efforts must focus around their protection and removal.
3. *Resources.* Don't wait too long to get the answers from #1 and #2 before you request additional help. For some chief officers, this should be an automatic thought based on the limited amount of firefighters initially dispatched. Don't hesitate, request additional help early.

4. *Incident management and accountability.* It is very easy for firefighters to become disoriented or wander into an area where they shouldn't be in these types of buildings. This primarily happens for two reasons. First is the complexity of the structure. Its height, area, and stock arrangement will make for a difficult operation. Second is the individual firefighter. Firefighters will often confuse taking initiative with freelancing on the fireground. Upon arrival, divisions/sectors must be established. Having individuals responsible for the operations on the different floors, the roof, exposures, etc., not only allows for coordinated efforts, but also greatly assists with the assigning and tracking of your people.
5. *Beware of conflicting radio reports.* In large area, one-story warehouses, the floor height of these particular buildings may not show their true conditions to the engine company making their way in at ground level. These high ceiling occupancies can produce light to moderate smoke conditions at floor level, when actually there may be a raging fire overhead. If the engine company stretching into the building radios they have a light smoke condition with no heat, and the ladder company on the roof says they are encountering heavy smoke under pressure from the roof openings, base your immediate decisions from the roof report!
6. *Supply the building systems.* When a factory or a warehouse has a sprinkler or standpipe system, take advantage of them. Augmenting them may help keep a fire in check, eliminate excessive hose stretches, as well as protect exposures. Ignoring them, or becoming complacent about their location and usage, will only prove for a more difficult fight.
7. *Big fire = big water.* When considering the fire potential these buildings can present, officers will need to establish and deliver large water supplies. Knowing the location of hydrants that are served by large mains will be beneficial. During large-scale fires, incident commanders can request that the local water company boost water supply to the operational area. With water supply and delivery being critical at these fires, officers will need to prioritize the use and placement of their hose streams. Small diameter supply lines and poorly placed handlines are a waste of time and resources.
8. *Exposures.* Check exposures as soon as possible. Buildings that are attached or closely spaced will allow fire spread to them early, or they themselves may become a secondary collapse concern later into the incident. Even those buildings that seem separated by adequate distances may require protection at some time during the operation. Plan ahead!
9. *Salvage.* As we stated in a previous section, this area of concern warrants early consideration. Building stock can far outweigh the value of the building where it is housed. Confining the fire involvement to a thousand dollar loss, only to find a million dollar loss on the floor below from water runoff can seem defeating.
10. *Large area searches.* These buildings are too big to enter into without the use of a navigational aid. Anyone going in without a deployed hoseline must use a search

rope to help them get home. Whether it is from a staircase onto the floor, or from the front door into the building, this reference from point A to point B has to be established. In addition, there should also be serious consideration given to the period of time firefighters could operate in buildings of this size. Having the ability to reference firefighters' operating time will allow them enough air to return.

When I was given the opportunity to develop the *Team Search Rope* program for Jersey City, one of the items I ensured it included was the ability for time management. After I presented a clipboard, a tactical worksheet, and timer to a group of firefighters, the disgusted look on their faces quickly changed after the first training evolution when a 200' rope search had firefighters attempting to return with *no* air.

11. *Create additional means of getting in, as well as out.* As we have stated earlier, firefighters should not hesitate forcing more than the one or two doors needed for initial entry. Due to the square footage, stock arrangements, and possible maze configurations presented, there may be a need for a firefighter to exit, or a rapid intervention team to enter, from a different exit/entrance point than the one or two initially created. Although it may be time consuming and task demanding, chief officers should not hesitate to call and assign additional resources for this task. Once additional exits have been created, advise your interior firefighters of the locations of these new exit points.
12. *Thermal imaging, thermal imaging, and thermal imaging.* What else can I say other than don't leave home without it.
13. *Collapse hazards and collapse zones.* Officers should look for and report collapse hazards that can cause concerns to firefighters operating within or near the building. Hazards such as heavy machinery, heavy content loads, water absorbent stock, and roof mounted water tanks and hoppers will add to the collapse concern. If available, a surveyors transit could be placed on specific items on the building, or specific areas of the building to detect early movement.

Fig. 8-22

Questions For Discussion

1. List the different types of construction associated with the factory, loft, and warehouse. In what class of construction, and with what type of construction materials used, is there a concern of the structure's walls being dependent upon the roof for their support?

2. List and describe the occupancy load and status concerns of the factory, loft, and warehouse buildings.

3. What factors should be considered in hoseline selection at a fire in these buildings?

4. What is the most common firefighter life hazard concern in these buildings? What options are available to deal with the concern?

5. What is a Blitz Attack? What are its considerations and limitations?

6. What are the considerations in protecting the exposures from a factory, loft, and warehouse fire? What tasks should be considered to protect those buildings?

7. What information should be included in a roof report to command for large area buildings?

8. List and describe the location and extent concerns of a fire in a factory, loft, and warehouse.

9. What are the concerns of high ceiling occupancies?

10. How does the time of the year affect conditions within factory, loft, and warehouse buildings?

9
High-Rise

Fires in high-rise buildings pose the most unique and greatest challenge to today's fire service. The building's design, its internal systems, the height and area of the building, its water supply, as well as the occupant load are just a few of the many elements that will greatly influence a fire department's operations. When operations are not able to be ground based, all resources and their assigned tactics must proceed up to the designated area and operate from within. This extended time, often referred to as lead or reflex time, can become quite significant. It is from these listed factors, that we arrive at the definition of a high-rise as being any building that is in excess of 75′ where fire department operations cannot be considered ground based, and where the height of the building creates a significant stack effect.

In this chapter we will reference both the residential and the commercial office high-rise. Each type of high-rise occupancy will pose unique and demanding challenges every time firefighters respond to them.

Construction: High-Rise

COAL TWAS WEALTHS

When we reference the class of construction for a high-rise building, we think of a Class 1 fire-resistive structure. By definition, we know that a building of Class 1 design is a structure whose components are designed and protected to resist the maximum severity of fire within the building. What the fire service has seen through the years, is a major change in the design and construction of what we still refer to as a fire-resistive building.

Older Heavyweights

As building architects started to design and erect buildings well above street level, their uncertainty forced them to plan and build structures with fortified designs. Buildings that were built during the period prior to 1945 were referred to as the heavyweights. They acquired this name from the amount of building materials used in their design and construction. By reviewing the amount of materials used, building designers and engineers noted accumulated weights of 20 to 23 pounds per cu. ft. of structural material throughout the design and installation of the building. The concept associated with the mass of these structures was to erect a fortress that could resist the effects of fire, the wind, and gravity.

Fig. 9-1 *Heavyweight high-rise*

In addition to their weight, their internal design and layout during this same time frame was also viewed as an asset. The high-rises of yesteryear did not have core construction, heating, ventilating and air conditioning (HVAC) systems, access stairs, curtain walls, etc. When a fire erupted in one of these buildings it generally remained confined. This luxury was soon to change.

Newer Lightweights

As building codes changed, construction and labor costs increased, and new and lighter building materials became available. Building designers and engineers soon learned that they could build more with less.

The lightweight high-rise that we see in the building industry today averages around 8 – 10 pounds per cu. ft. of building material—a big

Fig. 9-2 *Lightweight high-rise*

difference from its ancestor. This transition however, was not an overnight process. It seemed after World War II and up until the late 1960s engineers and designers actually went through a period that many in the fire service refer to as the middleweight high-rise. Weights per cubic foot found in buildings built during this era hovered around the 12 – 18 pound range. Although not as strong as the heavyweight, they and their cousins the lightweights introduced a number of structural features and materials that changed the way we conduct business in these buildings.

The Older vs. The Newer Generation

Structural protection

In older buildings of Class 1 construction, the structural steel members within the design are protected by encasing the steel in either concrete, plaster and metal lath, gypsum/sheet rock, or (in original installations) sprayed-on asbestos. Each one of the materials listed was installed to protect and insulate the steel from exposure to fire.

The more modern high-rise also provides protection to its structural steel, but through newer, safer, and more cost-effective applications. Although we still see the steel of many newer high-rise buildings being protected by fire-rated dry wall, a more economical method used by designers and installers today is to coat the exposed steel surfaces with a sprayed on insulation of non-combustible fibers. Many of the non-combustible fibers that are used in the process today will swell and then char like a giant marshmallow when exposed to fire. The insulating qualities of the material can prevent damaging heat from affecting the structural steel.

Fig. 9-2a *Sprayed-on insulation of non-combustible fibers*

Each method of protecting the steel had tremendous value, as long as it was applied and maintained correctly. The concrete encasement method of protecting the steel provides the best protection simply due to its inability to be easily removed once it cures. This was and is still by far the best, but the most costly method of protecting the steel. The modern and cheaper method of protecting the steel with a sprayed on non-combustible membrane can also be considered effective. This holds true until a section is scraped away by a welder attempting to achieve a ground connection, is found to be loosely or thinly applied in some areas, or during a fire when the protective coating is struck by hose stream dislodging and washing some of the material away. With buildings whose entire steel frame is designed to support the loads of the structure, it becomes critical that all areas be protected to the same level and standard. This level and standard must be maintained throughout all parts of the structure for the life of the building.

Curtain walls

In the modern high-rise with steel frame construction, the buildings exterior skin of walls and windows may consist of a series of framing, which will be mounted to the struc-

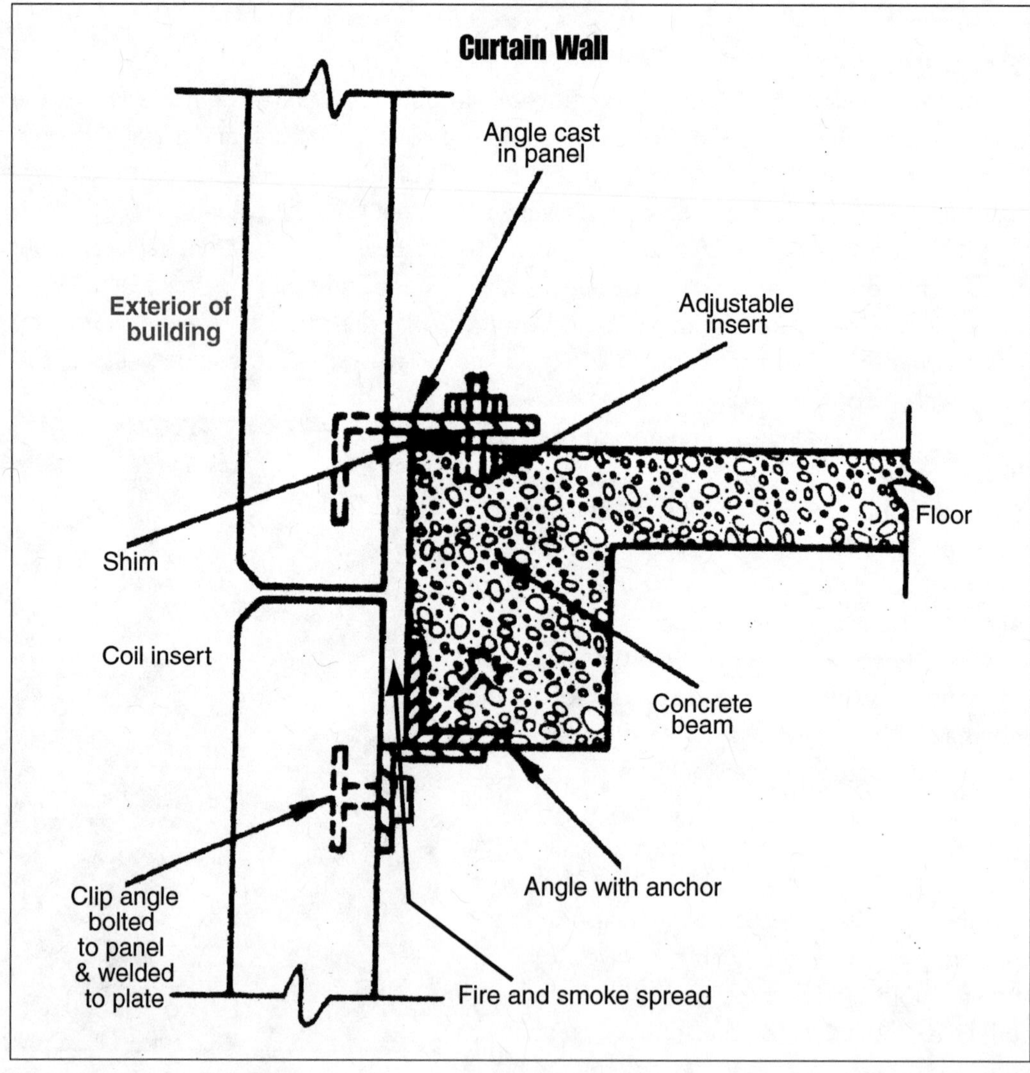

Fig. 9-3 *Curtain wall design*

ture to enclose each floor from the elements. These non load-bearing, prefabricated walls of steel, aluminum, and glass are referred to as curtain walls. The name is derived from the method used to attach the frame to the end of each floor.

The framing for the curtain wall will be bolted to pre-mounted clips at the end of each floor assembly. At the attachment point, there is often a gap or air space between the floor slab and the exterior wall that must be insulated and protected to prevent smoke and fire travel from the interior. In earlier installations, within the space you will find a metal flashing over fiberglass insulation. If the material was not installed properly, or if it was compromised in any way, it could allow fire to move to the floor above through this space. Fiberglass itself, even when installed properly, rarely prevented fire from extending up through the space. Today's codes are very specific on the materials used to protect this space. They must be fire rated.

Once windows fail, fire extending from a window can also affect the integrity of the curtain walls connection points. The intense heat and direct exposure from a well-involved fire can jeopardize the steel or aluminum frame that holds the large pieces of plate glass in place. Depending upon the design of the assembly, large granite or marble sections that are also hung at the ends of the floor assembly may fail, dropping to the street or sidewalk below. Failing curtain walls will easily allow fire to re-enter on the floor above. This, with the probability of large shards of glass followed by pieces of masonry falling to the street, only adds to the inherent dangers from the building's exterior skin.

Compartmentalization

Compartmentalization is the concept of dividing up a floor area to limit a fire's spread. It is an essential ally in a fire department's attempt to control a fire. In the older high-rise building, large, open areas were, for the most part, nonexistent. In the newer, more modern high-rise office building, their presence is a common site. Within the large, open floor spaces of the modern high-rise, cubicles or partition walls may be erected to provide some privacy for employees. Cubicles are nothing more than 4' – 6' dividers extending up from the floor, offering nothing more than an obstacle to firefighters. Partition walls are erected within the floor space to create an office or conference room for staff or management to conduct their business. It is important to note, as their visual appearance may indicate a divided or sectioned floor, the partition does very little to add to the compartmentalization of the floor. The reasons are directly related to their construction. First, their ability to retard fire is generally very limited. Some may be constructed of gypsum board, a reliable retardant of fire, but the thickness of the walls, presence of hollow core doors, and any poke-through accesses generally negate any significant fire resistance. Second, as partition walls are erected in an open floor space for the construction of offices, conference rooms, etc., the walls are generally only constructed up to the height of the suspended ceiling. This will leave the plenum undivided over the entire floor area.

Plenums

The plenum is the space between the top of the suspended ceiling and the underside of the floor above (a modern version of a cockloft). This space is often used for the horizontal distribution of the building's utilities, HVAC ductwork, phone, cable, fiber optic wires, as well as for the common return for the floors HVAC system. To extend a fire rated wall up through this area would add significant construction costs, and is often not required by code.

Core construction

Core construction, is the concept of containing all the building services, utilities, elevators, and stairs within a designated or core area. Core construction for the

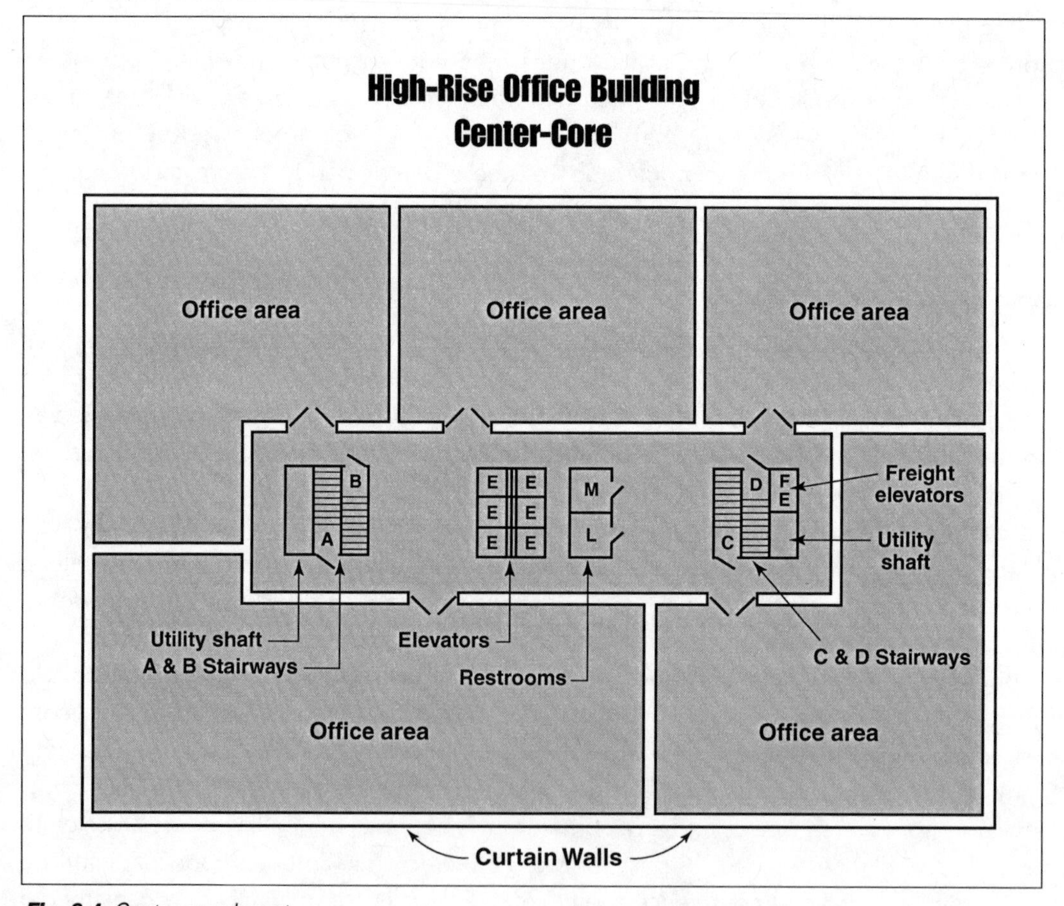

Fig. 9-4 *Center-core layout*

high-rise will generally consist of either a side or center core design, with the latter being more common. The concept behind the core design of a building was that it not only allowed the utilities and services to be placed within a designated area, but it also allowed more rent space in the building's exterior shell—a large open area with an unobstructed view. With the building's stairs, elevators, bathrooms, and utility shafts concentrated in this area, building owners could receive a better return on their investment.

HVAC systems

Probably the most influencing building system that will plague the fire service in the modern high-rise building is the HVAC system. The modern high-rise is a self-enclosed environment, which means that the building will be required to provide a clean, warm, and cool air environment for the building's tenants. The HVAC system in a building is designed to process and treat air to specified temperatures, humidity, and cleanliness, before distributing it throughout the intended spaces of the building. After the air is delivered to the designed areas within the building, it is collected and returned for processing and reuse through the same designed areas. The system type, design, and its serviceable areas are the predominant concerns for the incident commanders at a high-rise fire.

In the residential high-rise, the concern is much more limited than when compared to the commercial office high-rise. In the residential high-rise you will find that each living unit will have its own individual system of providing heat and air conditioning, with a central system generally found servicing only the common areas. In the modern office high-rise, the design will be much different. In this type building, the HVAC systems falls into the following two general categories.

1. *Non-central air system.* This is a very small and limited system. It will only provide air to one floor.
2. *Central air system.* This is a more common system. This system will supply air to a number of floors within the building. The central air system will be the more troublesome to the fire department.

To effectively operate and manage a fire in a modern high-rise office building with a HVAC system, the incident commander must have a working knowledge of the system, its design and use, and how it may affect the fire department's efforts. Later in this chapter, we will discuss a strategic plan for operation of a HVAC system at a high-rise fire.

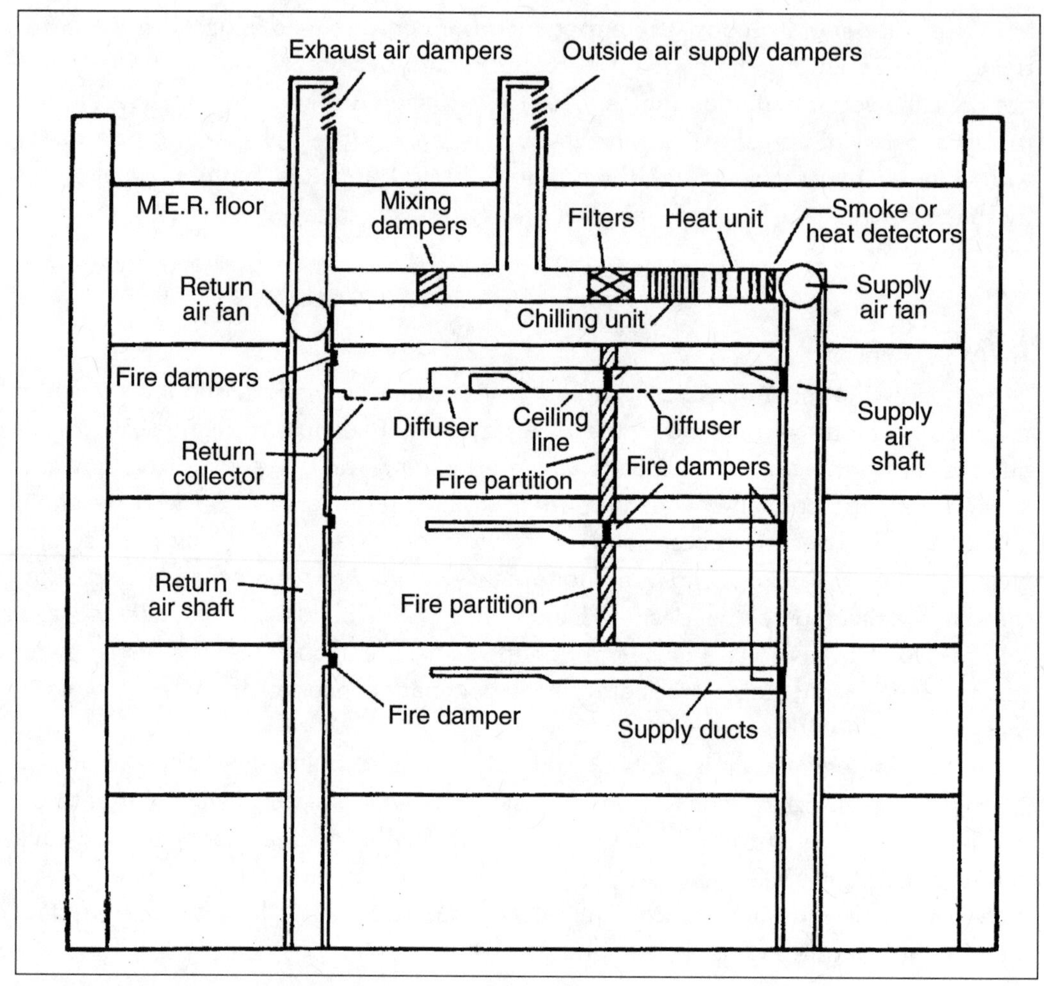

Fig. 9-5 *HVAC design and layout*

Occupancy: High-Rise

COAL TWAS WEALTHS

Occupant Load and Status

Residential

The occupant load and status in a high-rise can vary greatly depending upon the type of occupancy, the time of the day, and the day of the week. The residential high-rise building by its very nature will have a large occupant load 24 hours a day, 7 days a week.

With numerous tenants, varied work schedules, and the uncertainty that accompanies them, we must always expect large numbers of individuals in a building of this type.

Commercial/office

The office high-rise building on the other hand, will come with a different set of concerns on occupant load and status that the fire service must be prepared to deal with. First, simply by the nature of their design and use, the fire service must anticipate that these buildings will literally occupy hundreds, if not thousands, of employees during the business day, from Monday through Friday. We must also anticipate that the thousands of office workers who occupy these buildings five days a week, will, for the most part, probably know very little about the building they work in. Their normal routine will usually take them in through the building lobby, up the elevator to their office or work area, to an occasional trip to the restroom or cafeteria, and back down again to the lobby and parking garage. Not only can the mere numbers of people within the building prevent total evacuation, but also their lack of knowledge on any alternate means of egress, as well as any evacuation plans that might have been implemented, will compound your attempts when their evacuation becomes necessary.

The other occupant concerns within the office high-rise that we must address are those people who are associated with the property during the off-business hours, namely the members of the cleaning and maintenance staff. Depending upon the size of the building, they can number in the single digits to a hundred or more. These individuals that occupy the building during the night and early morning hours, can literally be anywhere within the building.

The concerns associated with this particular group of building occupants are noted for a number of additional reasons. First, many of these individuals may be non-English speaking, which will cause a communications gap, especially if you are trying to voice direction to them over the building's public address system. Second, few, if any, will have been subject to fire drills or evacuation plans. In the past, the fire service concentrated on practices and procedures for the building tenants during the height of the workday. This must change.

Occupant evacuation procedures

Firefighters all too often will depend on prompt, organized, and timely evacuation by the building's occupants so they can concentrate on rescuing the disabled, the young, the elderly, or anyone in the immediate fire area. Unfortunately, building occupants do not always follow evacuation plans. Many times they will delay exiting only later to become disoriented and trapped. Others will ignore communications altogether and self-evacuate, clogging stairwells and elevators. In many instances, their best course of action may have been to stay in place.

Fire departments must make extensive efforts to instruct building managers, fire safety directors, or building security to conduct occupant evacuation training for the building's occupants regardless of whether the building is a residential or office high-rise. Programs should include direct involvement of all the buildings occupants with establishment and review of the following:

1. How to notify the fire department.
2. The establishment and use of fire wardens. (Fire wardens are designated individuals who are given the responsibility of ensuring an orderly and prompt evacuation from a floor or area when evacuation becomes necessary.)
3. Scheduled fire drills.
4. Familiarity with building/floor layout and exits.
5. When to stay.
6. When to go.
7. Where to go.
8. Information relating to fire department public address messages.

Apparatus and Staffing: High-Rise

*CO**A**L TWAS WEALTHS*

Resource requirements for a fire in a high-rise building will be by far, the most demanding a fire department will ever encounter. What has often been referred to as the "100 man fire" becomes very evident when you need to get equipment and people well above street level.

The complement of equipment and firefighters to a reported fire in these buildings will vary from department to department. Our case history from these types of fires indicates quite clearly that you are going to need a lot of firefighters. This fact cannot be ignored. Ideally, they should be dispatched and on their way as soon as the first phone or fire alarm is received. If we wait to call the necessary resources until we actually need them, then the incident dictates the outcome.

Many fire departments researched will assign large complements of firefighters and equipment to buildings of this type simply because of the known difficulties they can present. Some fire departments researched with minimal on-duty or volunteer staffing, will establish mutual aid responses with neighboring departments for these same reasons. This proactive approach to a high-rise building will at the very least provide an increased level of firefighters responding into the incident at the receipt of the alarm. Others unable or unwilling to implement these options will be forced to deal with compounded and possibly overwhelming factors upon their arrival

The bottom line with this concept is simple. If you can muster them, start them in. If you don't need them, you could always send them home.

In the previous chapters' apparatus and staffing sections, we have given you researched, used, and suggested procedures for both the initial arriving engine and ladder company at a particular type of building. What follows is a reference list of assignments for initial arriving companies that I have reviewed from a number of different departments, large and small. As in previous chapters, the following is brief in text, and is primarily meant to inspire additional thought.

First due engine and ladder company – initial assignments

Prior to the arrival of a chief officer, obtain as much information as possible from the fire safety director, building security, building engineer, or fire alarm panel.

From the building staff

1. Location of the fire, if known.
2. If any access stairs serve or leave from the reported fire floor.
3. Evacuation procedures implemented, if any.
4. Current status of HVAC system.
5. Elevator status.
6. Elevator banks and the floors they service.
7. If necessary, location of fire service keys for the elevator.

From the alarm panel

1. *Type of alarm.* Is it a smoke detector, water flow, manual pull station?
2. *Multiple devices.* Are there multiple devices activated? (i.e., smoke and water flow in the same area; water flow and no smoke in the same area.)
3. *Multiple locations.* Are there multiple devices sounding on multiple levels? (i.e., smoke detection on floors 10, 11, and 12.)
4. *Alarm location.* If alarm is indicating the elevator machine room, you might need to walk up, or if available, use an unaffected bank of elevators. Also, if the alarm location is below the lobby, a fire below grade may affect the establishment of a lobby command post.
5. *Is the building fire pump activated?* This information may or may not be available from this location.

Note: Upon your arrival if the alarm panel gives limited information, as companies work their way up periodically view the panel for any changing or new information. There may be valuable size-up information here!

First due engine and ladder company initial assignments, continued:

1. Prior to the arrival of a chief officer, the officer of the first arriving company shall establish the Command Post. Some departments prefer to establish their command post outside the building 200′ away; others will prefer the lobby of the building.
2. With the arrival of a chief officer, members of the first due engine and ladder company are to use a fire service elevator and proceed to "two floors" below the reported fire floor. If the fire is within the first eight floors, companies going to work on the fire floor and floor above are walking up.
3. A minimum of one engine company and one ladder company should proceed together to the reported fire floor. Anything less is a dangerous practice. If your department does not use an engine and ladder concept, then a minimum number of firefighters should be assigned to handle the initial tasks.
4. Have members determine if the reported fire floor is the correct location, if so, return the elevator to the lobby with a firefighter. Having a firefighter return with the elevator car ensures control from within the car.
5. Have members report heat and smoke conditions on the fire floor.
6. Based on conditions and tasks to be assigned, prior to the arrival and assignment of a chief officer to the fire floor, additional arriving resources may need to be assigned. This may require establishment of an operations officer to coordinate operations.
7. As companies go to work in a specific geographical location, their management and accountability must be maintained. Depending on department policy and procedure, as well as the span of control of the area, this should involve the establishment of a division/sector or operations post.
8. Have members determine life hazard on the fire floor, and initiate primary search and evacuation procedures if necessary. This will require that a stairway be designated as an Evacuation Stair, and announced to Operations and Command.
9. Have members determine location of the fire on the floor, and select a stairway with a standpipe that will provide the best attack on the fire. This will require that a stairway be designated as an Attack Stair, and a stairway designated as a Ventilation Stair. Both locations must be announced to Operations and Command.
10. Have members initiate a hookup to the standpipe outlet on the floor below the fire floor and stretch up to the fire.
11. Stretch and operate on the fire floor, report progress and conditions to the Operations Officer.

First due engine company chauffeur/engineer

1. Establish a water supply and augment the standpipe or sprinkler system as per standard operational guidelines.
2. Determine if fire pump is activated. When feasible, determine pump-operating pressure.

Sprinkler operations for chauffeur/engineer, first due engine company

1. Establish a water supply
2. Stretch a minimum of two 3" hose lines into F.D.C. (When an option, larger hoselines may be used.)
3. Operate in volume at approximately 150 psi

Standpipe operations for chauffeur/engineer, first due engine company

1. Establish a water supply
2. Stretch a minimum of two 3" hose lines into F.D.C. (When an option, larger hoselines may be used.)
3. Operate in volume, starting at 100 psi plus 5 psi for each story above grade. Fire pump operating pressures may require a higher starting pressure. Note: The above is based on three lengths of 2½" attack hose with a smooth bore nozzle. Smaller diameter hose will require higher pressures.

Second due engine company – initial assignments

1. Report to the Lobby Command Post for accountability and assignment.
2. Determine if the fire floor has been verified by the first due engine and ladder company. This is critical; the fire's location must be identified.
3. Proceed to the location of the first due engine company via a fire service elevator, or if the fire is within the first eight floors, walk up.
4. Assist first due engine company with initial hose stretch. This is also critical. In many situations large hoselines may be required for their volume, reach, and penetration capabilities. Additional members may be required to ensure that the first hoseline is stretched to its intended location.
5. Second due engine company officer is to ensure adequate flow to initial hoseline from standpipe outlet. If additional pressure is needed, advise the Operations Officer. Water flows from standpipe outlets could be restricted from a number of reasons to be mentioned in later sections within the chapter. If the first hoseline stretched to the fire is not receiving what it is supposed to, don't think that the second or the third hoseline will.
6. If not needed with initial hose stretch, stretch backup hoseline. The second stretched hoseline must ensure the safety of those members operating on the fire floor. If they are making progress and do not require the stretching of a backup hoseline, then the second hoseline should be stretched to the floor above to prevent any fire from extending.
7. Report conditions and progress to the Operations Officer.

Second due engine company chauffeur/engineer

When feasible, determine if fire pump is activated and determine pump-operating pressure. Assist first due engine company driver/chauffeur with hook up to FDC connections. Establish a water supply and augment any additional standpipe or sprinkler systems as per department Standard Operational Guidelines.

Second due ladder company – initial assignments

1. Report into the lobby command post.
2. Proceed to the floor below the fire via the fire service elevator.
3. Confer with first due ladder company on areas to be searched, location of the fire, ventilation options.
4. Coordinate operations on the fire floor with the first due ladder company.
5. Report core type, other stair types, and their locations.
6. Primary search on floor above the fire, when possible.
7. Check for fire extension on the floor above the fire, when possible.
8. Report conditions and progress to Operations Officer.

Life Hazard: High-Rise

*COA**L** TWAS WEALTHS*

Firefighters

Elevators

Life hazard concerns for the firefighter are numerous when a fire occurs in a high-rise building. Probably one of the most dangerous is the use of the building's elevators. Elevators, when working, can bring large numbers of resources to the upper floors of building, eliminating the fatigue factor associated with climbing long, and numerous flights of stairs. The problem associated with elevator usage in a high-rise building is with elevator reliability, and the fire's location or reported location. Statistics show that 75% of the elevators in the affected areas of a high-rise fire are going to malfunction and shut down on average 15 to 20 minutes into the operation. This is a minimal amount of usage time. Some believe elevators equipped with firemen service will protect the firefighter from these concerns. This is not so. All elevators, regardless of a firemen service feature are subject to malfunction or shut down from water runoff, heat, moisture, or electrical failure. It could be from one, or a number of these concerns that could render an elevator bank inoperative, trapping members inside as they attempt to use the elevator. It is for this reason that some fire departments will not use elevators at all, while others will only use elevators that have a bank remote from the fire's location. The only thing firemen service equipped elevators specifically do for those members using it, is to allow them to override any previously made calls from floors that serve that bank, as well as allow members to control the cars placement at a specific floor from within the car. This procedure is only considered reliable until it is obviously affected by the elements of the incident.

Elevator operations

1. *Use fire service elevators.* Attempt to determine from building security or the building engineer, which, if any, are serviced by an emergency generator. This may buy you additional time in the event of a power loss.
2. *Access fire service keys and place elevator into phase 1 operation.* Phase 1 returns the car to the lobby or lowest floor. Depending upon code and the type of alarm system, elevator cars may have already been returned to the lobby or the lowest floor they serve.
3. *Prior to entering the car.* Scan the shaft between the car and the shaft and look for smoke in the shaft, fire in the shaft, or sprinkler water in the shaft. Unfavorable conditions presented within the shaft will determine its usage.
4. *Upon entering the car.* Place elevator into phase 2 operation – Phase 2 places control from within the car. Press the Call Cancel button, to eliminate any previously made selections. Next, place the car in fire service position.
5. *Car operations.* During the ascent, make a series of stops every 5 to 10 floors (depending upon floor destination) to ensure the elevator is responding to commands from within the car. Also during the ascent, engage the emergency stop button to ensure control of car. (Note: forcing the car doors may stop run-away elevators.) Proceed to two floors below the lowest reported floor. To open elevator car door, press and hold Door Open button. Verify floor and stair access location, then return car to lobby with member. (Note: Firefighter returning with the car should have own his/her own SCBA, tools, flashlight, and radio in the event the car becomes inoperative during descent.)
6. *Fire floor location.* Take elevator to two floors below the lowest reported floor. This is *critical*! Even with multiple sources reporting a fire on a specific floor, access stairs or convenience stairs associated with the commercial high-rise, as well as duplex and triplex apartments associated with the residential high-rise building, may not give the actual true fire location from a smoke condition or activated alarm. It is critical that members take the initial elevator to two floors below the lowest reported floor and then walk up.
7. *Elevator bank usage.* Members should determine blind elevator shafts and the floors they serve. Sky lobbies may need to be used in order to gain access to a certain bank of elevators.

8. *Firefighter accountability.* Each company or group of firefighters that enters the elevator should have their car number, company designation, in addition to their destination recorded at the lobby command post. This is required for tracking members in case of an elevator malfunction or failure.
9. Firefighter readiness. All members entering the car must have properly donned PPE, SCBA with cylinder valves turned on, portable radios, and assigned equipment. It is important that the car is not overloaded. In the case of an emergency, members must be able to move within, as well as out of, the car.

Heat

Another life hazard concern that affects the firefighter within the high-rise building is the heat conditions that fire in a structure of this design can produce. Class 1 fire-resistive buildings are constructed, arranged, and protected to resist fire. By the very nature of this type of design, buildings referred to as Class 1, will hold and retain their heat.

Depending upon the occupancy type, floor areas may present an added concern to already difficult heat conditions. The large open floor spaces commonly associated with the office high-rise will allow fire and heat to accumulate throughout the entire floor area. Residential high-rise buildings are more compartmentalized. Fire travel in these types of buildings will be more limited throughout the floor space, but will present extreme difficulty from heat conditions within the fire apartment itself. The compartmentalization, or lack of it, combined with the inability to achieve any effective ventilation, will create intense heat conditions for firefighters attempting to work in the area.

Fig. 9-5a *Residential high-rises may initially confine the fire to the original apartment, but they will be hot and difficult fires.*

Disorientation

Another serious consideration for the firefighter is disorientation. In any large area buildings, firefighters must be well organized, equipped, and accounted for to eliminate a disaster. Search techniques and patterns for large, open floor spaces in the office types of occupancies must follow a formal and systematic approach. In large, open areas, search ropes and thermal imaging continue to work well together. In more compartmentalized, and confined occupancies as displayed in the residential high-rise, thermal imaging cameras in addition to patterned searches will also play a vital role in the organized search. Obtaining a look at the floor below the actual fire floor may provide useful information about the hallway layout, the number of doors needed to reach the fire apartment, and any obstacles that may be waiting for you on the floor above.

Plenums/suspended ceilings

Plenums, a construction feature that is found in the commercial office high-rise, is a great concern to advancing firefighters. The large spaces created by a suspended ceiling are useful for many of the building's utilities and services. It is in this space that we find the return air system of the central air conditioning system, as well as a raceway for miles of insulated phone and cable wires.

The concerns from the plenum area are going to be twofold. First, is the possibility of fire extending into the plenum and igniting any combustible insulation or plastic coated wiring above the heads of advancing firefighters. Although current building codes prohibit the use of combustibles within the space, many of the materials that cause concern have been there before codes were enforced. Second, is the entrapment and entanglement nightmare associated with the suspended ceiling. When fire is suspected of being overhead in these areas, members should periodically remove ceiling panels ahead of their advance, checking for any fire. The thin metal wire and grid system that holds these ceilings in place can fail, not only dropping the ceiling tiles and its grid work onto firefighters, but also the lighting fixtures, duct work, and miles of entangled phone and cable wires. If the ceiling drops and entangles a member, firefighters must be well rehearsed in reduced profile and quick release procedures associated with their SCBA, in addition to the mentioned self rescue techniques. A small, but effective tool that helps with firefighter entanglement is a good pair of wire snips or wire cutters. This tool seems to be becoming more and more a commonly carried personal item, and for good reason.

Fig. 9-5b *Plenum space*

Falling glass

Falling glass from windows ten, twenty, or thirty-plus floors above street level can strike anyone on or nearby the fireground, to say nothing of what it will do to a hoseline. The premature and uncontrolled removal of glass in a high-rise building is an extremely dangerous tactic that in all cases must be controlled.

In the modern commercial office high-rise building, the windows are often sealed, and cannot be opened. Premature and uncontrolled breaking of glass will not only rain down large, heavy, and lethal pieces of glass, but their removal may also affect fire and smoke conditions within the building. Those that can be opened, or require their removal, must only be done under the direct control of the incident commander or operations officer. Those that need to be broken or removed, must be done under the added control and coordination of an individual referred to as a street vent coordinator. This individual through a coordinated effort with a police department liaison must evacuate all individuals from possible drop zone(s) before the glass can fly.

Collapse

Collapse is a real concern to firefighters even in a building that is referred to as fire-resistive. Depending upon the construction design, some high-rises will be constructed as a steel skeleton in a gridiron design. Buildings of this design will contain floors of corrugated steel sheets covered by 2" – 3" of concrete. When an advanced and prolonged fire heats up the underside of a floor's corrugated steel, the sections of concrete may crack and buckle the concrete floor upward. If this process is allowed to continue, the heating, warping, twisting, and buckling of the steel may loosen and drop sections of the floor anytime during fire department operations.

Collapse concerns can also take place on the exterior of the building as fire vents out of windows attacking the building's exterior facade. As fire exits a window, heat from direct flame impingement, the convection column, and the fire's radiant heat could split the concrete, granite, or marble exterior, dropping large sections of material onto the roofs of neighboring buildings, or to the street below.

***Fig. 9-5c** Once fire exists to the exterior, anticipate falling building materials.*

Smoke behavior

One of the last life hazard concerns that must be referenced in a high-rise building for the firefighter, as well as for any of the building's occupants is smoke stratification. Stratification is the ultimate cooling and layering of smoke within a building. As the smoke from a fire is heated, it moves vertically through the building's arteries. As the smoke cools and loses buoyancy, it will layer or stratify on floors remote from the actual fire floor. These conditions can trap firefighters or civilians in these layered pockets of smoke.

The stack effect, a condition that is caused by the temperature differences between the inside and the outside of a sealed building, will aid the movement of smoke away from the fire's origin. Depending upon the temperature differences between the inside and outside of the building, the height of the building, and the temperature being generated by the fire, smoke can travel extended distances vertically within the building.

Smoke within a structure can also be pulled down and stratify below the actual fire floor. This condition, referred to as reverse stack effect is more predominate in the summer months due to the cooler temperatures often associated within the inside of the building.

Finally, smoke movement can become further complicated from an activated HVAC system. HVAC systems still operating within the fire area can move smoke to unaffected floors within its zoned area.

Occupants

The life hazard concerns for the occupants in a high-rise fire can be extremely overwhelming. Depending upon the height, area, and occupancy type, the number of people in these buildings could be well into the thousands. The mere size of the building, accompanied with the number of people and the number of usable stairways, suggests that total evacuation of these buildings is often impractical.

In these situations it becomes the responsibility of the incident commander that he/she ascertain and verify the occupant status within the building. Information relating to the following questions will help determine if additional resources will need to be assigned to handle the civilian concerns.

Occupant considerations

1. Has the fire floor been determined by the initial companies? If so, what is the fire extension horizontally within the floor space, as well as vertically to the floor above.
2. What is the current extent of evacuation?
3. What evacuation procedures have been implemented?
4. If floor wardens have been established, what are their reports?
5. Are there any reports of severe life hazards within the immediate fire area?
6. Are there any handicapped individuals located on the fire floor or floors above?
7. What is the status of the HVAC system, and what floors are served in that zone?
8. What is the status of elevators—their number, location, and floors served?

9. Are there access or convenience stairs in the reported fire area?
10. What communications, if any, have been established between the fire department and the occupants of the fire building?

In order to control what can be a very overwhelming concern, it may be advisable to establish an evacuation officer to initiate communications with the building's occupants, as well as control the evacuation process.

When able to access a hard wire communication system in a high-rise building, the incident commander or evacuation officer should arrange for an announcement, if one has not been already been made from a canned message. Canned messages are pre-recorded information that automatically announces instructions to building occupants at the sounding of an alarm.

Most will agree that specific information and directions from a fire department representative is what everyone wants to hear. In most situations a calming voice can eliminate some, if not most of the *chaos*. Below are samples that fire departments can use to inform the buildings occupants.

Occupant voice evacuation directions

Sample #1: The ___________ Fire Department is on the scene to extinguish a fire on floor ____. Please remain calm within your apartment or at your place of employment, and await further information and direction from the fire department. Thank you for your cooperation.
Sample # 2: The ___________Fire Department is on the scene to extinguish a fire within the building. Occupants on floors ____, ____, and ____ are to exit those areas via the ____ staircase and proceed to________. Please remain calm and orderly. Thank you for your cooperation.

Evacuation efforts should focus on removal of the building's occupants that are in the immediate fire area first, with any other threatened occupants controlled and evacuated in three floor intervals if necessary. Having a controlled and organized approach for evacuating the building's occupants will not only eliminate stairwell overcrowding, but any premature or uncontrolled evacuation of the building will hinder fire control efforts, and add to the confusion on the fire scene.

Terrain: High-Rise

COAL TWAS WEALTHS

Accessibility

Topography concerns for the high-rise building are referenced in two specific areas: building setbacks and property accessibility. In the latter, property accessibility is categorized as those areas that are remote and difficult to access with large fire apparatus. It seems that in many departments, the size of the apparatus is getting bigger and bigger, and in some cases, bigger is not always better. When attempting to access a parking garage or rear loading dock, apparatus may not be able to get near the area in question. Narrow access driveways, overhanging marquees, multi-level garages with limited height and weight restrictions, may require longer hose stretches or use of auxiliary appliances within the area to get water on the fire (i.e., dry standpipe systems in parking garage areas).

Fig. 9-6 *Underground or lower level parking garages may cause accessibility concerns.*

Setbacks

High-rises that are set back from the street level on one or more sides of the building will create limitations and concerns on ladder reach, hose stretches, and general accessibility. Decorative landscaping, waterscapes, or pavilions will cause the engine company to stretch additional lengths of hose to hook up to the building's sprinkler or standpipe siamese, as well as limit the reach of an aerial or tower ladder.

Fig. 9-7 *Building setback*

Fig. 9-7a *Landscaped and sloped terrain must be part of your pre-incident information.*

Concerns in this size-up factor are taken for granted more so than any other size-up factor. It is important within this size-up factor that we reference the specific properties or areas prior to the receipt of the alarm. This will quickly tell you what you can, and cannot, do.

Water Supply: High-Rise

COAL TWAS WEALTHS

Of the reoccurring problems that plague firefighters at high-rise building fires, water supply emerges as a major concern. All effective high-rise fire protection systems involve the use of water. In order for the fire department to successfully combat and extinguish a fire in a building of this type, adequate water must be delivered to the floor(s) where it is needed.

Availability

Water supply for a sprinkler or standpipe system in a high-rise building will be available to the fire service from only a few sources. Those sources being from building water tanks specifically designed for that purpose, the city's public water supply, or from fire department supplied apparatus and hoses. It is due to the uncertainty from mechanical failure, power failure, or improper design, that many fire departments consider the primary water supply in a high-rise building to be from a fire department pumper.

In order for us to supply a building's sprinkler and/or standpipe system, we must be prepared to deliver an adequate and continuous supply to meet the needs of the incident. It is the fire department's responsibility to gather information about the water supply concerns for the building if they are going to be able to use the system effectively, efficiently, as well as be able to operate safely. There is no doubt that no one can remember all the information about an individual building, let alone the specifics relating to the building's water supply. The amount of information can be overwhelming. To help identify the areas of concern, firefighters must gather information from on-site inspections and code enforcement personnel to aid them in this quest.

Detailed information relating to the building's water supply can be listed within a pre-incident form specific to the building, with specific and more refined information listed within an alarm printout sheet. Information that could be quickly viewed to put water supply operations into motion, as well as information that may be used for incidents that can escalate into multiple alarms, is information that must be researched and documented.

Information gathered about the water supply concerns of a high-rise building should include the following:

1. The size and capacity of the building's standpipe system.
2. How is the system supplied?
   - Public water?
   - Fire Pumps?
   - Pressure tanks?
   - Gravity tanks?
   - Fire Department?
3. Are there connections on every floor?
4. In what stairs are the risers located?
5. What class standpipe system is it—1, 2, or 3?
6. Does the building use zoned systems? If so, what floors are served in each zone?
7. Are there any pressure reducing or pressure restricting devices within the system? If so what are our limitations?
8. When the building uses fire pumps, do the pumps directly supply the system, or are they augmented by the building's water tanks?
9. If the building contains water gravity tanks, do we know their capacities and locations? Is there dedicated water for firefighting?
10. What are the locations of fire department connections and the systems and zones they serve?
11. What are the hydrant locations and main sizes surrounding the building?

This may seem like a lot of information to gather, but if there is a fire in a high-rise building, you will want the answers to these questions. It would be best, if they can be researched and documented before any incident occurs.

Delivery

Fire pumps. Fire pumps in a high-rise building are generally located on a lower floor, or cellar within the building. They may be started manually, or operate automatically upon a drop in pressure within the system they supply. Some fire pumps are of the combined type that may supply both the sprinkler and the standpipe system for the building. With any type of installation, it is important to know the type of pump system you are utilizing. With pumps that have to be started manually, it becomes the responsibility of the building engineers to start the building fire pumps, and to maintain control of the system when an incident occurs. As one can anticipate, engineers (if present) are often assigned or involved in another task when a fire incident occurs. Nevertheless, this leaves the operation and control of the system with the fire department. It is important for an educated, radio/phone equipped firefighter or fire officer to be assigned to this job pending the arrival and assistance of a building engineer. Even with a qualified building engineer, his/her lack of knowledge with required fire department flows warrants us to assign a member to this location.

Fig. 9-8 *Building fire pump*

The building's fire pumps are a valuable asset in structures of this design and type, but they too have limitations that we in the fire service must be prepared to deal with. The building's fire pumps should be capable of delivering their rated capacity at a pressure of 50 psi at the highest floor hose outlet. A relief valve placed within the system is required at the fire pump, limiting that pressure to 15 psi above that required to deliver its rated capacity at 50 psi. This allows for a maximum total of 65 psi to the highest floor outlet. In newer buildings with newer standpipe installations, we are starting to see 100 psi requirements at the most remote outlet. But is this a true figure that can be counted on at the end of the nozzle? Namely because of the above listed restrictions, the building's fire pumps may not be capable of supplying adequate flows to hoselines, especially if inadequate and improper hose diameters and nozzles are being used. In addition, it is also important to note that the building's primary water supply to the standpipe system may be a gravity tank. In this situation there may be inadequate head pressures supplied to the upper floor outlets, which are supplied by the gravity tank. In newer and more recent installations, gravity tanks are required to have a booster pump to supply the floor in which they are located, as well as a number of floors immediately below.

Pressure reducing devices/valves. Pressure Reducing Devices (PRDs) and Pressure Reducing Valves (PRVs) could be placed within a standpipe system to prevent dangerously high discharge pressures from hose outlets. There are two major types of devices used for reducing pressure within a standpipe system; flow restricting devices and pressure reducing valves.

Flow restricting devices control the discharge pressure by restricting the flow through a reduced opening. Pressure reducing *valves* limit the pressure and volume by being set to a specific pressure that will not be exceeded under any flow.

The concept behind pressure reducing valves and devices is the concern that firefighters would be exposed to dangerous operating pressures and forces if they connect their hoselines to standpipe outlets near the base of risers that are of substantial height. In cases just the opposite, fire departments have also encountered additional problems where the pressure reducing valves (PRVs) were not set for the required discharge pressure. This would limit a sufficient flow to a fire department hoseline, especially of a smaller diameter. When indicated and when possible, devices and valves should be removed or adjusted to deliver adequate flows to the size and length of hose deployed.

Water delivery to the fire is vital, and there is a critical set of tactics that must be understood when water supply and the building itself hinder effective flows. It becomes the engine company's responsibility in transferring the available water efficiently, effectively, and safely to the area(s) in question. All of the tactics and procedures associated with these concerns will revolve around selecting and operating the right hose and nozzle to meet the demands of the job. So much has been written about pros and cons of different diameter hose, as well as the use of smooth bore nozzles and automatic nozzles, that furthering the debate would not be beneficial here. The common concern for all, has to be the required pressure to ensure an efficient and effective flow for the hoseline and nozzle you have selected. Think about it.

As we previously stated, pressures at the standpipe outlets in high-rise buildings that use older systems will be limited to an outlet pressure of 65 psi. Newer systems are required to deliver 100psi at the most remote outlet. With this in mind, you must take into consideration the length of your hose stretch, the diameter of the hose, the nozzle selected, and the possible presence of pressure reducing devices or pressure reducing valves if you want to ensure adequate flows to extinguish a fire.

Excluding pre-incident specifics of a particular building, many departments researched use as a standard 2½" hose with smooth bore nozzles for fighting fires in high-rise buildings. The only exception noted to this procedure was with the selection of 1¼" or 2" hose with smooth bore nozzles for some fires in a Class 1 housing project high-rise. This option, notably in Jersey City was added for this particular type of structure due to the construction and compartmentalization design of the building, and the added mobility requirements for the floor and apartment layout. Additional factors that would alter the smaller hose selection would be inadequate outlet pressures to support larger hoselines, as well as the weather considerations, specifically the wind. Wind at the upper levels of a high-rise fire can overwhelm even a well-placed, and well-supplied 2½" hoseline.

Taking all of the above into consideration, knowing the particular building and its limitations becomes your best resource for making the determination for what you can and cannot use.

Auxiliary Appliances and Aides: High-Rise

*COAL TW**A**S WEALTHS*

Past experiences in high-rise building fires reinforce the fact that automatic sprinkler systems are the most effective way to prevent a major high-rise fire. Their presence and operation will become the fire department's best friend when a fire occurs. Many major incidents have been prevented and controlled by well-placed and sufficiently operating sprinkler heads. As with the standpipe system, it becomes the responsibility of the fire department to gather as much information as possible about the system within the building.

Information that should be gathered about a sprinkler system in a high-rise building should include the following:

1. Does the building have a sprinkler system? Depending upon when and where the building was built, there may not be a sprinkler system.
2. If a sprinkler system is present, what area(s) of the building does it protect?
    - Public hallways?
    - Loading docks?
    - Storage areas?
    - Compactor areas?
    - All areas?
3. Are there multiple sprinkler zones within the building? If so, where are the sectional valves located?
4. What fire department connection will serve what zones/floors?
5. How is the system supplied?
    - Public water?
    - Gravity tanks?
    - Pressure tanks?
    - Fire pumps?
    - Fire Department?
6. Is it a combined system (i.e., Class III standpipe and automatic sprinkler system)?
7. What type of system is it?
    - Wet?
    - Dry?
    - Pre-action?
    - Special?

Fig. 9-9 *Not only is it important to know the type of system you can utilize, but also the specific areas/floors they serve.*

Detection systems

Information received from the building's alarm system must also be referenced and considered within our auxiliary size-up. Simply due to size of the building, identifying and interpreting useful and relevant information can aid the fire department's efforts. Often one of the first items the incident commander will look for is the alarm annunciation panel. Without prior knowledge, or at least someone there to guide you, its location may actually be difficult to find. Although the educated fire officer would expect to find the alarm panel to be located in the fire command center or in the lobby area of the building, a recent story told to me about a high-rise building, included the fact that the location of the alarm panel was in a closet down at the end of the hall, because it was unattractive.

Once you find the panel, determine the type of alarm, and the location and number of devices activated. Alarms that are referenced to as being trouble signals are those identified water flow or manual pull stations. Others referred to as smoke or heat detectors, will be of the indicating type. If the panel also shows that individual detections are increasing in number, or there are activations on more than one floor, this may be a reliable sign that you have an advancing fire.

The gathering of information about the type of alarm system should include whether or not its activation also affects any of the building's systems or features. Systems of the indicating or detection type may be tied into the HVAC system, the elevators, or the stair tower doors. Systems may be designed to shut down an HVAC system. They may also be designed to bring all the elevators down to the lobby or sky lobby. Some alarm systems may also unlock and open doors to specific areas. Knowing this prior to the receipt of the alarm at a specific address will let you know well before you get there what the system can and can't do for you, as well as how it may help or hinder your operations.

Newer and more advanced building alarm systems are controlled and referenced via a network of computers. Computerized alarm systems can assist a fire department by monitoring the building's systems during an incident, as well as give a printout or history of what has occurred within the system prior to your arrival. A printout or history reference may allow you to determine the seat of the fire as the smoke originating from one area traveled to another area, tripping alarms along the way.

Pressurized stairways

Pressurized stairways involve the concept of placing a positive pressure within a stairwell shaft to prevent smoke from entering. In most installations, fans used to pressurize the stairwell are not activated until an alarm activation is received within the building. When activated, the constant pressure within the shaft is designed to keep any contaminants from entering into the shaft if a door is opened on the fire floor. This is the same principle of the positive pressure within the mask on your breathing apparatus. When an opening is created around the seal of the mask, the positive pressure forces the smoke away from the opening until the face mask can be readjusted. This type design usually works well within the stairwell of a building until a number of doors are opened and the pressurization of the shaft becomes ineffective.

Fire tower stairway

A further-advanced design of a smoke-free stairwell is referred to as the fire tower stairway, or by many in the East Coast as the Philadelphia Tower. This stairwell does not rely on the concept of forced air within an enclosed shaft, but with a design that has the stair shaft separate from the actual floor. It achieves a smoke-free atmosphere by having an open air, outside balcony between the floor space and the building's stair shaft. In this design, building occupants will traverse from the floor area, outside across an open-air balcony into a separate stair shaft. Any smoke that exits out the door of the involved floor will automatically exit into the open atmosphere. Building occupants will then be able to exit down a separate, and smoke-free shaft.

This design, although outstanding, is rarely found due to the added costs in design and construction.

Buildings communication system

The building's communication system in a high-rise must also be viewed and referenced as a auxiliary appliance because of its ability to aid the efforts of the fire department. Depending upon the radio system, or systems that are being used by the fire department, high-rise building construction may shield radio waves and hamper communication within the building, as well as to those to the outside of the building. Department procedures should reflect the use of simplex (non-repeater) channels in addition to attempting transmissions near the building's exterior walls when communications become difficult. Depending upon code and jurisdiction, newer high-rise buildings are being required to install internal communications equipment that will enhance fire department radio usage.

Fire department use with a building's communication system will depend on what is available to them. Public address type systems work well for occupant notification, and occupant direction. They may also be used for giving special or urgent information to fire department units. Simply the fact that the system could communicate to large areas at one time, makes the public address system a consideration for this purpose.

Fig. 9-10 *Portable phones should be stored within the buildings Fire Command/Security Center.*

What we find most often is that most fire departments will use the hard wire systems that can be plugged into, and referenced to key areas around the building. Portable phone/jack systems allow fire officers to communicate with each other from different locations throughout the building. Phones can be plugged into lobby areas, elevator cars, elevator vestibules, standpipe cabinets, fire pump rooms, as well as the building's fire command center. In addition to the above, use of public telephones within the building is a viable option, in addition to the use of cell phones and even sound powered phones when available.

Whatever devices are available, and whatever means we can create, it is vital to establish a communications link throughout the building.

One note, if you're going to attempt use of the building's public phone system, obtain the phone number of the lobby or command center before you go up—it will save time, trust me.

Assistants/aides

The last reference within our auxiliary appliance and aide section is the people within the building who can aid the fire department in their efforts. This can be anyone from the fire safety director and their assistants, the building engineer and their assistants, building security, to the possible presence of fire wardens. Each will play a vital role in the fire department's efforts.

Building engineers are critical in the use and control of the building's systems. From control of the HVAC system, the starting and stopping of the building's fire pumps, to obtaining copies of floor plans for the fire floor and floors above, their knowledge and our access to it is critical.

Fire safety directors, building managers, and building security should be able to reference occupancy loads, evacuation implementation and its phases, occupant notification, the whereabouts of anyone needing special assistance, as well as occupancy content concerns.

Fire wardens can aid with evacuation notification, evacuation direction, and possibly occupant accountability.

We need to add another group of individuals that will greatly assist us with our efforts before a fire involves a high-rise building. Individuals associated with the fire department's Code Enforcement Bureau or Fire Prevention Bureau must become a vital part of your information gathering process. With the amount of information that should be gathered and reviewed before the receipt of an alarm, in addition to ensuring that building systems and appliances are being maintained to code, individuals must be delegated to ensure that these concerns are being meet. These buildings are too big, too complicated, and (in some cities) too numerous to attempt it by ourselves.

Street Conditions: High-Rise

*COAL TWA**S** WEALTHS*

Safety and control

When we focus our attention with street conditions at a high-rise fire, we don't think of problems necessarily with street width, street direction, or street surface conditions. What will often become a concern to the fire department is the conditions or hazards that the fire, or the fire department's operations can create. Glass, as well as large pieces of spalling marble and concrete can come raining down the side of the building into the street, and on top of adjoining buildings. The dangers that these objects can create to occupants and firefighters can be lethal, if the street is not controlled.

The first, and the most obvious concern is to shut down all pedestrian and vehicle traffic surrounding, or at the very least to the side of, the building in question. Very few, if any, departments have the initial staffing to delegate to this task. Quick and early establishment with the police department becomes your best course of action to meet this objective. Next is control of the exiting occupants to the street(s) in question. Here, you must gain control and direction through the use of another agency, in this case building security is the option that should be initially considered.

With this all in mind, just as we must direct and control those who want to leave the building, we must also direct and control those who want to enter, namely, the firefighters. Running through falling debris is not the best option. Seeking and establishing alternate approaches to the building through rear entrances, underground parking garages, and the unaffected side(s), will add to the safety of members who are attempting to go to work.

The controlling of the street from added hazards that we, or the fire might create, can be directed to an individual called the street vent coordinator. If during fire department operations it becomes necessary to remove glass from certain areas to assist with operations, the untimely, and uncontrolled removal of glass could injure people who are still evacuating, firefighters who are entering from this area, as well as damage hoselines that have been stretched to fire department connections.

If this position can be delegated, the following is a list of responsibilities of the street vent coordinator.

Street vent coordinator

1. Establish communications with Command and or Fire Operations.
2. If possible, advise Command and Operations of wind conditions on the side of the building in question, prior to venting.
3. Ensure total evacuation and closure of the street on the questioned side. This must include control of any exiting civilians from adjacent and nearby buildings into this area.
4. Establish a liaison contact with the police department to assist with this operation.

5. If possible, attempt to cover and protect any stretched hoselines into fire department connections on the side to be vented.
6. Upon completion of the ventilation, report observations and conditions to Command and/or Operations.
7. Maintain control of the street throughout the entire incident.

Weather: High-Rise

*COAL TWAS **W**EALTHS*

When we consider the weather and its effect on fighting a fire in a high-rise building, two significant factors must be referenced. First, is the wind and its role on fighting a fire in a high-rise. Second, is the ambient temperatures and their effect on the smoke and heat movement on the inside of the building.

Wind

Wind is a critical concern when firefighters go to work in a high-rise building. There seems to be no lack of this weather element, especially when responding and operating within a high-rise district. Wind velocities will, for the most part, be much higher and occur more often in a high-rise district than in any other part of a city. As taller buildings accumulate in a particular part of a city or area, they put a barrier, or wall of resistance up as the wind currents attempt to move around them. As the wind currents fight to realign and move with their intended flow, their velocity increases as they find a path of least resistance around the sides of the building. If there are a number of high-rise buildings grouped together, wind currents will increase as they fight their way around any obstacles. Wind currents can increase and become turbulent to the point where they can create eddies between buildings.

Even on those days when you exit the cab of the apparatus or chief's vehicle, wind conditions that seem minor or nonexistent at street level, can actually be very significant at levels as low as ten stories above grade. This initial observation can give a false interpretation of factors, which can significantly influence your operations.

When fire vents through windows on an upper floor of the windward side of the building, conditions will be quickly noticed as the wind creates blast furnace fire conditions within the building. In those situations where fire has not vented itself and the question of horizontal ventilation comes into play, serious consideration must be given to simulating the ventilation operation, to discover how the opening may affect fire and smoke conditions within the building. If ventilation of the fire floor is being considered, it is highly advisable to recreate the ventilation pattern on a lower floor in an attempt to simulate wind direction and conditions, prior to actually performing this tactic on the fire floor.

One to two floors below the actual considered floor will give you your best readings on whether you should, or should not attempt this option.

Temperature

Atmospheric temperatures outside of a high-rise building will greatly affect the building's stack effect. Stack effect is defined as the natural vertical movement of air within a building due to pressure differentials, which are caused by the difference in temperature between the inside and outside of the building. This movement of air is normal in tall structures, and seems to increase in speed and intensity, as buildings get taller. Normal stack effect is more significant during the colder temperatures of the season because of greater temperature differences between the inside and outside of the building. Stack effect in a high-rise building is responsible for the wide distribution of smoke and gases throughout the building in areas well above the actual fire floor.

In the summer months when the temperature outside the building is warmer than the inside temperature, there is the possibility of reverse stack effect. The cooler dense air within the building can pull smoke and its gases down below the actual fire floor causing additional operating concerns for the fire department, as well as an added evacuation concern for some of the building's occupants. With all of this in mind, we must also consider the core design of the building. Those built of the center or side core design will have a tendency to draw the products of combustion toward the vertical arteries within the core. Remember, it is also the building's core where we will make our attack from, as well as where the building's occupants will attempt to leave.

Exposures: High-Rise

COAL TWAS WEALTHS

The surrounding properties may be a vital concern when fire occurs in a high-rise building. The concern is not primarily from a fire extension, but from the possibility of glass and masonry raining down onto rooftops of adjoining or nearby buildings. Earlier, we discussed the need to control entry and exit points to all sides of the fire building. In addition, we discussed the establishment of a street vent coordinator with the added help of a police liaison to eliminate the movement of individuals on the questionable side of the building. However, when the exposure is a building that is nearby and lower in height than the involved floor of the fire building, the fire department must give added concern to the evacuation and removal of occupants in this particular building, specifically those who are occupying the top floor. When fire extends out a window or series of windows, the probability of masonry spalling off the side of the building, and then falling on top of the roof deck of the attached and lower exposure building is great. Large sections of marble and concrete weighing hundreds to thousands of pounds, falling ten, twenty or thirty plus stories, will have no problem finding their way through the roof deck and onto the top floor of a neighboring building.

The only plus from a lower exposed building, is that when a window fails, or is removed for a ventilation operation, the roof deck of the exposure building may catch some of the glass, preventing it from raining down directly into the street.

Area: High-Rise

COAL TWAS WEALTHS

Square footage

Area square footage concerns for a high-rise building will differ based on the occupancy, and the use of the building. In the residential high-rise, our concerns are somewhat lessened due to the compartmentalized design of the building and its apartments. We say somewhat lessened, because of the possible added difficulty associated with an apartment being a duplex, or possibly triplex design. As more and more newer residential high-rises spring up, we are finding more and more multi-floor arrangements within the apartments of the more affluent areas of our cities and towns. These type apartments, will double, and even triple initial square footage concerns, to say nothing of how fire will travel vertically within them.

The commercial office high-rise buildings of today are being built much larger and taller than ever before. Open floor spaces exceeding 20,000 to 30,000 sq. ft. are becoming more of the norm within a center- or side-core design. The center-core, being the more common design, will have all of its utilities and accesses within this vertical artery in the middle of the building. The obvious disadvantage of the open floor design is the unimpeded, horizontal fire travel throughout the floor. The advantage, if we want to use that word at all, is the unobstructed reach of our hose streams once we attempt to regain possession of the floor.

Convenience/access stairs

If it wasn't bad enough having large, open floor spaces within the office high-rise building, add the possibility of a convenience or access stair between floors of the building. Depending upon the building tenant, and the number of employees associated with company, many companies who lease within the building will occupy more than one floor. For convenience they may add stairs within the floor space in an attempt to eliminate company employees from going out into the core of the building and taking an elevator or public stair to the floor above or below. The convenience or access type stair is becoming very common within a multi-floor occupied space. There should be no doubt of their ability to allow fire, heat, and smoke to move vertically into the floor space above.

Location and Extent of Fire: High-Rise

*COAL TWAS WEA**L**THS*

Occupancy type

Being able to anticipate how a fire will spread in a high-rise building will greatly aid in the direction and management of your resources. The first of many considerations for the fire department will be the occupancy of the building, and its related construction design. We start by associating residential high-rise buildings as being compartmentalized and defined. We generally consider buildings of this type to limit fire spread to the apartment of origin. The commercial office high-rise building on the other hand will have large open floor spaces. This will allow smoke and possibly fire to have immediate access to the entire floor space. It is from these associations that we can start to anticipate and direct our concerns with the fire's location and its extension probabilities and possibilities. However, before separating and comparing the differences between both buildings, we find that there are many similarities, simply because they are both considered high-rise structures.

Fig. 9-10a *Compartmentalization of residential high-rises will slow the fire's spread.*

Elevator shafts

A concern that will be universal to both types of buildings is the elevator shaft. These vertical arteries can become an avenue for smoke spread for the entire height of the shaft. The fire officer's initial thought in this size-up category must start with the fact that the shaft that encloses the elevator car and its mechanics will not be airtight. Often you can stand in an elevator lobby and feel the movement of the air, as well as hear the whistle of air coming from the shaft doors as the elevator car moves up or down. With the movement of the car in the shaft, the air within the shaft will become pressurized, moving smoke and gases much further than we would initially expect, most notably in the direction of the traveling elevator car. In addition to the shaft's vertical alignment within the building and its location within the core design, it will also be subject to the building's stack effect, which could further influence any smoke movement.

Stairwells

Stairwells within a high-rise building are often considered another universal extension concern. Depending upon the age, the type of building, and its occupancy usage, you may find more than one type of stairway represented in the high-rise building that can cause havoc to fire department operations. Unless they are of the pressurized design, their presence within a core will draw smoke and gases toward the shaft. These vertical arteries, like the elevator shaft will be subject to the building's stack effect, allowing smoke to move throughout when given the chance.

Auto extension

Auto exposure or auto extension is often referred to as the most common means of fire extension in a high-rise building. Fire lapping out of a window and rolling upward, will eventually weaken the integrity of the window, or windows above allowing fire to enter into the floor area. Removing combustibles away from the area in question, as well as the stretching and operating a hoseline on the floor above will prevent serious extension via this means.

Commercial/Office High-Rises

Curtain walls

Fire extension concerns from curtain wall design and poke-through construction focus within the office high-rise building. The curtain wall design, as we mentioned earlier in the chapter is nothing more than a exterior skin, attached to the shell of the building. In older designed high-rises, this space received little attention. As we stated in an earlier section, there was often a gap of an inch or two that received fiberglass insulation. It was the original intent of filling this space with insulation to preserve the R-value of the floor. At the time of installation, no one knew that fire could find its way into this space and travel from floor to floor. Originally used fiberglass insulation was later found to go through repeated phases of filling and then drying out with moisture from constant temperature changes of the exterior skin of the building. Depending upon the moisture content of the fiberglass at the time of the fire, the space could easily allow fire to extend from floor to floor. In other types of insulation used, some were found to actually be combustible adding to the fire's ability to spread. New codes have eliminated both concerns.

Poke throughs

Fire can also spread through poke-through holes. These holes are created between floors in a high-rise for building utilities to pass through. Depending upon the attention to detail, and any additional renovation work, the fire stopping around the piping, wiring, and ductwork, can either be non-existent, or improperly installed. The vertical alignment of these utilities will most often be found within the building's core. The utility shafts are a constant concern for the checking of smoke and fire penetration. Any significant fire below this area may allow extension to the floor above through these openings.

Plenums

The plenum space is another extension possibility that lends itself more to the office high-rise building when compared to the residential high-rise. This space between the top of the suspended ceiling and the underside of the floor above is generally open for the entire occupancy floor area. Partition walls will rarely penetrate up through the dropped ceiling space. This horizontal space is not only an additional raceway for phone and computer cable, but it may also serve as a large return duct for the HVAC system. If the HVAC system is not shut down, it can draw smoke and fire through the plenum area in the direction of the return shaft opening.

HVAC system

The buildings HVAC system can play a major role in the fire's ability to spread smoke from one area of the building to another. The modern office high-rise with sealed windows is a totally self-contained environment. The interior of the building is designed to rely on a self-contained air handling system. Knowledge of the workings of a building's HVAC system is critical to the management of a fire in a high-rise building. Preventing a system from intensifying a fire, limiting further smoke spread, and possibly venting smoke or purging unaffected areas are real possibilities, if the system can be controlled and managed.

Time: High-Rise

*COAL TWAS WEAL**T**HS*

Occupant loads

Initially, the thoughts of the occupant load and the time of the day will be the concerns when responding to a reported fire in a high-rise building. With residential high-rise buildings, our concerns will be for the entire 24 hours. Without doubt the concerns will increase during the late evening and early morning hours; however, due to the large number of occupants within the building and the varied work schedules and sleep patterns, the concerns will be high at all times, day or night.

The office high-rise building will present a greater increase in the number of building occupants during the day hours, notably business days (Monday through Friday). As we mentioned in the life hazard profile, after the business workers have gone home there will still be a group of individuals within the building for cleaning, repairs, or after-hours work that must be factored into our life hazard concerns.

Building systems

Another important consideration within our time size-up factor that must be noted, is that the status of the building's systems in the office high-rise building may change, most notably after normal business hours. The building's HVAC systems may no longer be on recycling times due to less of a demand in the floor spaces. Elevators and the building's fire pumps may be shut down for repair or preventive maintenance. A number of similar or related repairs may be in progress during the after business hours due to the drop in occupancy within the building. Information on the status and operations of the building's systems are critical pieces of information that must be referenced.

Lead/reflex time

Other information that should be factored when arriving at a high-rise incident is the lead/reflex time. With the lead/reflex time of an incident being defined as the timeframe from when a fire company or group of firefighters is assigned a task or objective, until the time they arrive and go to work on the objective, the timeframe can become critical. The average lead reflex/time in a high-rise building can be 5, 10, or 15-plus minutes depending upon the assigned location and their access to it. The higher the floor, generally the longer the time.

Fig. 9-11 *Even when incidents are within reach of fire department ladders, you still have to expect extended times before resources can commit to the fire floor.*

Time of the year

The time of year and its weather affects can affect smoke movement throughout a building. As we have cross-referenced in other size-up factors, hot weather during summer months as compared to cold weather during the winter months will affect the building's stack effect, possibly affecting smoke movement in a high-rise building.

Height: High-Rise

*COAL TWAS WEALT**HS***

Accessibility

Elevators

Height concerns and considerations for a fire in a high-rise building are of significant concern as a fire starts and extends beyond the reach of ladders and hose streams. When fires are above our ground-based reach, generally our first question is, how do we get up there? The answer to this question is usually within a fire department's standard operating guideline. In my research, I found that most departments will reference the reported fire floor, and depending upon its height within the building, they will take a fire service elevator or climb the stairs. Few departments I found in my research stated that they will use the stairs regardless of what floor the fire is on.

There are two main reasons why the majority of fire departments use elevators in a high-rise building. One, is to reduce the lead/reflex time it takes to go to work at a fire well above street level. Fire, smoke, toxic gases, and their affect on the building and its occupants will not wait for us to get there. The second reason is that in order for people to go to work at this fire, we need to get them to areas within the building with some energy left so they can mount an initial, as well as a sustained attack. Walking or running up a number of flights of stairs with the amount of gear we need to carry, will completely exhaust the most physically fit firefighters. Elevators therefore become the necessary evil. In many cities, the protocol is that if a fire is reported within eight floors of the street lobby or sky lobby, firefighters will walk up. Beyond that, firefighters should only use elevators that can be controlled by fire service operations.

Fig. 9-12 *Walk or ride?*

Stairs

When it comes to the question of stairs within a high-rise building, there are a number of concerns that need to be answered in our attempt at an efficient, effective, and safe operation.

Consider these questions:

1. What is the lowest reported floor of the fire? With multiple floors reported, proceed to the lowest floor first.
2. When factoring the floors within the building, is there a designated floor 13? Floor designations may go right from 12 to14.
3. How many staircases are within the building? There is generally a minimum of two.
4. Do all the staircases serve every floor of the building?
5. Does any one of them terminate at roof level? At least one will.
6. Are they of the return or scissor type?
7. Do we have access to pressurized stairs, a smoke-proof tower, or a fire tower?

When attempting to get answers to these questions, it is important, especially when relating to #2, to determine if the building identifies a 13th floor. For superstitious reasons, some buildings will not label a 13th floor. What you will find is floor 12, followed by 14. Although in the actual counting of floors within the building there will be one factored as 13; the identification and how we factor it into our management of the incident is important. Complacency with this concern may cause confusion not only with climbing stairs, but with the setting up of different posts within the building.

Stratification

As we have already discussed, height considerations can play a significant role on the building's stack effect. The higher the building, generally, the greater the stack effect. But just as conditions can move smoke and deadly gas to the upper floors of a high-rise building, others may limit the smoke and gas and move smoke up and out of a vertical opening.

When the fires, smoke, and gases leave the initial fire area, they are for the most part, hot, buoyant, and under pressure. As these products leave the fire area, depending upon the artery they seek and the building's stack effect, the products of combustion could lose their buoyancy and cool as they come in contact with the entrained air, concrete walls, and the building ceilings. This cooling, and ultimate layering, of the smoke is referred to as stratification. These pockets of smoke could consume entire floors remote from the actual fire floor. Many civilians and firefighters have been caught in thick smoke conditions ten, fifteen, to twenty-plus floors above the actual fire floor.

Special Considerations: High-Rise

*COAL TWAS WEALTH**S***

Due to the complexity and demands that a fire in a high-rise building can create, the fire department must categorize, reference, and assign duties and responsibilities to its members in order to provide an efficient, effective, and safe operation.

In order to develop a standardized approach to the demands that these buildings present, here is a list of guidelines and concerns that will aid fire officers in their assigned area. Different departments may reference or further delegate some of the duties listed. This is an acceptable and, at times, advisable practice. What follows is an orderly approach to put the process in motion.

High-rise strategic operating guidelines

1. *Establish the lobby command post.* This becomes the responsibility of one of the first arriving fire department units. The incident commander must secure control and operation of the lobby area, and the building's fire command center. It is here where access to the building's fire safety director, the building engineer, as well as to the building's systems will be. Due to the responsibilities of this assignment, some departments will add the establishment of a lobby control officer. This individual will have the added responsibility of firefighter accountability.
2. *Determining and verifying the fire floor.* Knowing the actual floor where the fire is located, is initially, the most important piece of information the fireground commander must receive. This piece of information will influence all other decisions from the command post. Upon arrival, companies should seek preliminary information from the alarm activation, the fire safety director, building management, or building security. Initially, this information may be sketchy. It is critical from a safety standpoint that companies report to the lowest reported floor when multiple areas are reported in the building to verify the true fire location, and to report those findings back to the incident commander.
3. *Gain control of the building systems.* It is vital when fire department operations can no longer be considered ground based, that we gain control of the building systems that could influence fire and smoke conditions, as well as fire department operations. Some departments will assign this duty to a building systems officer who will report directly to the incident commander. Building systems that must be controlled include the following:
    - *HVAC system(s)* – The fire department must determine the status of all the HVAC systems in the building. This is to include Mechanical Equipment Room (MER)

floor(s) and zones they supply. Any systems that have not been automatically shut down must be manually shut down. This is to include supply and return fans of each affected area.

- *Elevators* – The fire department must ensure that all of the building's elevators are recalled to the lobby or sky lobby through Phase 1 operations. It also critical to note and identify which elevators serve what floors. Fire department members should attempt to identify which elevator bank(s) are covered by emergency generator service in the event of a power loss. Only elevators with fire service operations should be used.
- *Fire pumps/standpipe systems/sprinkler systems* – Determine the location of the fire pump room. A fire department member must accompany a building engineer to the fire pump room to assist with their operation, as well as establish communications with the operations or command post. Once communication has been established, determine and inform of the pump stages and floors served. The fire pumps must be operated in coordination with the operations or incident commanders requests. (Fire department must be prepared to augment both systems.)
- *Communications* – Communications at any fire scene is critical for an effective, efficient, and safe outcome. In a building of great height, it becomes more of a concern. It is important to gain access to all means available, whether they are fire department radios, hard wire communications, in-house telephones, cell phones, or sound powered phones. Establishing a primary and secondary communications link for all designated areas is critical.

4. *Establish the fire operations post.* The operations officer will establish this position generally one or two floors below the fire floor. The variation between one or two floors below is dictated by tenable conditions, which may be affected by a reverse stack effect within the building. It becomes the responsibility of the Operations Officer to establish a communications link with the incident command post, to initially report the size of the fire, the attack stair being used, the progress of the first deployed hoseline, and additional resources needed. The fire attack should begin with a minimum of an engine and ladder company, preferably with two engines and a ladder company. As conditions progress, the chief in charge of operations must control and coordinate all units operating on the fire floor and all floors above. He/she must also advise command of any problems encountered as well as anticipated, and establish communications with the staging area once it has been established.
5. *Begin the evacuation process.* Begin by evacuating those people who are in immediate danger. Those priority areas will include the fire floor and the floor above. In many instances occupants will have begun self-initiated evacuation by leaving an area by any means possible. Until control can be obtained, and people properly directed, members must be alert to civilians in the attack stair as doors are opened

in an attempt to advance onto the fire floor. Due to the demands of this assignment, some departments will assign an evacuation officer. This individual will report to either the command post or the operations officer.

6. *Establish an interior staging area post.* As operations will dictate, a staging area should be established one floor below the operations post. This area should be of adequate space to hold units and equipment in reserve, have direct accessibility to the operations floor, as well as have direct access to companies to report into staging.
7. *Establish a logistics officer.* Due to the demand that these buildings will present, a logistics officer should be assigned and become directly responsible for supporting all the incident's needs. Those areas should include the following.
   - *Communications* – Ensuring that multiple communications links have been established. Coordinate this operation with the building systems officer if one has been established.
   - *Stairwell/elevator support* – Ensuring that all the necessary equipment and firefighters are getting to the designated staging area. Coordinate this operation with the building systems officer if one has been established.
   - *Establishment of a medical and rehabilitation post* – This area is assigned to delegated member and established at least one floor below the interior staging post. Depending upon the access and the spaces provided, it is generally not advisable to place this same post within the staging area. Confusion will be eliminated and responsibilities will be more defined if they are separate from one another.
8. *Establish a safety officer(s) and a rapid intervention company(s).* Depending upon the size and complexity of the incident, a minimum of one, preferably two, safety officers with a number of assistants may need to be established. Initially, interior safety should be established at the operation post to assess fire conditions and operations on the fire floor and floor above. An exterior safety officer should also be established to monitor exterior hazards to firefighters and to escaping building occupants. Some departments may substitute this individual with a street vent coordinator or use these individuals in coordination with each other. It is also critical that *at least one* company be established as a rapid intervention company. Their location should be at the operations post in the event of a lost, or trapped firefighter in need of assistance. The primary purpose of their location is to eliminate lead/reflex time concerns.
9. *Establish a street vent coordinator.* This is one of those positions where you wish you had the extra people in order to establish the position. Their assignment and responsibilities are directly related to the life safety concerns of the incident. Being able to control the evacuation to the exposed side, as well as observe and report wind and fire conditions could greatly affect the outcome of the incident.
10. *Establish a planning officer.* As incidents escalate to multiple alarm assignments, Incident Commanders will need to delegate responsibilities to this individual to aid with the following:

- The tracking and assigning of resources.
- The maintaining and displaying of situation status.
- The reviewing, evaluation, and revision of the action plan.
- When necessary, the preparation and implementation of a demobilization plan.

High-Rise

HVAC Tactical Guidelines

These guidelines describe a combined effort between the incident commander, the operations officer and the building engineer. If any one of these individuals is missing from the equation, it should not be attempted.

1. *Determine the status of the HVAC system.* All HVAC systems that have not been shut down automatically, must be shut down manually. This should be an automatic step for the first arriving fire department units.
2. *HVAC use/phase 1 operations.* Once the fire floor has been determined and verified through the operations officer, all HVAC zones that do not serve the fire floor can have their supply fans activated. This procedure will supply fresh outside air to these areas, which will pressurize the zones, limiting smoke into these floors. It is critical that the fire floor be determined before this procedure is implemented.
3. *HVAC use/phase 2 operations.* Once this Phase 1 is completed and is reported as working well, the incident commander and operations officer can evaluate the possible evacuation of smoke from the non-pressurized zone. Prior to using this tactic the Incident commander and operations officer must gather the following information from the building engineer, the HVAC floor diagram for the area in question, and the fire operations post:
   - The exact location of the fire on the floor.
   - The location of the return airshafts.
   - The location of the attack stair.
   - The location of the evacuation stair.
   - The location of the ventilation stair.

Gathering this information will help the incident commander and operations officer factor if this procedure will accomplish the following:

1. Draw the fire toward or away from advancing crews.
2. Intensify the fire by use of the system return fans.

3. Cause the fire dampers to close from the temperature of the fire, which will negate the operation altogether.

If Phase 2 is to be considered, all companies operating on the fire floor must be backed to the attack stairs before this operation can be implemented. Once all units have been accounted for, the incident commander can have the building engineer activate the return fans for the fire zone. Conditions within that zone must be monitored and evaluated for their effectiveness. All findings must be reported to command and operations for further evaluation.

Questions For Discussion

1. Identify and review the differences between the older and a newer generations of high-rise buildings. How will these differences affect fire department operations?

2. Review the listed duties of the first due engine and ladder company when arriving at a high-rise. Compare the listed duties with your own department's procedures, and discuss how they may differ and why.

3. List and discuss the hazards to the firefighter when arriving and going to work at a fire in a high-rise building.

4. What concerns need to be addressed in the street, when a fire involves the upper floor of a high-rise building?

5. What aides may be in a high-rise building to assist with the fire department's efforts?

6. What are the concerns associated with the wind?

7. What is a convenience or access stair, and how will their presence affect fire department operations?

8. What are the possible fire extension concerns within a high-rise building?

9. What building systems need to be controlled within a high-rise building?

10. List and describe the strategic plan for the use of an HVAC system.

10 Vacant Buildings

Fires in vacant structures require fire officers and firefighters to take a well-calculated and cautious approach before going to work in these types of buildings. It doesn't take very long for a building that has been unoccupied or abandoned to become a death trap for neighborhood firefighters. Once abandoned, these structures will become quick preys for vandalism. Anything of value in the building (namely the lighting, plumbing fixtures, appliances, wood trim, floor boards, copper pipes, and cast iron tubs) will have been stripped and salvaged to be sold or used for their inherent aesthetic value. Their method of removal, and the additional openings that they create will definitely increase the fire and collapse concerns within the building.

In addition to the salvaging of the building, we must remember that these buildings may have also been subject to years of neglect and disrepair that could have been further compounded by their exposure to the weather. Buildings exposed to rain, sleet, snow, and ice will create additional structural concerns, which will again increase the collapse potential of the building.

Additional considerations could also come from any previous fires within the building. Many of these buildings become easy targets for arson. A fire of any size could have done considerable damage to a building whose structural integrity is already questioned.

Whether you have one, or thousands of these types of buildings in your town or city, fire departments must be prepared to deal with the difficulties that these buildings may present. When going to work in or near a vacant building, the safety of the firefighter is the primary concern.

Construction: Vacant Buildings

COAL TWAS WEALTHS

Vacant buildings will not be subject to one particular type of construction. They can encompass buildings of fire-resistive design to buildings built of wood frame. It is important to initially note in this size-up category that the class of construction firefighters are responding to and going to work in, will still have the same fire spread and collapse concerns when the building was occupied, but now those initial concerns can be greatly magnified.

Abandoned

Buildings abandoned and neglected by their owners will fall into disrepair quickly. From this you must expect that windows will be broken or removed, glass in skylights will be missing, and the bulkhead doors and scuttle hatches will be damaged or gone. From this association, it must be further anticipated that weather in any form will have been allowed to accumulate on floors, walls, and stairs. Water entering the building will not only affect the interior surfaces of the building, but prolonged exposure will eventually penetrate surface areas, rotting and destroying the integrity of structural members. These dangers are further enhanced when the damage extends to load-bearing members. Buildings that have visually appeared to be structurally sound and attainable, have totally collapsed, killing and injuring firefighters.

Buildings subject to structural rot, will present the least visible sign to the imminent dangers of structural collapse.

Fig. 10-1 *An example of an extremely dangerous building.*

Salvaging

Salvaging is a significant concern in any vacant building regardless of its class of construction. However, if we were to fine-tune our thoughts on where this will cause the greatest concern for the firefighter, we need to start with those buildings built of Class 3 or Class 5 construction. I say this from the obvious ease associated with the removal of wooden members when compared to masonry or steel.

Buildings, once abandoned, may contain hundreds, if not thousands, of dollars of salvageable items; some of the most notable being decorative wood trim, wood flooring, marble, and cast iron. Depending upon the condition of the salvaged material, these items can gather large sums of money at a local antique dealer. From the removal of copper piping originally associated with the building's plumbing, aluminum and copper from the building's wiring, iron from the old gas fixture piping within the building, to any marble that can be found in the hallways, building vestibules or staircases, all will be worth money to someone.

One of the specific concerns that the fire department has with some of the more heavier items, is the methods used for their removal. Sinks don't present too much of a problem for the individual attempting to remove them. They simply can be pried from their setting, and either carried down a flight of stairs, or dropped onto a pile of mattresses to a floor or courtyard below. Furnaces still within the building are going to obviously be a very heavy item, which will require that they either be broken up by a sledgehammer, or that they be cut up by a torch in order to be removed from a basement or cellar area.

The technique used to remove a toilet or cast iron bathtub however, is going to be much different. These obviously heavy objects are going to be worth money if they're in good shape, and if they can be removed intact. The easiest way to achieve this objective is for the *salvage team* to remove the structural members below the tub, in an attempt to drop it to a floor or level where it can be easily removed to the exterior of the building. The concept associated with cutting, prying, and removing floor joists from below, allows gravity to drop the fixture to yet another pile of mattresses. This technique of removing the floor joists and allowing the fixtures to fall in this manner is a technique used for all toilets and tubs, in all bathrooms, on all floors throughout the building.

With this in mind, we know from our previous discussions concerning the locations of bathrooms and kitchens within a multiple dwelling, that they will be stacked in a vertical line throughout the building. If this particular group of individuals removes and drops all fixtures as previously described, these new openings within the building will not only affect the structural integrity of the building, but their presence will greatly enhance the fire growth in the building.

Previous fires

Structural damage from previous fires is a serious problem in vacant buildings in all areas, not only the inner city. The damage done from one fire, let alone a number of fires within a single building will greatly impact the structural integrity of the building, which will directly affect the firefighter's ability to safely operate within the building.

During the heavy fire duty of recent years, the fire activity in the inner cities was at a staggering pace. Going from one vacant building fire to another was a common occurrence. It seemed that knowing where you were operating was not as important as the number of fires you responded to that tour. The danger that developed from this complacency was that firefighters were responding to numerous fires over and over again in the same building, and not passing the concerns on to other shifts of firefighters. Depending upon the size and area of the building, as well as the location of any previous fires within the building, signs of structural damage may not easily be evident to a new group of firefighters.

One incident that leaves a lasting impression on me as a new firefighter was responding on a second alarm to a fire in a vacant, six-story, Class 3 constructed multiple dwelling. Approximately 15 minutes into the operation, the entire back wall of the building collapsed into the rear yard, bringing six firefighters down with it. Luckily, all six firefighters survived, but three were forced into early retirement from their injuries. Later we found out that this particular building had six previous multiple alarm fires in it. This was totally unacceptable. Risking the lives of firefighters for what was really a six-story piece of garbage should have never happened. From this lesson, it became obvious that a vacant building marking system to safeguard the firefighter was a must.

Fig. 10-2 *Previous fire damage must be identified.*

Vacant building marking system

Many cities throughout the country use a simple marking system of drawing on the side building with a can of spray paint to give information relating to the integrity of the building. A series of spray painted boxes with additional symbols is designed to give the next arriving compliment of firefighters vital information about their ability to operate safely in, or around the building. As simple as it may seem, the key to its use is that it remains current, and that all members strictly follow it.

Even with what appears to be a small fire, an X in the box indicates that firefighting forces must

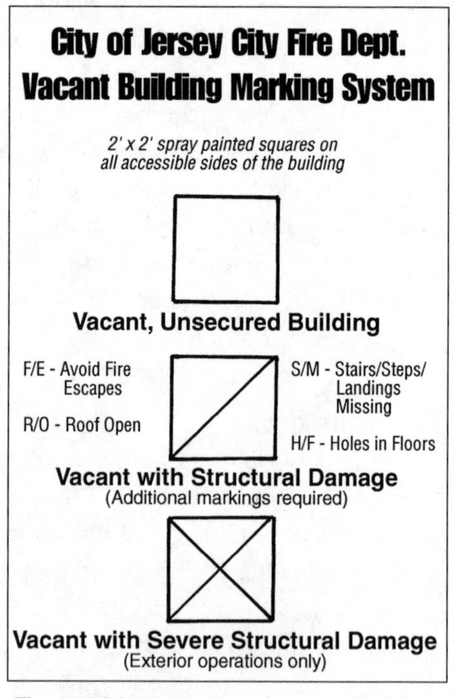

Fig. 10-3 *Marking system for vacant buildings.*

stay outside and extinguish the fire defensively, regardless of what is showing. Placing an X in the box states the building has suffered enough structural damage from the weather, rot, vandalism, or a previous fire, that any attempt to enter the building would be extremely dangerous.

To keep the information relevant and useful, it is also critical that the markings on the building be updated periodically. This is advisable for a number of reasons. First, the building's conditions can change. A building that was originally marked as structurally sound with no concerns, could have been salvaged only weeks, days, or even hours after your review, creating openings in floors, walls, and ceilings. Those buildings that have been left unattended and unseen for an extended period of time could also have additional areas now subject to rot and decay. Other buildings may have had a fire in them while your shift was off duty ultimately affecting the integrity of this building and its features. Marking the building once and thinking conditions can't change for the worse is unrealistic. Any building left unattended for any considerable length of time is not going to improve.

Fig. 10-4 *Markings must include any damaged and dangerous building features.*

One of the easiest ways of updating the markings is to do it right after extinguishing a fire in the building. If additional damage by the fire, or from fire department operations has rendered a portion of the building, or all of the building to be dangerous, then it should become the responsibility of the incident commander to ensure the marking system is updated with the latest information before the last company leaves the scene. Ignorance or complacency with this system could allow a different shift of firefighters to crawl in the same building at great risk.

Fig. 10-5 *This marking advises firefighters that the building has structural damage. Most notably to the fire escapes and the roof.*

An additional concern for reviewing and updating the markings on a building comes from the possibility of faded, or altered markings. Over a period of time markings will fade. In an attempt to make them as visible and lasting as possible, many departments use fluorescent orange or yellow spray paint as the standard marking colors. These two particular colors seem to stay visible the longest.

Fig. 10-6 *Place your markings high on the building.*

Altered markings were an additional concern that was initially a problem because of our own doing. We were finding in some sections of Jersey City that juveniles were spray painting over, or adding markings to the fire department's, which made the symbol at times difficult to read. This was quickly remedied by placing the markings high enough on the structure so they could not be easily reached.

If you have one, or thousands of vacant structures in your city, you should have a marking system. It may not be a foolproof system, but it will definitely help with the safety of your firefighting operations at these structures.

Occupancy: Vacant Buildings

COAL TWAS WEALTHS

If for a moment you thought that your occupancy concerns when arriving at a fire in a vacant building are not a relevant consideration, think again. The same occupancy concerns we have been discussing throughout the book, specifically the people, content and the occupancy/construction concerns, are still going to be a significant list of considerations that the fire officer must factor into his or her decision making when arriving at a fire in a vacant building.

Load and status

On more than one occasion I can remember responding to what appeared to be a vacant apartment building from the street address side, only to quickly find in our search efforts fully occupied apartments in the rear of the building. Large area buildings, generally those that are deep off the street and attached on both exposed sides, will not allow for a view of the structure other than the street side. Windows boarded up in the front possibly from a previous apartment fire, vandalism, or renovation, may initially give the appearance that the entire building is vacant. Even those companies assigned and going to work in the front of the building to fight a fire may still not realize that people inhabit apartments in the rear. It is for this reason that buildings, even if viewed as abandoned, must be searched for apartments still in use or for any homeless who have reoccupied the building.

Firefighters and fire officers who have vacant buildings in their response districts should be alert to signs that a building has been taken over, or re-inhabited by squatters and the homeless.

Occupancy indications: vacant buildings

1. Electrical extension cords stretched from building to building. These can be stretched from window to window, through alleys, or laid in rear yards.
2. Lights or candles in windows/rooms.
3. Curtains, shades, cardboard, or newspaper over windows.
4. Flowers or plants in windows.
5. Openings in previously sealed areas.
6. Large amounts of activity around entrance openings.

There is also the possibility that vacant structures are occupied by workers renovating the building. What initially appears to be an unoccupied building may actually turn out to be a building with trapped construction workers. When any of the following are present, the building can possibly be occupied by construction workers.

Additional occupancy indications: vacant buildings

1. A presence of construction vehicles in the street.
2. Any construction debris/dumpster near the building(s) in question.
3. Construction materials near the building. (i.e., piles of lumber, sand, stone, block, etc.)
4. Any newly constructed fences around the property to safeguard the storing of construction materials and the building itself.
5. Signs of newly installed material or recent work. Examples are power-washed brick, repointed brick, trimmed window openings, new windows, and demolition tubes.

During normal district inspection, or when returning from alarms, sometimes all it takes is a glance to the left or right side of the street to gather this additional information.

Previous occupancies

Buildings that are abandoned and vacant may also lend some useful information from their previous listed occupancy. Buildings whose occupancy name still appears on the facade of the structure may assist a fire officer in making decisions that may prevent firefighters from becoming trapped, or lost from unsuspected conditions. Buildings with names like, "Drennan Dye and Wax," "Acme Cardboard," "Blanchard's Brewery," "City Roofing," and "Foy Paints" should be a clue to responding firefighters of the possibility of any inherent building hazards and features that still may be within the building.

Fig. 10-7 *City Chemical Corp.*

It's not uncommon to find wax-coated walls, ceilings, and floors still within an abandoned candle factory. When present, these type conditions will *still* add to a rapid fire growth. Buildings that were a former cold storage or brewery may still contain insulated floors and rooms lined with cork. Although abandoned, buildings of this type will still be able to produce heavy smoke conditions and extreme fire loads, all within a maze of rooms. Old abandoned manufacturing mills will still contain oil soaked floors and walls. From the added possibility of abandoned stock and product residue, accelerated fire conditions can trap unsuspecting firefighters operating within this type of building.

Although the building is abandoned, its former use will still present information that should affect fireground decision-making.

Occupancy/construction associations

From the same practice of gathering information from the building's previous occupancy, we can also put this same information to use in determining any inherent construction features that were once necessary to meet that former occupancy's needs. Abandoned bowling alleys, movie theaters, and supermarkets (just to name a few) all required at one time large, open areas to accommodate their business needs. Large, open floor spaces that require large spans can only be accommodated from a truss design. Truss designs within the building can be in the form of large wooden timbers, smaller dimensional lumber, or unprotected steel. In these cases, exposure to the weather could have rotted wooden members, moisture may have corroded gusset plates, and steel could have lost any protection previously associated with its installation.

What this information boils down to is this, what construction features we feared when the building was occupied are still in place, but now, in a deteriorated and unstable condition. Simply, when left abandoned and vacant, nothing changes for the better, it only gets worse!

Apparatus and Staffing: Vacant Buildings

COAL TWAS WEALTHS

The amount of resources assigned to a fire reported in a vacant structure should not lessen by any stretch of the imagination. Because the building is referred to as vacant and abandoned, the only concerns that might lessen are those in the category of the occupant life hazard, and even that is not a guarantee. Fires in vacant structures require the same complement of apparatus and firefighters to respond as when the building was fully occupied, and in many instances, you are probably going to need more.

Engine Company Operations

Hoseline selection, stretch, and placement

Fires in vacant structures for the most part will be engine company oriented. This means well-established water supplies with intensive efforts on proper hoseline selection, stretch, and placement. When interior operations allow, hoseline selection must be predicted on existing, as well as developing, conditions.

Big fires will still require big water. However, fires that are small to moderate in size may prompt fire officers to order larger size hoselines to be stretched. The reason behind this consideration is not only dictated by the fire load and the potential fire area of the building, but again from the fact that the building may be open, which in turn can enhance the fire's ability to take possession of the building. Fires that initially appear small and confined on arrival can quickly become large and overwhelming for the smaller stretched hoseline. Officers should never hesitate to order a larger hoseline stretched, or for that matter, a backup hoseline even when the initial observation shows a small fire. The stretch, placement, and operation of any hoseline must be done with strict attention towards member safety.

An incident commander should focus on hoseline stretch and placement around the concept that if one hoseline is stretched and working, you should probably order three; one to the fire, one to back it up, and one to the floor above. Additionally with a fire in a vacant building, we must add the possibility that fire can drop down through openings in the floors. This may warrant redirection of a hoseline, or depending upon conditions, the ordering of an additional hoseline.

Fig. 10-8 *Plan on rapid fire growth.*

To further this thought, anytime the building's accessibility is blocked, damaged, or questionable in any way, officers should consider hose stretches via ground ladders, a safe and secure fire escape (if there is such a thing), or by rope in an attempt to not only lessen the impact friction loss may have on the stream, but to provide the safest possible route for firefighters. As simple as this may sound in print, hose stretches of the types mentioned may be staffing intensive. As in all cases, long stretches may require the grouping of engine companies, or if there is a lack of truck work, the assistance of a ladder company to ensure that the hoselines get to where they are supposed to go. The speed at which the hoseline needs to be stretched, the distance and route it may need to travel, accompanied with the high probability of kinks from building debris, warrant additional help with hoseline stretches in vacant buildings.

Ladder Company Operations

Forcible entry

As we had mentioned earlier, operations in vacant buildings will be primarily engine company intensive. This is not to say that the ladder company doesn't have specific responsibilities when arriving at a fire in a vacant building, they will. However, depending upon the specific building, many of the building's obstacles may have already been removed. If this is the case, some of their initial responsibilities may have lessened. However, in other buildings, forcible entry difficulties may have to be overcome before the engine company can attempt their hose stretch. With cinderblock doors and windows, plywood covered windows and entranceways, or roll-down steel gates and doors, the tasks may be more difficult and time consuming than when the building was occupied.

Fig. 10-9 *Closed and boarded-up structures will create added difficulties.*

Searches

Searches in vacant structures still must be conducted to ensure that there are no occupants in the building. Buildings, as uninhabitable as they might seem, could very well have a number of homeless or squatters living inside. Be conscious of this thought, especially with large area buildings. Whatever the situation, incident commanders must assign personnel to conduct searches in both the primary and secondary modes when the building and fire conditions allow.

Ventilation

The one area of responsibility of the ladder company that will often require the least attention is ventilation. For the most part, these buildings will be open. Glass in window openings will be gone, scuttles and skylights are probably missing, void spaces will have been stripped for their copper and iron, and the roof may already be open from a previous fire. All of this eliminates the need for the firefighters to open up.

Fig. 10-10 *Advanced conditions will require large streams.*

Apparatus Placement

When conditions are well advanced, apparatus must be placed with elevated streams in mind. Tower ladders and ladder towers are probably the fireground commander's best tool for combating this type of fire. Their versatility provides the incident commander with the capability to extinguish a tremendous amount of fire, not available from ground-based streams. When feasible, they should be positioned to cover two sides of the building with exposure protection being the initial consideration. These pieces of equipment are often the key to an effective outcome in many large volume fires. Elevated streams from these apparatus can deliver between five and six tons of water per minute, all from safe distances.

Life Hazard: Vacant Buildings

*COA**L** TWAS WEALTHS*

This is an important, as well as lengthy, section of text, and for a valid reason. Many firefighters have been severely injured and killed fighting fires in vacant buildings. These are the most dangerous types of buildings that firefighters will encounter. There aren't too many cities across the country that don't have at least one vacant structure within their boundaries. To be aware and prepared for the hazards these buildings present could very well prevent serious death and injury. It's critical that we remember the concerns and share the information when and where we can.

Firefighters

The primary mission of the fire department is to extinguish fires in structures with the most efficient, effective, and safest procedures available. This mission obviously includes fires that involve vacant structures. At fires in vacant buildings, first arriving officers must consider the attack mode to be used with its primary emphasis on firefighter safety. Modes at times are often easy to identify. Buildings that are heavily involved will obviously require a defensive mode to stabilize the incident. Buildings that are less involved may require an offensive procedure, with hoselines and members going to work in the building's interior. However, fires in vacant structures will require additional information other than fire conditions to determine if an offensive attack is the safest procedure for members to attempt. In urban areas, many fire departments are confronted with hundreds, if not thousands of vacant structures within their jurisdictions. Attempting to maintain a mental note on the conditions of this many buildings is an impossible task. It is for the reasons previously mentioned that departments use, and continually update a vacant building marking system for safeguarding their firefighters. This information will aid the incident commander in determining an offensive or defensive mode. Light smoke showing from a window of a building that has an X painted on the front wall equates to putting the tower ladder up and hitting it from the window.

As disciplined and accurate as a fire department can be with a comprehensive marking system, there will be many times when we must go to work inside. When this occurs, firefighters should always be alert to specific areas within a vacant structure that are known trouble areas.

Firefighter Hazards: Vacant Buildings

Fig. 10-11 Underside of open steel pan interior staircase

Interior stair cases – This area seems to suffer the most damage, and cause the most concern. We initially focus on this thought from the fact that the interior case will have been subjected to the most abuse over the years, simply due to it being the main artery of the building. Neglect from years of abandonment, disrepair, abuse, and salvaged items initiate the concerns.

Added concerns will come from the fact that staircases in most multiple story buildings will have windows at their half landings, as well as skylights, scuttles, and bulkheads over them. In a vacant structure, all of these openings will be broken, missing, or if present at all, leaking. Rain, snow, and ice constantly penetrating these areas could turn the building's most vulnerable artery, into the most dangerous.

Window openings, top floors, and roofs – For the same reasons mentioned with the interior staircase and its vulnerability to the weather's elements, we must also place the same concerns to areas just inside window openings, and depending upon the condition of the roof, the entire top floor. Anytime weather is permitted, repeated and prolonged contact with structural items that were not designed and protected from this exposure, rot has to be anticipated.

Just inside window openings, water, ice, and snow will be allowed to accumulate jeopardizing the integrity of the floor. Roofs in disrepair, or those that have previous holes from a fire or from fire department operations, will also jeopardize the integrity of the roof deck, its supporting members, as well as the top floor.

Fire escapes – What once served as a secondary means of egress for escaping occupants, can now become a steel death trap hanging from the side of a building. They can become a major collapse concern to members attempting to climb or descend them. Years of neglect and disrepair will cause the steel to corrode and rot, which will affect the integrity of the fire escape steps, landings, handrails, and ladders.

Their stability could be further questioned at the point where their supporting bolts and plates secure the entire assembly to the exterior wall. A structurally damaged wall may allow the bolt and plate assembly to pull through once a firefighter places weight on it. They should be avoided at all vacant building fires!

Bathrooms and kitchens – Certain rooms within a vacant structure will cause concern from their previous use, as well as from their anticipated salvage. Bathrooms and kitchens are known difficult areas within a structure. We know that the many utilities that service these areas have allowed fire to extend from one floor to another when the building was occupied. With the building now being abandoned, these are generally the first two areas that will have their piping, wires, and fixtures removed as salvageable items. With their removal, will come additional openings further enhancing the fire's ability to spread.

These same areas when in use, were also known as damage prone areas. Water leaks, even those that were minor and left neglected, may have rotted floorboards and floor joists. Household fixtures and appliances when left in place, increase the hazard from their added weight on the deteriorated areas. One of the most notably troubled fixtures is the bathtub that is full of water. Water accumulation from exposure to the weather or from fire department operations, could give the bathtub the added weight it needs to drop through the floor(s) on its way to the basement. Firefighters should never stand in a bathroom, or obviously below a bathroom when heavy fixtures are still in place. Overhaul should be done from a door opening, not from within the room.

Vulnerability to fire spread, salvaged items, rot, as well as any heavy appliances and fixtures that may still be present, make both of these rooms two definite trouble spots that firefighters need to be aware of.

Rapid fire spread – Rapid fire spread within a vacant building must be anticipated. From the numerous openings created within the building, to deliberately set fires, firefighters and fire officers must use calculated and cautious decisions when going to work at a fire involving a vacant structure.

Firefighter accessibility and egress – After a building is taken over by a bank, city, or landlord, an attempt may be made to safeguard the structure until the building is resold, reoccupied,

***Fig. 10-12** This conflagration must be planned for.*

or torn down. The most common way of securing the building is with plywood over its door and window openings. The plywood may be simply nailed or screwed over the opening, or it can be further secured by installing stainless steel bolts through the plywood cover into a series of lumber that extends across the outside and inside of the window opening. This method of securing window openings was first employed in buildings that were being protected by the Department of Housing and Urban Development. HUD windows, as they are still referred, will add an extra dimension for their removal.

Fig. 10-13 *Sample of a plywood fortified window.*

In less frequented situations, buildings can become completely fortified structures. Building owners may close all the openings in the structure by having masonry block installed. Concrete block placed in window and door openings will turn the building into an actual fortress. The inability, or at least the significant delay in creating any type of substantial opening creates difficulty for firefighters attempting to enter and exit the building, in addition to allowing a fire to spread throughout the building. These now windowless buildings, will significantly compound all phases of your operations.

Firefighter life hazard concerns related to vacant building *accessibility and egress* can be summed up from the following:

1. *Delayed discovery of fire.* Buildings that are closed and secured may not show signs of smoke and flame until companies open up the building and commit themselves to the interior.
2. *Delay in extinguishing the fire.* Even if we catch the fire in its early stage, the delay in removing the secured openings may allow the fire to rapidly progress.
3. *Increased risk from flashovers and backdrafts.* Depending upon the method of securing the building's openings, the compartmentalization of the involved area, and the building's fire load, firefighters can be at greater risk from flashovers and backdrafts.
4. *Lack of secondary egress openings.* With boarded up or concrete blocked window openings, firefighters lose any alternate means of egress if they become trapped inside the building. Re-establishing those openings must become a priority if firefighters are going to work inside.
5. *Disorientation.* Boarded up or blocked up openings may cause members to lose their bearings within a room. The ability to establish a window opening for ventilation and

egress purposes will allow firefighters to orient themselves within the room layout. Not having that opportunity may further enhance any disorientation as members attempt to leave the room or area.

6. *Deliberately set traps.* As sad as it may sound, many firefighters have encountered traps deliberately set to injure a firefighter. As rare as these instances occur today, firefighters should still be alert to their possible presence and plan to protect themselves. A number of these situations that have been experienced and documented include:
   - *Containers of gasoline* – attached to a ceiling, or placed in overhead areas. Members advancing in to extinguish a fire may find a fireball of liquid flash over their head, in addition to having flaming liquid rain down on them.
   - *Multiple fires* – Sometimes arsonists will light a significant fire on an upper floor to initially draw the immediate attention of the firefighters to stretch and operate on that level. While this is taking place, a smaller fire set on a lower floor with diesel fuel, which initially burns slow and then rapidly, traps firefighters operating above it.
   - *Holes in floors below window openings* – Holes have been deliberately cut in a floor just inside the window opening to drop a firefighter to a lower floor. These openings may have carpet, linoleum, or cardboard placed over them to give a member a false feeling of security when probing for a floor. These items will easily collapse under the weight of a firefighter.
   - *Stairs, steps, or landings removed* – As we have already stated, these are known trouble areas simply from their location and neglect. Additionally, steps or landings may have been deliberately removed to catch an unsuspecting firefighter, dropping him or her to the floor below.
   - *Large pieces of furniture or building materials used to block entrance and exit openings* – This is done for a number reasons. It could be from squatters attempting some type of security for their building or apartment. It may also be done to slow down advancing firefighters from getting water on the fire. In either case, it causes an accessibility and egress concern in an emergency situation.
   - *Refrigerators or large furniture positioned so they could fall on advancing firefighters* – A common place for these items will be on a stair landing, or at the top of a staircase. Any large item coming down a staircase may crash through the staircase bringing everyone down with it, or the item may slide down the staircase like an out-of-control bobsled.
   - *Buildings used as drug dens* – These buildings could very well be booby-trapped to keep individuals out to protect their inventory or privacy. This may include piano wire across doors and window openings, plywood boards containing exposed nails (giant pin cushion) laid across the floors just inside door and window openings, or fire escapes that have been partially dismantled so they can pull away, or collapse under an individual's weight.

Prevention tactic

Any time that engine companies are advancing into a vacant building fire, especially one that is known to be a crack house or drug den, it should be a policy for the nozzle man to sweep the floor with the hose stream to not only cool the floor from embers and debris, but to also to wash away any needles or broken vials that may penetrate a member's turnout pants. We used this policy for years in all types of buildings to prevent burns to the knees, most notably before the introduction of turnout pants. Washing the floor in front of you as you work your way in will not only help to eliminate red knee caps, but may also eliminate items that may cause additional harm.

Terrain: Vacant Buildings

COAL TWAS WEALTHS

Your size-up concerns relating to building setbacks, and structures built grade-level do not change because the building is vacant. The same concerns that are noted when the building was occupied, will still be there. Apparatus placement and operations pertaining to laddering, hose line selection and stretch, and all their related concerns will still continue. Officers must keep in mind the delays associated with this particular size-up concern and the rapid fire conditions associated with these buildings. This will greatly affect your decision-making.

Water Supply: Vacant Buildings

COAL TWAS WEALTHS

Availability

Initial water supply concerns for fires in vacant structures will depend upon the availability in the area. With the vast majority of these types of structures being in the urban areas, water supply will most notably be from the public water system. Water availability must be given further consideration from the possibility of an inoperative, or vandalized water supply. In many of the inner city areas where vacant buildings accumulate, fire hydrants are often tampered with, or vandalized. This could mean anything from missing hydrant caps, damaged threads, to barrels filled with debris. This concern becomes further complicated when an arsonist takes a couple of swings with a sledgehammer to the threads of all the hydrants in the immediate

Fig. 10-14 *Vandalized/neglected hydrant*

area. Vandalizing a number of nearby hydrants will obviously cause a major delay in establishing an early and sustained water supply.

Auxiliary Appliances and Aides: Vacant Buildings

*COAL TW**A**S WEALTHS*

To sum up this section in a nutshell—forget about it! If the building hasn't become vacant within the last 24 hours, then anything of value has most likely been stripped from the building. Just the fact that the building has become abandoned will render the building's alarm system and suppression system unusable. In those rare cases where there is an attempt to protect the building, the property owner may maintain the building's systems. This information, when available, must be noted into the department's pre-incident information. Knowing that you have fire detection and suppression equipment in service at a particular vacant structure not only aids in protecting the property, but also aids in protecting the firefighter.

Street Conditions: Vacant Buildings

*COAL TWA**S** WEALTHS*

The information and concerns related to street conditions will not have changed because a building on that block has become abandoned. Any previous concerns with the street's width, grade, surface, and traffic flow will generally be the same when the building(s) in question were occupied. Concerns, however, will increase with street conditions when entire blocks or entire neighborhoods are abandoned. Areas that have been forgotten often have their streets forgotten as well. From potholes, collapsed surfaces, or abandoned autos may be just some of the few concerns that you may be presented with when you must go to work. Any one of these problems could render sections of the street impassable, unstable, and difficult to place apparatus on.

Knowing the difficult blocks in your district is a plus, especially before you need to drive down them at night. Chances are any street lighting in these areas is long gone also.

Weather: Vacant Buildings

*COAL TWAS **W**EALTHS*

Wind

Of the weather concerns—wind, temperature, humidity and precipitation—the wind again seems to stand out as the predominant concern. On more than one occasion firefighters have stretched and gone to work on what started as a small manageable fire only to have it fanned out of control by a gusting wind. Firefighters should never underestimate the wind, especially in a vacant structure.

Vacant buildings will offer no resistance to the wind. With the building's windows missing, wind can turn what appeared to be a controllable fire into an inferno in seconds. This becomes a critical element in your fireground size-up that cannot be ignored. Being observant to the weather before an alarm is received, knowing wind conducive areas in your district or city, as well as being observant of wind conditions upon your arrival, will greatly aid in your decision-making. Anticipation continues to be the key.

Exposures: Vacant Buildings

COAL TWAS WEALTHS

The potential fire involvement at any incident will come from a number of size-up factors that we need to gather upon our arrival. Factors most notably will be from the height and area of the building involved, but we must also add the concern of any nearby or attached exposures. This becomes a real concern for the fire officer when confronted with a vacant building fire. We say this simply from the fact that more times than not we are going to be encountering an advanced fire upon our arrival. As stated, fires in vacant structures just don't happen, they are mostly incendiary.

With the protection of any exposure building, the protection of life still remains as the governing factor. This concept doesn't change with a fire in a vacant building, even when the building is heavily involved. When fire officers are met on arrival with a vacant building that is heavily involved, the first consideration for the placement of the first deployed hoseline is between the fire and the most severely threatened life exposure, not necessarily the most severely threatened exposure. This previously discussed concept continues with your exposure concerns when confronted with a fire in a vacant building.

Fig. 10-15 *Prioritize exposure protection.*

When no life is endangered nearby and the fire building is threatening both exposures, initial consideration has to be toward protecting the more substantial piece of real estate. Considerations for this placement are the:

- Exposure proximities – Generally the closer it is, the more trouble you'll have.
- Exterior sheathings – Brick may buy you more time than wood.
- Height and area of the exposures – Taller and larger buildings may increase the fire's spread potential. Protect the big one first.
- Fire load of the exposures – A fully stocked warehouse provides more of a initial threat than a vacant one.
- Value content – High value properties should be given early consideration. The building's stock may outweigh the value of other exposed buildings.

Area: Vacant Buildings

*COAL TWAS WE**A**LTHS*

Area size-up concerns for vacant structures will still follow the same listed concerns for fires that involve occupied structures. The building's square footage, shape (regular or irregular), and whether it is interconnected or not, will still cause concern relating to the potential area of involvement. What fireground commanders must continue to remember is the fire extension probability and possibilities of these types of structures. Large square footage buildings, irregularly shaped structures, or structures that may wrap around a nearby exposure can produce large fire conditions that will threaten nearby exposures, as well as entire neighborhoods. Obtain square footage information as soon as possible.

Location and Extent of Fire: Vacant Buildings

*COAL TWAS WEA**L**THS*

Until this point, we have been stating concerns about the stability of the building from vandalism/salvage, weather damage, previous fires, or incendiary devices. These same concerns (when present) will add serious concerns to our fire location and extent concerns. Initially, remember that when the building was occupied the same location and extent concerns that existed then will still be there. From open interior stairs, light and air shafts, to pipe chases and plumbing voids, these arteries will still allow fire to travel within the building. In vacant structures, however, additional openings for fire to travel will be numerous and throughout the building. These openings, which will vary in size, will create additional drafts of air with a ready supply of oxygen, further enhancing the fire's ability to spread. Incendiary fires will further this concern.

Firefighters must be reminded of this each time they go to work within the building's interior. Attempting to search above an involved area will have to be delayed. Larger than normal hoselines may be ordered for anticipated fire size, and multiple hoselines may be required to handle numerous as well as altered fire spread.

Anticipation, combined with the knowledge and experience of the firefighter and fire officer has always been critical to the successful outcome of an incident, but with vacant buildings thinking well ahead of present conditions is critical when going to work at these types of structures.

Fig. 10-17 *If you find an opening, fill it up with water.*

Time: Vacant Buildings

*COAL TWAS WEAL**T**HS*

When considering time and how it may impact fire department operations at vacant structures, we find very few specific areas for us to focus our attention on. Time of the day will have no direct bearing on an incident in a vacant structure. If the building is inhabited by vagrants or homeless people, it can be occupied any time of the day or night.

Time of the year

Time of the year will present a concern, especially if it is impacted by the weather. In the cold weather months of the year, more of the area's homeless will be looking for shelter from the cold and the snow. Vacant structures during these periods may house more individuals than we might encounter in the warmer months. This probability combined with primitive methods of keeping warm will add to the difficulties.

If not given consideration, the number of people you may find in a particular building may not only be surprising, but also overwhelming.

Height: Vacant Buildings

*COAL TWAS WEALT**HS***

Just as vacant buildings can come in any class of construction and area size, they can also come in any height. Many of us when thinking of vacant buildings within our town or city, consider the lower-rise, residential buildings. Wood frame or ordinary designed buildings of three-, four-, and five-story heights generally don't cause us too many concerns within the height size-up category. Within this mix of three-, four-, and five-story buildings, there maybe the eight, ten, or higher story structures that will.

Accessibility

Elevator access and usage may be extremely limited to non-existent in some buildings. If building security and maintenance is maintaining the building, you could expect to find some type of upper floor access. However, if there is any present, it may be limited to only one elevator that will often be a freight elevator that does not have fire department controls. This may force firefighters to climb up. Without a safe and controlled elevator, firefighters will be forced to walk up. With this will come delays in reaching the fire, extended hose stretches, long flights with tools and equipment, and the need to provide alternate means of egress. Any time we are operating at an upper-level fire in a vacant structure, we have to always be cautious about any fire that may drop below us. Chief officers should not hesitate ordering a number of aerial devices to be placed all around the building. This tactic with members observing conditions on the floor(s) below may prevent us from getting into trouble.

Special Considerations: Vacant Buildings

*COAL TWAS WEALTH**S***

Fire officers must constantly review procedures for these types of buildings. Emphasis has to continually revolve around a slow, cautious approach, with the safety of the firefighter being the overriding concern.

Vacant building fires will demand that extra time be afforded to building size-up. No officer should attempt decisions without an extensive review.

Vacant Buildings

1. *A review of any vacant building markings.* A system that identifies any structural concerns within the building will be one of the most useful pieces of information that an

officer could use to safeguard his members. Markings indicating that the building has holes in the roof, stairs missing, or that it is totally unsafe to enter, provides information that is critical to a safe outcome.

2. *The location and extent of the fire.* This observation along with information about the building's stability will determine how, where, and if, we will assign forces. Surveys on all six sides of the main fire area are a requirement to ensure members do not become trapped from rapidly changing conditions. This survey may also prove to show additional structural concerns that can be relayed to operating members.
3. *The class of construction/the building's height and area.* The class of construction will give inherent information relevant to that type. The building's height and area will lend information about the structure's potential involvement.
4. *Exposure problems.* This will dictate priority concerns based on a number of factors, notably the most severely threatened life exposure, as compared to the most severely threatened exposure. One is different from the other. Otherwise, review the factors presented and assign the resources.
5. *Any indication of occupants within the building.* This may come from a number of sources previously mentioned. Their indication will require decisions to be made for their search and the protection of those searching.
6. *Indications of the previous occupancy.* This may allow information about the fire load, content loads, hazardous materials, structural features, etc.
7. *Area/building accessibility.* Obtain information pertaining to additional routes from court yards, alternate building entrances, parking lots, rear building lots, and adjoining buildings to aid in the suppression effort.
8. *Any additional structural defects.* Cracked walls, out-of-line exterior walls, out-of-line window frames, sagging lintels, bowed ceilings, etc.

When structural observations from either the marking system or your actual view show that the stability of the building is questioned, officers should prepare members for exterior operations.

With operations in a vacant structure focusing around a slow and cautious approach, take advantage of the additional time that you have allowed to do a thorough building survey, there is no reason not to.

Questions For Discussion

1. List and describe the reasons why vacant buildings are more vulnerable to fire spread and collapse as compared to when the same building was occupied.

2. List and describe all the possible occupancy indications of a vacant building.

3. Of all the ladder company functions, which will receive the least attention, and why?

4. List and describe the egress and accessibility concerns that you may be confronted with from a fire in a vacant building.

5. List and describe the firefighter life hazard concerns that may be presented in these buildings.

6. What areas within a vacant structure are most vulnerable to rot? List the reasons why.

7. What is the governing factor when protecting an exposure building from a fire involving a vacant structure?

8. How may your time size-up factor influence your decision-making?

9. How can the previous occupancy of the building influence your decision-making on the fireground?

10. List the concerns relating to the use of auxiliary appliances within the building.

Index

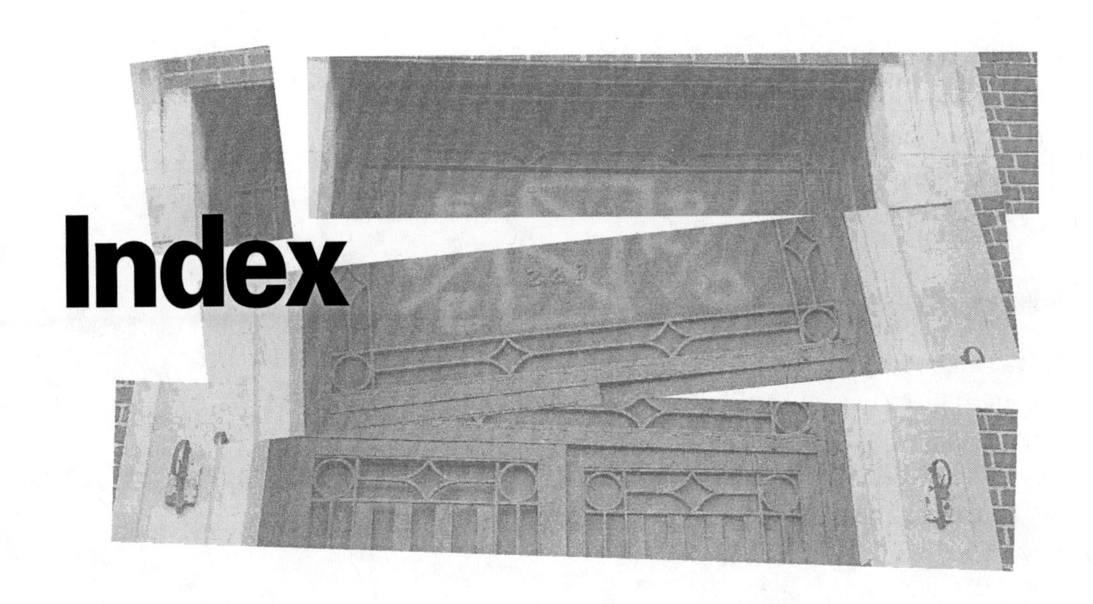

A

B

C

D

E

F

G

H

I

K

L

M

N

O

P

Q

R

S

T

U

V

W

Y